Gurus' BODYGUARD

Darrell McDowall

"Wow, what a journey! I literally could not put this book down. I was absolutely enthralled and received lessons, reminders and updates on how I should act, think and react. Darrell's wisdom is deep, and I cannot thank him enough for the knowledge he has passd on. I love it !!"

Shane Robins

"This man is not ordinary. His thoughts are golden and his experiences are something we can all learn from. This bookchanges the way I feel about life".

Paul Innes

"This is the best book I've read in many, many years. It is a must for anyone on a spiritual journey".

Paul Laufenberg

Copyright © 2022 by Darrell McDowall

All rights reserved. No part of this publication may be reproduced, distributed, or transmitted in any form or by any means, including photocopying, recording, or other electronic or mechanical methods, without the prior written permission of the publisher, except in the case of brief quotations embodied in critical reviews and certain other noncommercial uses permitted by copyright law. For permission requests, write to the author, addressed "Attention: Permissions " at gurusbodyguard@gmail.com.

Darrell McDowall

Ordering Information:
For details, contact: gurusbodyguard@gmail.com

ISBN: 9798474183718

Design & Production: Lance Innes
Editor: Lois Farrington Hunt
Cover Design & Production: Stephanie Crane

Imprint: Independently published
Printed in the United States of America

First Edition

CONTENTS

Introduction — *i*
Prologue — *ii*

Part 1
The Early Years

1	Growing Up on The Station	1
2	South Vietnam 1967	7
3	A Homer	25
4	Saigon	37
5	Special Services	51
6	Terrorism in Vietnam	59
7	The Minefield	79
8	Napalm	85
9	The Medal	89
10	The Long Hai Mountains	95
11	The Fight	103
12	Death of Another Friend	113
13	Ending	121
14	Home	127
15	The Hotel	141
16	The Psychiatrist	145

Part 2
The Search

17	Mary	153
18	Malaysia and the Muslim Religion	169
19	Becoming a Buddhist Monk	179
20	The Dream Destroyed	201
21	The Cuckoo's Nest	211
22	The Spiritual Test	227
23	The Ashram	239
24	A Miracle	265
25	Back to India	271
26	Home Again	305
27	The United States	307

28	*Linda*	325
29	*Other Amazing Teachers*	329
30	*Australia Again*	335
31	*India*	341
32	*Esalen*	357
33	*Motorbikes & Deserts*	365
34	*India Again*	373
35	*Another Master*	395
36	*The Familiar Road*	401
37	*The Unexpected McKenna*	407
38	*Garimo*	417
39	*Desire*	433
40	*Thailand*	453
41	*Karl Renz ...The Authentic Truth*	461
42	*Belief*	465
43	*Journey's End*	471

INTRODUCTION

One of the things I lost in Vietnam was open-hearted trust. But today I accidently left the back door open and was lying down watching television.

I did not hear him enter, and there he was standing at the foot of the couch looking at me with wild and angry eyes. Before I could move, he was on me, wrestling furiously as he punched.

I finally managed to roll him off and onto the floor but he jumped up, gun in hand, and shot me. I felt liquid running down my chest. Then he pulled out a knife and stabbed me. The gun in his hand was a water pistol, the knife was plastic.

The intruder was my handsome little grandson and his obsession of the moment is guns, knives and fighting me or his dad. I love him deeply but his mum and I are worried that his obsession with guns and fighting may encourage him to join the army when he is old enough. He thinks I'm a soldier hero.

He is almost five years old and has a loving heart. He shares everything with his friends when they come to visit. His adoring schoolteacher mother has raised him with lots of love and has never hit him.

This story is for him and children everywhere, to show them what war is really like. It is not about heroes and valour. War is not a place of love and respect. It is a breeding ground for hate, fear and revenge.

Prologue

Jay Zee was dying. She only had hours to live. Her real name was Janet Zuckerman but those of us close to her called her Jay Zee. She was one of the most renowned and respected teachers at Esalen Institute, the celebrated New Age learning centre situated beside the ocean in Big Sur, California, and this was the early 1980s. Here I walked and talked with famous scholars and teachers such as Joseph Campbell, Gabrielle Roth, Fritz Pearls, Dick Price, Jenny and the Nine, Feldenkrais, Traeger, Lily, John Soper and Nancy Lunney.

I was fortunate to be there as a "work scholar", a masseur, a gardener and sometimes an assistant group leader. People called me the "curious Australian", because I asked lots of questions, usually of a spiritual nature.

This particular day everyone was eating in the large dining hall, but I was alone with Jay Zee, at her home, holding her hand as she lay close to death. She was awake and quite lucid, looking off into the vast boundless sky and I wondered what she was looking at.

When she turned back to me, I quietly asked: "Jay Zee, what is prayer?"

She turned lovingly towards me and answered: "Prayer is listening to the silence." I did not really know what she was pointing to on that day, yet after a lifetime of searching, one of the things I have discovered is that in the silence, in the stillness of a mind freed from desire for anything in the world, therein lies true peace. To be in the world but not of it.

The shock of Vietnam wiped my mind clean of all my previously held beliefs. To replace what I lost on the battlefields I had to find new answers

Now, in my last years, my desire is to "give back", and share with you the many teachings, words and wisdom passed on to me by the remarkable people I have had the good fortune to meet and serve.

It is high time for change, especially in myself.

Chapter 1 - Growing Up On The Station

The men in my family, my father and uncles, were all big and strong, weighing up to 18 stone (about 114 kg), whereas I was born with what the doctors call a "hole in the heart". I was the runt and by the age of eleven only weighed 6 stone, which is 74 pounds (about 30 kg). My cousins were all bigger than me, as were the boys at the local school, so I had to learn to move quickly to avoid their punches and beatings. Luckily my dad had been training me in the art of boxing since I was four while waiting for me to "grow up", as he put it, so I at least knew how to duck and defend myself.

On the vast family cattle-stations I had been riding horses since an early age. By the time I was a teenager I knew how to break in all kinds of horses. This was my first job even though I still only weighed about 6 stone.

During the days on the 55-square-mile station, we chased the herds of wild brumbies that roamed freely, and brought some of them back to the homestead to be broken in. We then rode them for weeks at a time on the long musters into the seemingly never-ending Australian bush.

My mouth sometimes became so dry, when riding for days on end, that I would chew grass or leaves just to be able to swallow. I remember often imagining an ice-cold soft drink or a frozen ice block soothing my dry mouth.

When we spotted the brumbies, my two part-Chinese uncles, Jimmy and Boof, my father, and the Aboriginal men who were also cousins and part of my family, would all take off in pursuit, riding at a flat-gallop through the thick Australian bush, jumping logs and dodging trees. These men were some of the best riders in the world, certainly the best in this tough environment, and they knew this vast country like the back of their hand. They could find their way anywhere in the thick Australian bush. This is quite an art which they slowly taught me, while my black "Uncle Harry" and his son Stanley taught me how to find "bush tucker". I was amazed at the amount of food that could be found in the bush, such as honey, yams which taste like a sugary potato, chinky apples, wild bananas, flowers, weeds and much more. I soon realised that this wonderful earth feeds all the birds and animals, so why wouldn't it feed humans too?

My Uncle Boof was a famous fighter (so earning him the name Boof). He also gave me lessons in fighting and shooting. He was the best shot on the station and an excellent teacher. By the time I was twelve, only he could best me with a rifle.

I was often sent out to hunt the wild dogs that kill our stock, plus the feral pigs that destroy our land by digging large holes. These holes can break our horses' legs when we are riding at a full gallop.

We don't shoot the purebred dingoes, only the ones that have cross-bred with an Alsatian or any other large domesticated dog.

Once there was a huge black dog that was part German shepherd and that had been killing a lot of our stock. I was sent out to hunt it down with a 270-bore rifle. The adults had already failed to find and kill it. A 270 was one of my favourite long-distance weapons. It's smaller than a 303 rifle, but the bullet will fly straight through a large gum tree trunk, whereas a 303 bullet will not.

I picked up the dog's tracks near the Mitchell River which ran through the middle of our property. I soon came across the cattle it had brought down. Most were still alive, but crippled and lying on the ground. This large dog attacked and pulled them down, then chewed a hind leg until they could no longer walk. He would leave them bellowing in pain and then, in his greedy lust for blood, move on and bring down another poor beast.

He was an efficient and compulsive killer.

I passed six of these hopelessly crippled creatures before I finally sighted the black dog. He was busily mauling another beast. He didn't know I was there. I had made it my business to always be downwind from him and to move silently through the bush.

I knew that I would not get close to this dog, as my uncles and father had already tried. He had quickly escaped as soon as he sensed or smelt them approaching. His survival sense was acute.

He had a cattle dog bitch with him that he had coaxed from a neighbouring station. My father had told me to shoot her too as she was probably carrying his pups.

I had been tracking this black dog all day and I was tired and thirsty. I knew I had to hit him with my first shot even though I was at least 300 yards away. I adjusted the sights on my rifle and took aim.

Remembering my Uncle Boof's lessons, I pulled the rifle tightly into my shoulder and took aim while I softly drew my breath half-in and held it. The trigger on this fine rifle had a first pressure, which I softly squeezed. I then relaxed my body and slowly let out my breath. Second pressure, then an awfully loud bang. The dog jumped up and bolted. Had I missed? No, I hit him all right but not the heart shot I was trying for. I had hit a back leg, high up, and he was dragging it as he ran.

Thinking about it now, you could say it was Karma in action.

Anyway, I took off after him and soon caught up. With a fearsome growl he turned and came at me with large fangs bared. A brave dog, I thought, and wished he were mine so that I could retrain him. But he was a killer now and sorely wounded, so I had to shoot him before he got to me. He was a magnificent specimen, as cross-bred dogs and cross-bred humans often are.

I reluctantly put a bullet through his head.

I also had my razor-sharp knife strapped to my arm in case my rifle misfired. In the bush one must be prepared for anything, as the law is often kill or be killed.

Close by, the bitch was whimpering and cringing in fear. She had no blood on her so she obviously hadn't taken any part in the killings.

In my short life I had hunted and killed too many pigs and other animals. I decided that I could not do it anymore. I suddenly realised that I hated killing.

Looking at her lying on her back, legs up in surrender, I

couldn't bring myself to kill the poor little bugger so I tied my belt around her neck and led her on the long walk back to the homestead. When I arrived, my father ordered me to shoot her. I begged him to let me keep her. He flatly refused.

Then good old Uncle Boof stepped in and took my side, as he always did, and finally convinced my father to let me keep her.

She became one of our best dogs. She loved me as I loved her and we worked well together. Still, Dad killed all her pups when they were born.

I was twelve years old when Dad sent me to a Christian boarding school. It was out west and a long, long way from my home. Here I had to continue ducking the big boys' punches as well as some of the Christian Brothers who were molesting the little boys. I had to say a strong "NO" to the Christian Brothers, as I was still a pretty little boy back then. For my refusals they would often give me the cane on my already bruised bottom. I once held the record, at this school, for the most beatings in a week. I suppose they hoped I would give in when I had suffered enough.

But the endless beatings only made me tougher, though for those first few months I would hide in the toilets and quietly cry where no one could see me.

Then out of nowhere something remarkable happened just after I turned thirteen.

My mum came down to the school and took me to a heart surgeon. He told us that he would have to operate in three weeks or I would die, and he could not guarantee that the operation would be a success.

As we walked out of that hospital, hand in hand, my mother could not stop crying. I turned to her and said, "Mum, I can tell you one thing for certain, I am not having that operation and I won't be dying." She stopped crying as she looked deep into my eyes.

When she took me back to that surgeon the hole in my heart had miraculously closed and the murmurs had ceased. The doctors were amazed.

It changed my life and in an incredibly short time I put on 50 pounds and grew 6 inches. Soon I was holding my own with the big boys and no longer needing to duck punches. All Dad and Boof's training now gave me an edge and after I flattened one of the school bullies, everyone, including the teachers, knew to leave me alone.

I won some swimming and boxing tournaments, and was later picked in the first thirteen-rugby league team. I was 6 feet tall, with dark hair, and good-looking enough to be noticed by the girls at the nearby schools.

During the school holidays I took the long train journey home and returned to our latest cattle station near Cooktown. This was wild, unfenced country and we often spent weeks at a time riding through the endless bush, never seeing another human or even the trace of one. We rounded up wild cattle to sell back in the city. Sometimes we ended up with hundreds of cleanskins – that is, cattle which had no brands or ear marks.

By the end of three long hard-working weeks, we only had potatoes and fresh killed meat left to eat unless I managed to find wild honey and yams when I had the time.

We rode everywhere, and in the oppressive heat of summer would sometimes have to navigate through huge thunder and lightning storms. Once, during a bad storm, I sought cover under a big tree because my horse was spooking and bucking in fear. Luckily for me my father screamed out to get away from the tree. I reluctantly did as I was told. Moments later a bolt of lightning hit that tree and it split in two. I would almost certainly have been killed, along with my horse, if Dad hadn't yelled out. The Australian bush is a dangerous place if you don't know what you are doing.

Back at school I finally graduated, and then went to Teachers' College down in Brisbane.

Mustering on the station.

Chapter 2 - South Vietnam 1967

"War is organised murder and nothing else."
Harry Patch – last surviving soldier of World War I.

At the age of nineteen the Vietnam War was definitely not my chosen path but it sometimes seems that destiny has a mind of its own. Military conscription was no longer a choice.

Let me first mention the shock of recruit training after I was reluctantly conscripted into this man's army at the age of nineteen. A letter had come in the mail telling me where and when I was to report for two years of service.

In Singleton NSW I quickly realised that army training was being carried out using pretty much the same techniques as are used in brainwashing. Fear and tension were quickly thrust upon the young recruits, plus physical training to the point of collapse from total exhaustion. This was accompanied by loud and threatening abuse every day. Emotional abuse and a total lack of privacy were just some of the techniques that the army used on us. This was done so that we would supposedly obey all orders immediately, without thinking. "Soldiers are not paid to think, you dickheads," the officers and NCOs often yelled during exhausting training drills.

Locked in the army barracks at Singleton, seven days a week, created a prison environment for the duration of this difficult recruit training. Often, we were forced to humiliate ourselves, answering the same question, shouted in our ears: "What are you?" with always the same demanded answer: "I'm a dickhead, sir."

We were finally given two days' leave in the middle of our forced training, and twelve of us got a tattoo of an eagle on our left arms as a bonding thing. We knew by then that we were eventually going to Vietnam, like it or not.

When we arrived back at the army camp, we were lined up to receive our smallpox injections. Our overweight sergeant, a mean bastard of a man named Wally, happened to be standing next to me when I offered my right arm for the injection. He ordered me to turn and take the needle in my left arm. I told him the tattoo was new and would get infected. His ruthless reply was that if it did, then that was my lesson for getting a tattoo.

Of course, the whole damn tattoo became infected so badly that I was sent in an emergency ambulance to the navy hospital in Sydney. My entire left side swelled up and I was in agony and I was placed in isolation. When I returned to camp after three weeks, I had missed a lot of important training and was still too weak and underweight to train properly for another two long months. I wish I had punched that fat, sadistic, foolish sergeant, who did not approve of harmless tattoos.

Finally, in my weakened condition, I was sent to the Infantry Corp, even though I had asked to be sent to three other corps on my application. I knew that I wanted to be in anything but the infantry. I made a foolish mistake when I had demonstrated that I was one of the top three shooters in the battalion. I was also on the boxing team, before the smallpox injection, and had beaten all comers, including a man who had returned from representing Australia in the Olympic Games. I realised too late that fighters, and good shots, are always sent to the infantry.

After infantry training at Singleton, we were sent off for some more hurried training at the jungle-training centre in Canungra, which is said to be the hardest jungle-training camp in the world. The Americans who trained there certainly thought so.

I really do not know how I struggled through that place, as I was still very weak and underweight. I was in agony on the forced marches and together with the physical training, I was always exhausted. I chose to persevere because I did not wish to let my mates down.

After such shortened training time, I was sent off to

Vietnam, even though I did not want to go. I was told to either go or spend the next eighteen months in jail, with a light always on, being woken every half hour, and with water dripping on me in an enclosed space. "Your choice, soldier," the officer shouted.

The much-publicised political story that we were all given a choice of whether we wanted to go to war was definitely not the whole truth. The beginning of political lies and bullshit.

Of course, I saluted and went, thinking, unfortunately at the time, that Vietnam was the easiest way out.

We were hurried onto a roaring, cumbersome Hercules troop carrier. We finally landed at Vung Tau military airport after an uncomfortable, cold, seemingly endless, vibrating, bone-shaking, noisy flight, all the while sitting on hard tiny web seats.

Stepping thankfully onto the tarmac, we were immediately wrapped in a blanket of humid, sweat-inducing tropical heat, as this was the height of summer in South-East Asia. Thirty-eight of us were National Servicemen conscripts, here against our free will, ironically sent to stop the Communist hordes from taking away our freedom. We were eventually to replace the killed and wounded men in the 2nd Battalion, which was already based here. Two tough regular army sergeants accompanied us. One of them, whom I met at Ingleburn infantry camp, after the jungle training at Canungra, was now one of my best mates.

We instinctively made for the only shade available, a 6 X 6 metre iron roof perched on 3 metre wooden posts. The airport was alive with the movement of American army personnel, vehicles and planes. This all must cost a fortune, I thought. Phantom jets, Hercules, helicopters, and the huge lumbering B52 bombers. These bombers were loaded with 1000-pound bombs that supposedly kill and destroy everything they are dropped on, yet, as I soon learned, they do not kill the wily enemy, the Vietcong, who live safely in their massive underground tunnels.

After a long wait, as is routine in this man's army, an officer drove up in a truck riddled with bullet holes. He was clearly excited, and quickly informed us that these bullet holes were the result of a Vietcong ambush he and his men drove through on their way to pick us up. "Welcome to South Vietnam," he shouted. "You are now at war, soldiers. Take up battle positions on the trucks, weapons cocked, safety catches on."

This order got my adrenaline pumping immediately, and with my standard issue SLR rifle I was quickly in battle position. And so the wild adrenaline rushes began.

A long army "what the fuck is going on?" hour passed. We were sitting soaked in sweat in this humid, sweltering hot sun before we finally lumbered off.

Our base at Nui Dat in Phuoc Tuy Province was a couple of hours' drive inland from Vung Tau, where we had landed. It was exciting to be in a foreign country for the first time in my life. The small villages we passed through were alive with people and activity. These strangely dressed people, mainly peasants and farmers, were hurrying about their business, hardly giving us a passing glance. It did not seem like we were welcome as saviours here.

The incredible smell of decay in these villages was quite unbelievable to my untrained nose, and I was surprised that everyone was not dead or dying from the plague or some other horrible disease. Most of the people were dressed in ragged clothing and appeared to be very poor. Groups of little children rushed to the side of the road when they saw us approaching and begged with outstretched hands. Because we had nothing to give them, they yelled, "Uc Dai Loi [Aussie], you want to fuck my sister?" Innocence took its first small blow. The officer screamed at me to shoot them if I saw one of them with a grenade. Innocence (shoot a child?) took another of many larger moral blows to come. I knew, without a shadow of a doubt, that I could never shoot a small child.

Leaving the villages, we often travelled through seemingly endless rolling green rice paddies and intermittent rubber plantations. Finally, we approached the cold steel gates, nicknamed "the pearly gates", of Nui Dat which was the entrance to the Australian base in Vietnam. Rows upon rows of curled barbed wire, sandbags, tents, and huge artillery guns filled my vision. Half a mile into the heart of this cold soldiers' nest we were ordered out of the trucks and lined up on parade at the Australian Reinforcement Unit (ARU). This was the place where we were to become acclimatised and trained for the next six weeks. The captain of this unit informed us that Nui Dat is where we would spend the next thirteen months, when we were not on patrol in the jungles. I glanced around at the dirty old four-man tents and muddy ground. The captain smilingly informed us that the cots, or beds, for our tents may or may not arrive in a couple of weeks. A sick, trapped feeling crawled through my stomach and I mumbled, "I don't believe this shit." But I mumbled too loudly and the sergeant major screamed at me to fall out. Discipline is strict in this man's army, and nobody ever dares speak on parade, especially not in a war zone, I found out.

The rest of the men were dismissed and I was marched into the "bosses" tent where I was summarily placed at attention, "on the mat", before the commanding officer, who holds the rank of a captain. Our two accompanying sergeants, Jonesy and Ballard, also came into the tent with me. The captain ignored me and asked them who the hell I think I am. Luckily, in Australia Jonesy and Ballard were my mates, and we'd had some fun times and a couple of fights together in the Sydney pubs. I was also honorary platoon commander for a while at Canungra, supposedly the toughest jungle-training camp in the world. My mates told the CO this, adding that I was the best bushman they had, plus a good soldier and also a Golden Gloves boxing champion. As the CO was a well-known boxer, he looked at me with new respect and laughed, saying, "Well, it is a bit rough here and it's been a long journey." Then, wonder of wonders, he saw on my file that I had worked in our family hotels, so instead of charging me with an offence, he offered me the job of running the bar and PX (shop) at ARU, effectively handing me the mantle of "King Rat".

The next day, when the rest of the platoon went on dangerous training patrols, learning what to carry while becoming acclimatised in Vietnam outside the wire, I stayed in camp and set up the PX shop and the bar.

Within a week, after a couple of fist fights and a power display for the NCOs (non-commissioned officers such as corporals) backed, of course, by my two sergeants (sergeants really run the army, unofficially) I was handling all the considerable gambling, the movie nights, and the black market. The profits from these sidelines were huge and I split them with Jonesy, Ballard, and my closest friend, Private Kim Cuddy. Within a short period of time, I was making so much American military money I started giving it back to the heaviest consistent losers, together with some free advice on gambling, which always seemed to fall on deaf ears. I was, obviously, not a good teacher, and never have been.

Apparently, Oscar Wilde once said: "The only thing to do with good advice is to pass it on. It is never of any use to oneself."

A person soon does strange things in a war zone. In our spare time Kim and I stood on a plank about 6 feet above the ground and, using our fists, tried to knock each other off the plank. We were in the army rugby team and were now very fit. Kim was one of those men who are always surrounded by friends, because he usually had a ready smile and a kind word for everyone. I enjoyed this game with him as he did not lose his temper when I hit him. Kim was 6 feet tall and weighed 200 pounds. When he hit me, it really stung, but I was faster than him and a more experienced fighter, so he was the one who was more often climbing back on the plank. A few strangers played with us but usually only once.

I remember learning to play this game as a kid while standing on a dead tree over a river on one of the family cattle properties. It was a large station of over 50,000 acres in the Australian bush and blessed with big flowing rivers.

The boredom and shock of war makes men play strange games. Russian roulette with a .38 pistol soon became fashionable with large amounts of money on offer. For some, death became a fair and sought-after option here. My turn to play this game would soon come, though I would never have believed such madness could be possible for me.

Each day I saw the patrols returning to base, sometimes carrying men on stretchers, usually suffering from heat exhaustion, as they struggled to carry their heavy loads and become acclimatised.

After a couple of contacts with the enemy, my mates began giving me their valuables and written last messages to their loved ones before they went on patrol. After a few weeks they were all looking gaunt and tired and it was obvious how quickly they were losing weight. This acclimatising, and training, is hard work it seemed.

One of the men, a quiet fellow from Melbourne, had been forced to leave his pregnant wife, and he was taking on a terrible, fearful, haunted look. The government obviously had no compassion, keeping him away from the birth of his first child. He must have been wondering if he would ever see his unborn baby or his wife again. (The war must go on, no matter the cost for individuals and the deaths or mutilations of young men.)

Anger welled up inside me as I looked at my friends. "What is this stupid war all about?" I saw soldiers being forced to kill men, women, and even children, simply because they were called Communists and may be carrying a hidden bomb. Communism, capitalism, Catholicism – all the isms – will pass away as everything of man's "monkey mind" eventually does. Why then kill for thoughts about beliefs?

And huge guns in the Nui Dat compound were firing every sixty seconds onto "forbidden" areas. Bad luck if the Vietnamese kids were out playing in these areas of their homeland, which they sometimes were as I later found out.

These guns are far from accurate and the noise emanating from them was deafening. A vacuum of air hits you up the bum when you're sitting on the "thunder box" (a toilet dug in the ground). The first time this happens, the newcomers think a king cobra has bitten them, and they often came crashing through the wire mesh door with their pants still around their ankles. This reaction was seeded by a well-displayed photo of three men holding a huge cobra, which was pinned on the door of the toilet. The old hands smilingly told all newcomers that the snake was killed in the "thunder box" after it bit somebody.

It was an amusing pastime for the regulars here at ARU to

sit watching the thunder boxes, with a beer in the hand, waiting for a newcomer to be half frightened to death.

Occasionally our base was harassed by small ambushes during the night, and we all quickly hit the frontline foxholes while men fired at anything that resembled movement. We sometimes lay there for hours, battling with the shadows of the night. So, the nervous system suffered further and men changed gradually but surely on a deep, hidden level.

Meanwhile the only entertainment in camp was what I managed to organise and this eased some of the tension, yet fights still broke out. One night a solid bloke, a conscript named Paul from Melbourne, voiced what I was thinking. He said that he will not kill a man just because he is called a Communist. Paul was immediately set upon and beaten to the ground. Bleeding, he gamely climbed back on his feet and shouted that his father was a waterside worker in Melbourne and a member of the Communist Party. Shouts of "traitor" and "kill him" were taken up and Cuddy, who I've never seen angry, jumped forward and quickly pummelled Paul to the ground again. Sick at heart, I reluctantly stepped forward and stood in front of the mob and my best friend Cuddy.

I quietly said, "He has had enough. If any of you want a fight, then fight me." Kim's fists were up and he had a fierce look in his eyes, as the pack was behind him and howling for more blood. Paul once again staggered to his feet. At that particular moment I did not recognise these men and my blood boiled. My face must have brought back their sanity, as they recognised and were possibly afraid of the reflection of their own madness.

Nobody took up my challenge because they had all seen me fight. I took Paul's arm and put it around my shoulder, helping him to the sickbay. Within two days he was transferred from Nui-Dat back to Australia. A pity I had not spoken my truth as I could have joined him, although of course he now had to face the enemy in front him. I was told by my sergeant mates that poor Paul would be deprogrammed in a mental hospital, probably drugged. He would be forced to dodge army psychiatrists who would be ready to give him shock treatment at the slightest hint of non-conformity. I wondered how this treatment could possibly exist in the world today. Is war just blind belief fuelled by fear?

Six weeks finally dragged by and my fully trained platoon were given their posting orders to various units in the two infantry battalions that existed in Nam.

The captain had already told me that he was giving me my corporal stripes making me an NCO. I would be staying on as permanent staff at ARU to continue running the PX and also train men for boxing tournaments he wished to run. This suited me just fine as I had no wish to be a Vietnam hero. Also, being an NCO would get me out of all menial duties, such as dish washing and late-night sentry duties.

Then I made one of the worst mistakes of my life that cost me dearly and changed me forever. A newly posted, big tattooed corporal, with two of his own "heavies", came into the bar and informed me that they were taking over my gambling table and commandeering my "Crown and Anchor" set. Unfortunately, Jonesy and Ballard, my heavyweight sergeant backups, had both been posted and were no longer here to give me protection. Cuddy was also gone and I did not have time to find a new sergeant who wanted to get rich quick, or find a strong backup man.

The corporal and his heavies made the unfortunate mistake of physically threatening me with their fists. About a minute later the tattooed corporal and one of his heavies were carried out of my bar. "Better learn to fight if you want to take over my bar and the Crown and Anchor table," I said, as they were being helped out.

The next day the boss called me up, and I was "on the mat" once again. He told me that a company in the 2nd Infantry Battalion had had another contact with the VC and lost a few more men. They urgently needed reinforcements and, as I was the only one available, my orders were to proceed immediately.

"But, sir," I said, "I have not had any training outside the wire and I'm not acclimatised for action in the field." Sadly, he pointed out that a private soldier cannot punch non-commissioned officers and his hands were tied. It was the 2nd Battalion or jail for me. My choice, he told me.

I foolishly chose the battalion because I did not have the slightest inkling of what lay ahead. I thought that I could get by with the help of some mates who were probably posted there, and they would show me what to do and what to pack when going on patrol. Anyway, I thought, I had not liked taking money from men on gambling tables.

This turned out to be a terrible choice, for which I was soon to pay dearly. Alone, I was taken to a strange land of veterans who looked like tough extras in a Rambo movie. I soon realised that I was in trouble because I was not "blooded in battle", as they were, nor did I know the current terminology for things only found in Vietnam outside the wire and in the field of combat. It was like reporting for duty on a ship and calling the portside the left side. Well, I thought, I'll keep quiet, tough it out and edge my way in because, unfortunately, none of my old mates had been posted here in this platoon to help me. These men were all strangers to me.

As the platoon corporal was escorting me to my tent, we passed the company bar, which was a large tent with a man's bare skull hanging over the entrance. I asked the corporal who the skull belonged to and he said, "A Vietcong that we shot." I spontaneously said, "I thought Australians respected the dead and gave them a proper burial."

In an instant, with that bare skull went "respect", another of the many morals taken from men and sacrificed in a war zone.

This vengeful corporal told the entire platoon what I had said, and my nightmarish time began. Not only had I replaced their dead mates, I had also dared to criticise their skull placement of the hated enemy they had proudly killed. The army taught us to hate and kill, well and quickly. It did not teach us about the shocking consequences of killing, as I would later find out.

I again noticed that men were moving around the company area as if they were dressed like extras in an American war movie, with bandoleers around their heads, large knives strapped to their sides, bullet belts around their bodies, feathers in their hats, grenades and various other deadly war paraphernalia hanging from them. Welcome to hunter's paradise, I thought. I did not realise that they would soon be hunting me.

Posted here alone, separated from all my mates, I was feeling lost and lonely as never before. These men did not even say "g'day" to me as they walked by. They treated me as if I were invisible, although a couple of them sneered at me and shook their heads as if to say, "what have they sent us".

The abrupt, unfriendly corporal immediately yelled at me, in front of the men, "There is your tent and tomorrow you go on patrol for a week. Be ready to move out at dawn."

He was obviously doing this very loud-mouthed number on me to show the rest of the platoon how tough he was. I think he was gay and he was compensating for his insecurity by using me to aggressively perform and emote in front of the men.

Wonderful – I was not even sure what gear and equipment to take outside the wire, and these guys were all fair dinkum war heroes, so I dared not show my ignorance by asking them. They would have only sneered at me as my all too obvious ignorance could get them killed. They soon told me just that.

After a humid, hot, restless night, which included two hours' sentry duty at 1 am, the first ray of the sun saw us lined up and ready to leave. The corporal took one look at me and asked where my second grenade was. "I've left it behind," I answered. I had done this purposely because this gear was extremely heavy and I was not sure I could carry it all with no training.

"Go and get it and don't forget it again, you dickhead," he yelled, with a smirk on his face, enjoying his performance.

As we were making our way out through the barbed-wire enclosure that surrounded our camp I was consciously holding my rifle pointed to the left towards the looming jungle. The abusive corporal again loudly barked at me, "Private, can't you see that the man in front of you is covering the left? Wake up and hold your weapon to the right."

I wanted to tell him that to my right is our camp and there is no enemy there. If we are shot at, it will be from my left. I was soon to learn that these regular army officers do everything by the book, even if it makes no sense.

Logic disappears when a man is hiding behind his rank and desperately needs to prove himself to his men. The corporal made sure that the entire platoon heard that I had made two mistakes before we even got outside the wire. I was given no chance to fit in to this platoon. Jesus, what a place to find myself stuck in with no mate to help me out or speak up for me. This isolation had never happened to me before and I would soon shockingly realise that I had enemies both in front of and behind me.

Quickly covering the cleared ground outside the wire, we stepped into the thick, foreboding virgin jungle where there were no paths and only hard swung machetes, wielded by tough, experienced soldiers, could cut a way through. I soon noticed that this was often a noisy, slow, and sometimes stupid way to get anywhere, as the Vietcong knew exactly where we were. But it is in the military rulebook, written by officers safely back in camp, and we were forced to fight by the book as "good" soldiers follow orders to the letter. Unfortunately for us the enemy did not have a rulebook and often set up an ambush, as they could easily figure out the direction we were slowly heading. It became quickly obvious to me that the Vietcong knew their jungle much better than we did. This rulebook mentality is how and why our platoon sergeant and others later died.

We continued to spend our days cutting through jungle, and crossing rivers, creeks and swamps with the men nervous and trigger-happy. They were often shooting at shadows or animals, mistaking them for an unseen enemy, further giving away our precarious position.

The Cong often ambushed us with one or two hastily fired shots and then quickly disappeared, but our return fire seldom hit anyone. These Vietnamese soldiers were too experienced.

Actually, the Viet Minh were the regular troops of the North

Vietnam Army and had been fighting foreign invaders like us all their lives. They were not the recruited Vietcong of the south, dressed in their cheap black pyjamas. Wars had been ongoing here for 150 years.

I was sorry for being a part of perpetuating their wars.

That night I heard news on the radio that two companies from the other battalion accidentally ran into each other in a forest and opened fire. Quite a few of the men were killed and many were wounded. Later I learned that this "accidental contact" did not get reported truthfully in the papers back home. The true story would not glorify our soldiers. "Killed in action", the papers read. Died like heroes, poor buggers!

Each night we set up ambushes, lying on the hard usually cold, wet, ground. During this rainy season I seemed to be wet all the time, either from sweat, rain, or from crossing swamps and creeks. Blood-sucking leeches covered my body and I used salt to get them off my skin as I had been taught on the family cattle stations. No one helped me get them off, though they helped each other. Alienation held no compassion.

On the third night in the jungle our platoon opened fire from our ambush position, but I could not see a target or any flashes from return fire, so I did not waste a shot. My father taught me, at a young age, never to fire a shot unless the target was firmly in my rifle sights. He also warned me, before I left Australia, never to shoot a man in the back.

The next morning there were no enemy bodies to be found and we moved on. During the day fresh orders were radioed in telling us to meet up ASAP with the other platoons of our battalion which were far to the north of our current position.

We began a forced march that lasted nearly all day. My feet were soon bleeding from the burst blisters caused by these cheap new hobnail boots that I had never worn before or had time to break in. My clothes were soaking wet from sweat. One of the men passed out, so we quickly built a stretcher to carry him. I was put on the carrying detail and my shoulder soon felt like it was going to pop out from the heavy weight. I was purposely given no rest spells, like the other carriers. I felt like I would soon pass out. Dizziness overcame me, but I gritted my teeth and made up my mind that I was not going to show any weakness or complain to these men. I was aware that I was in a very dangerous social position, and any sign of weakness would simply earn me further contempt.

The men already had names for me. One of these names was "recruit". A recruit is someone who first comes into the army to do basic training and recruits are also referred to as "dickheads". It is a belittling name and the abusive corporal loudly told me that recruits and "dickheads" like me do not belong in Vietnam.

"Fine," I replied, "send me home."

Once again, he relayed what I had said to the platoon. He was getting more bravo points with the men at my expense.

He must really be afraid of them and his position, I thought, to continue this unwarranted abuse.

I was no longer worried about the enemy as they were not always present, so they were not as dangerous to me as my own men. I knew, at this early stage, that I would not kill any man unless he was threatening my life and there was nothing that I could do to stop him in any other way. I had told the Intelligence Captain this very fact when he was trying to program our minds in Sydney. As a result, I was offered time in jail or a year here. If only I had known about this platoon, I would have taken the eighteen months in jail where I could have at least come out alive. In this situation it increasingly looked like the odds were against me to get out alive.

What concerned me the most was the fact that these men could kill me just as quickly as the so-called enemy. The men continued to remind me of the number of "dickheads" who had been "wasted" (killed) out here and threatened that I may be next.

Terror is a tool of this war. After days in this endless hell, I actually dared death to show itself so fear did not overwhelm me. Death now seemed like the only friend that could quickly get me out of this living hell. On second thought, if I died who would look after my family as I am the only son, and I would also like to see my fiancée again, who had travelled with me to Sydney and was patiently waiting for my return.

Choosing life over death, I continued on.

We passed through small villages that had been bombed by the Americans. I found myself surrounded by the true victims of war – the poor Vietnamese peasants who were clinging to wounded children and begging for a helping hand. Our military training taught us that anything or anyone could be a trap. If we stop to help, we may be ambushed or blown up by the people we are trying to help. I despaired at the realisation that man always has a reason for cruelty or extermination, either killing humans in a war zone or animals, such as kangaroos, back home.

I broke the rules when I saw a young baby screaming in pain.

She had fresh, horrible smallpox scabs all over her head and face. I gave the crying, pitiful, begging mother my tube of anti-itch cream to ease the baby's pain. Word was quickly passed down the line that I had stopped and done this. As a result, it was the sergeant who came and abused me wholeheartedly for being an idiot. In front of the other men, he informed me that if I got any itches, no one was going to give me some of their cream. The platoon laughed at me again. I could no longer tolerate this treatment. I needed to find a way to stop it.

As we were leaving the village, a man in cheap black rags, which the native peasants had been wearing for centuries but which was also the official uniform colour of the Vietcong, broke cover and was running for his life down a track. He was a harmless Vietnam Aboriginal, a Montagnard, who are looked down upon and treated badly by the Vietnamese, because they are black and very poor. Both the sergeant and the corporal loudly ordered me to shoot him. Would this killing make me one of them, I wondered, and will the platoon then accept me?

I aimed but could not shoot him in the back, especially as he was unarmed. I remembered yet again my father's teaching to never shoot a man in the back or kick a man when he is down. With the running man's back firmly in my rifle sights, his words came back to me strongly. Everyone was now screaming at me to fire, so I fired seven rounds, just over his head. I had been in the company shooting team in Australia and could have easily killed him with my first shot.

He finally disappeared around a corner, and the officer and men hurled abuse at me for being "a useless dickhead who cannot even shoot straight". A section of men then set off in pursuit of the scared, unarmed, fleeing man.

I was relieved when they returned empty-handed, though I was abused and disliked even more and respected even less. What an impossible situation. All my life, since my heart healed, I had been a leader and respected by my mates.

Here I was a failure. I was not a killer and I did not hate the Vietnamese people because I was the invader in their country.

We had no further contacts with the elusive Cong on this march. When we arrived at the other platoon's position, the fighting was over. The march back to base was filled with more pain as my feet were now infected from the burst blisters. Even with blisters under blisters, blood on blood, I was slowly becoming used to the heat and the marching. We at last made our way back through the minefield and the tons of barbed wire surrounding Nui Dat.

The retribution continued. After a week of dish washing, plus all the dirtiest and hardest jobs such as digging latrines, spraying an oily substance around the base (the true nature of which I was to discover later), and filling sandbags, together with the worst sentry duty at the oddest hours, I was feeling totally worn out and even more lost. These soldiers were doing everything that they could think of to break me. I was very much alone and not feeling so tough anymore. I was giving up. The men did not let me play cards or ball games with them and no one would talk to me. During the lonely nights I drank my two allowed beers alone in the boozer under the enemy skull.

I had a birthday, with no party or greetings. Slowly I was recognising my need to be accepted by these soldiers, perhaps even at the expense of becoming an angry, trigger-happy, merciless hunter of humans. A killer! This thought sickened me yet it was coming down to a choice of kill or be killed. What else could I do? Could I kill a running man by shooting him in the back, and also ignore a baby's distress? The answer was still a definite "no".

One day I had been given, as usual, extra duties and lots of sandbags to fill. After another long day I was lying with my back against my tent, enjoying a rest and watching four of the boys playing volley ball, wishing I could play with them. To my absolute surprise they invited me to have a game. The big redhead who invited me was one of the soldiers who had often told me that I was going to die by friendly fire.

On approaching them they all smiled and talked to me and a feeling of huge relief filled me. Had they at last accepted me because I worked hard and did not complain and I had learned the rules of how to be a soldier?

Before we began playing, the big redheaded man invited me to link arms back-to-back with him in a certain way to see if I could lift him. I stupidly did as he asked and then was pulled to the ground. I could not break free from his cunning hold. The other men quickly pulled my army trousers down around my ankles and a mixture of something black and sticky, combined with flour and feathers, was poured over my genitals. As they stood around jeering, laughing, and calling me a useless "dickhead", I dragged myself to my feet, gritted my teeth, and walked quickly back to my tent. An awful burning had begun and my skin was on fire. I bathed quickly, but all the skin soon peeled from my genitals.

I felt hopeless. I sat in my tent and played with two grenades and my rifle. I imagined going down there and forcing them to lie on the ground, while I poured their evil brew on their genitals. But someone else in the platoon would shoot me if I did that. Tears came to my eyes. I felt shamed and lost in this hellish world.

Daily this was getting to be too much for me and I wondered how much more I could take. I began to plan my only option. I would soon head into the jungle alone. But first I would have to survive my final inevitable explosive confrontation with these men. Even though death was looking much more acceptable each passing day, the only sensible option, my urge to live, was still slightly stronger. I had survived in the bush alone many times in the outback of Australia. I had been taught how to survive by my Aborigine and bushmen family. I now made the plan to leave this platoon when the first shot was fired at me, and I would be prepared in advance.

My Gradfather, Gerald Watson, ironicaly, an Infrantry soldier in 2nd Battalion WW1.

Chapter 3 - A Homer

Something strange happened that almost earned me a trip home – a "homer" as they call it here in the hell of this hopeless war in Vietnam. I was sent for a medical inspection and it was found that my front teeth, both top and bottom, were all chipped and broken. The doctor was shocked and spoke to the commander. He thought that the water probably caused it and I should be sent home before I lost all my teeth. The commander immediately signed the appropriate papers to send me home. A breathless happiness swamped me.

I was immediately sent to the dentist to have my teeth filed to make them smooth again. The dentist asked me if I had been chewing anything hard and I honestly told him, no. He then looked at my metal dog tags which hung from my neck. They were pitted and bent. He asked if I knew I had been chewing my dog tags. I definitely did not remember doing this, but told him my nerves had been on edge lately. If only I had known that I had been trying to bite through my twin metal tags, I would have taken them off before seeing the dentist.

The dentist told the commander, and my signed "homer" papers were shredded. He then filed my chipped teeth to make them smooth and told me to stop chewing my dog tags.

Just when I felt that I could stand no more of this hell, a message arrived to ease my troubled nervous system. It was from D Company, which was stationed about a quarter of a mile away. My best mate, Cuddy, was back in camp and he wanted me to come over for a visit that night. A few of the boys from my old platoon were also there, and they greeted me warmly with respect as their mate and passed me a beer. The feeling that flooded through me from this simple human acceptance lifted a great weight from my shoulders.

We spent our evening talking about old times in Australia and the good times we had together. Finally, as the evening was coming to a close, I told them of my plight with my platoon. They immediately stood up and took me to their company commander's tent and though it was late he received them warmly and openly. A lot different from my company and my situation as I was not allowed to approach any officer without being summoned. They repeated my story and asked their commander to arrange an immediate transfer to their platoon so I could be with them.

He said that he would, but it was not possible right now as he had just received new orders. The entire battalion was to go back into the field tomorrow. The enemy had attacked and burnt a friendly village up north. Smiling, he assured me that when the fighting settled down and we returned, he would get me a transfer to Cuddy's platoon.

I told the boys to take care of themselves and I reluctantly dragged myself back to my "hellhole" of a platoon, not suspecting for a moment that this night was to be the last time we would ever spend together.

The next day as my platoon moved slowly on foot through the thick jungle, I was placed in the position that most soldiers here tried to avoid – last man down at the very rear of the platoon. It is considered a lonely, dangerous place. The Cong often kill this man and then quickly melt off into the jungle, but I liked this position. The thought of death held no fear for me in these circumstances. Because I was separated from these men, I would not be "accidentally" shot in the back. Being at the rear, I could put quite a bit of distance between them and myself.

I realised, while plodding through the jungle in my sad predicament, that I was not really angry with the men in this platoon. I knew it was not their fault that they were given an untrained replacement. It was the fault of the officer in charge who put insignificant rules before reason.

Each day on this long trek I hung back a little further from the platoon. I am a bushman and had my own compass and map to lead me to our final destination at the end of each day. Because it took many hours for the platoon to travel a short distance while cutting through the thick jungle, I was soon heading out on my own little excursions, sometimes kilometres away from the rest of the platoon. I often saw signs of the so-called enemy but that did not concern me. I thought I would be able to survive here on my own, easily avoiding the noisy Americans, the slow Australians, and even the wily VC.

I always rejoined the platoon before they halted for the day and they had no idea that I had been doing my own "recon", which I had decided to do in order to prepare for the time when I would be forced to leave this platoon and survive on my own. Because the army would brand me a "deserter", even though a few of these heartless men had fired at me and tried to kill me, I would have to disappear.

After another restless night spent in the wet jungle the welcome sun rose once again, and we continued to hack a path through the dense foliage, supposedly to avoid ambushes or step on mines. Vietnamese children were running back and forth on a well-worn track just 50 metres away, yelling as loudly as they could, "Uc Dai Loi number 10", meaning "Australians are the worst". They were intent on revealing our not so secret position to everyone in the area. Once again, we were ignorantly put in a dangerous situation. Our officer was blindly following orders given by some general or brigadier, playing war games on a whiteboard, while seated safely back at Nui Dat in Battalion Headquarters. These commanders could not see or hear what was happening here and we were not allowed to change the predetermined tactics, which would soon get someone killed again.

The innocent civilians knew the sounds of war and armies moving through their land, suddenly killing them or destroying their villages. They were aware that some nervous, foreign, trigger-happy soldier would open fire at the first sign of movement. Their heart-rending experiences had taught them to avoid soldiers and warn all others of their approach.

Soon we were ambushed, but not from behind me, as expected. The snipers fired at the forward scout and then melted off into the surrounding jungle. Their quick escape route was already mapped out, and we had no chance of finding them. They had tunnels everywhere.

Angry at our inability to adjust to this dangerous situation, I strode away from the platoon and explored further than I usually did, meeting people and checking out the countryside, while looking for any snipers or VC.

My history protected me in this unfamiliar territory. My great grandfather was a cattleman as was my grandfather and father. They were among the first free white settlers to reach Australia's wild, vast and beautiful unfenced north-eastern interior. As a young boy I could still ride for days in a land where white men had never set foot, and there were no barbed-wire fences. My great grandfather had chosen an Aboriginal wife, so my forefathers, including whites, Chinese and Aboriginals, taught me to track and hunt. Because I was ordered to hunt and kill from a very early age, I was thoroughly sick of killing innocent beings. Killing tribal kangaroos because they ate too much grass, wild pigs because they were dirty and dug holes, wild dogs because they may eat a calf, killing birds to eat, or magnificent bulls because they had gone wild. There are so many logical reasons for killing, especially here in a war zone. I could no longer believe in the reasons. I was long ago sick of the bloodshed.

Fortunately, I had been taught well, and in these jungles, I could see that humans were easier than animals to track. When I suddenly appeared before some of these local people, I could not help but be amused at their surprise. The first person I made contact with on this particular day was a hill tribesman. His skin was a very dark black and he was dressed in raggedy black. He would most likely be shot on sight by the invading soldiers. When this poor man saw me, he started running, so I put down my rifle and waved at him with a smile on my face. Looking back, he was surprised and stopped running. I walked up to him and shook his hand. He could not speak any English, but by this time I could speak a little Vietnamese. I gave him a chocolate bar and smiling in innocent happiness he led me to a clearing where he proudly showed me his ancient clay ovens where he made bricks and from which he strived to eke out a miserly existence. I had a little Vietnamese money on me, which I gave to him. He then led me to his home nearby, which was an underground hovel with rags and plastic for a roof. He proudly introduced me to his wife, who carried a boy child in her loving arms. They too were both dressed in black rags. While we had a cup of tea together, I tried to explain to him not to wear black because it marks him as a VC, in the eyes of foreign soldiers, and is very dangerous.

I learned later that innocence is not allowed, even for this man. If the Vietnam armies decide to enlist this man and he is reluctant, they will demand allegiance. They will take his family hostage or even cut his wife's throat, just to convince him, and any others, to join whichever army is currently in control. In Australia, the military would only put a defector in jail, ruin his life, and bring shame on his family. This is the psychology of war. In the arena of war, the main overwhelming emotions are fear and hatred, not love and goodwill.

Bidding them a gracious farewell, the next two people I surprised were outside a small farmhouse. Their rifles were leaning against a post a few metres from their reach. I was kneeling about 50 metres away at the edge of a forest, next to a large tree. These men were obviously guerrillas, and with my rifle at my shoulder I thought about killing them. I quickly realised that this thought was foreign and ridiculous to me, because I did not really know whose side they were on. When they finally saw me, as I stepped away from the tree, they seemed to become frozen in shock and were certainly no threat. I did not know if these were the men who shot at my platoon so I lowered my rifle and waved to them, with a smile on my face. They couldn't quite comprehend my peaceful actions and looked at each other with their mouths hanging open.

They finally smiled stupidly and slowly picked up their weapons before slinking off into the forest. They even gave me a hesitant wave. I could not help but laugh ironically. These men were friendlier to me than the men in my platoon, even if they were the enemy.

On the return journey from my jungle exploration, I saw a refreshingly beautiful Vietnamese girl of about twenty, my age, working in a small field. I was drawn to her like a magnet, even though I knew it was very dangerous and foolhardy to be caught out in an open field.

I approached her with the sun at my back, as I have been taught to do with wild horses so they cannot see me approaching. She looked up, startled, when I touched her shoulder. I tried to calm her with the little Vietnamese I knew. Slowly she relaxed and offered me a small smile. I could not resist reaching out, just to touch her on the arm, but she jerked back with a look of horror on her face. I felt awful for frightening her in this way and apologised profusely as I backed away. I only wanted a human female touch and a little loving kindness that only women can give. I forgot for a moment that I was dressed in a uniform and was carrying deadly weapons. I quickly walked on.

Cutting across my platoon's trail, I soon caught up with them and followed their slow hacking wander through the jungle. Finally, we stopped for the night and I was forced to dig a latrine for the platoon before I was allowed to dig my bomb shelter. I was alone once again. This site, quite far from the rest of the platoon, was chosen especially for me by the crazy corporal. It existed in the rockiest ground he could find where it was almost impossible to dig a hole.

I was always the isolated outcast, while the rest of the platoon worked together. Everyone else dug, worked, and slept together in pairs.

I turned this adversity into an advantage. I had grown much stronger with the extra hard work and was now fully acclimatised. Finally, I knew the rules and terminology of their war and exactly what to carry while on patrol. I had learned the hard way, which made me more skilled.

Some of these boys still followed their favourite pastime, which was needling away at me with their comments and taunts. "You will be dead soon, dickhead," the dark machine gunner quipped. I simply ignored them.

The lieutenant announced that he wanted five men to investigate suspected enemy activity a couple of kilometres away from where we were dug in. He picked me as one of the five.

There were about twenty of us gathered in a circle. The usual corporal, the one who had given me up to the platoon for the skull comment, was picked to lead this patrol. He loudly announced, so everyone would hear him, that he refused to have me on his patrol. He was appealing to the men so they sided with him, and some of them obligingly reacted with scoffs and snide remarks about me.

I was amused at his antics. This fool still believed he was quite safe, baiting me like this. I knew I would not come off second best in any encounter with this bully.

I was grateful for this rejection. I was much better off not going on patrol with these guys. Even though these men may have doubted my ability as a "by-the-book" soldier, I no longer doubted my ability as a bushman or a fighter. I no longer needed to prove myself to them.

During the night our position was bombarded by a short mortar attack, but the shelters we dug were well designed and no one was injured. During the attack the ground shuddered, but I was not in the least bit afraid. I pondered my change in attitude. Death seemed to have become my friend, and even an injury was almost welcome if it could get me out of this hellhole alive.

I began to question my reality. Who put me in this waking nightmare? Is there a God? If so, how could he do this to me and most of all to the innocent children of Vietnam as he sits up there in heaven?

After breaking camp in the early dawn, we headed further north where signs of enemy activity became increasingly obvious. I chose not to wander off on this day because we were following a road for a change. The road narrowed into a little track through thick forest where the Vietcong finally ambushed us. One man screamed as he was shot. I was being attacked so I looked hard for the attackers. With my finger poised on the trigger of my rifle, I couldn't see any movement or sign of anyone. As usual they had hurriedly retreated after their initial burst of fire. They had simply disappeared into the dense jungle.

The rest of the platoon were firing their weapons into the surrounding foliage and the noise was deafening. I refused to waste my bullets on this useless folly, when a target cannot be seen.

My father's voice came to me and held me steady. "Do not pull the trigger until you have the target lined up in your sights, son. Save your ammunition, because you will not hit an unseen target."

Being nervous and trigger-happy in a war zone is common and very dangerous. A lot of casualties happen because of this lack of control. For this reason, men accidentally shoot each other. A previous sergeant of this platoon once threw a grenade into the dense jungle at an elusive, unseen Vietcong. The grenade hit a tree and bounced back to kill him. This avoidable tragedy was never reported back home. Instead, he died as a hero, killed by the enemy. The government created fake heroes in any way that they could to ensure that young men would continue to join their armed forces.

After this contact with the Cong, I was handed another cross to carry as my bullying corporal immediately inspected my rifle and vehemently announced to the entire platoon that I had not fired a shot. I was immediately branded a coward. This time I spoke up and explained why I did not shoot. Though my reasoning seemed to fall on deaf ears, at least the lieutenant did not formally charge me, yet threatened to if it happened again. It is a pretty serious charge, "cowardice in the face of the enemy". I wondered if they could still shoot you, as they used to be able to? Probably! How stupid, but the important thing to me is that I know I am not a coward and will quickly defend myself, wherever the threat comes from.

Some of the men approached me in private saying they knew I was speaking the truth. One man from the outback of Australia even told me, quietly, that he knew I am all right but nobody dared support me in public, so the corporal and his small pack kept at me.

Over the next week we had more fleeting contacts with the almost invisible Vietcong. We were like a large, slow lizard winding through the bush and they were like small dogs biting us before we could move.

A big country boy named Doug killed two unarmed men in black one dark night when he was on sentry duty. Apparently, they should not have been wandering around here at night as there is a strict curfew. They were probably Cong or at least poor villagers recruited by the cruel Cong, as I would later find out. Doug was a strong silent country boy who had never said anything against me.

In the morning he stood solemnly looking at the dead while everyone was congratulating him. I could see the effect this killing had on him. He looked confused and his eyes had a faraway look in them. He could see quite clearly that these men were unarmed.

I was concerned about Doug. Killing is something you can never forget and God help you if it is premeditated and unnecessary. Here the rule was kill or be killed. But I would soon see men killing the innocent and the damaging effect it had on them.

To save Doug's reputation, a grenade appeared in the pocket of one of the dead men and a claymore mine was found next to the other. The report was then announced, on the two-way radio, "Two armed enemy killed in action."

Truth is another sad casualty of war. Lying joins our forgotten honour.

Yet as the great Oscar Wilde once said: "The truth is rarely pure and never simple." Here in Nam, it was simply sacrificed.

However, these soldiers did not have much choice other than to place armaments on these dead men. If the dead are not armed in death, then there will be an inquiry and Doug would have to be charged and sentenced as this is what would be politically demanded.

Individuals, whether they are unthinking soldiers or

unfortunate civilians, caught in a war zone, can easily become innocent victims of an often-dark power present in any war forced on humanity for any reason. Thoughts of hatred, revenge and raging anger become the norm. Yet politicians would rather maintain the public belief that we, the war combatants, act with virtue, respect, honour and kindness. We are the goodies and the nominated enemy are the baddies. Wars, however, can quickly make baddies of everyone in the frontlines, as I would soon experience.

I was learning that turning the other cheek, or showing mercy to the enemy, had been strictly left out of all army manuals and training.

At last we were ordered back to camp at Nui Dat and this barbed-wired hellhole never looked so good, but not for long. The full force of this Eastern war with its accompanying gifts of intense burning hatred, a thirst for revenge, and an overwhelming desire to kill, was about to also overcome me like a huge tsunami. I would be neither neutral, thoughtful nor peaceful any longer, for quite a while.

Unexpectedly, with hardly time to change our stinking clothes, we were ordered back into the jungle to handle an emergency. A large force of Vietcong had been sighted near our base camp. We hurried towards their position. The brigadier and other armchair officers were giving orders over the radio, directing their exciting war game from the safety of their Nui Dat base camp.

Finally, we were ordered into complete silence with our radios off. Our personal weapons were cocked and loaded, with safety catches off.

This position was very dangerous because the adrenaline is pumping and a shaking finger can cause an accidental shot. We had been ordered to "shoot on sight" as the "enemy" are near and were all dressed in black. I wanted to tell the lieutenant that, from my experience, a lot of Vietnamese dressed in black, not just the Vietcong, but I knew he would laugh at me and tell me to shut up.

Because the enemy was to our front, I was placed behind

the forward scout from the other platoon. As we turned and advanced towards the enemy in an extended line, I dropped back a fraction because I had so often been threatened with death by some of my own men. It seemed stupid and careless to be in front of them if all hell broke loose, giving one of them the opportunity to "accidentally" shoot me in the back.

A mist was rising from a dirty smelling swamp as we crossed it. I could feel leeches on my body but had no time to extract them.

There were two Australian platoons made up of about twenty-eight men. The forward scout from the other platoon was now a long way from our group and he sighted the enemy before we saw them. Because the last order from the brigadier was to "shoot on sight", he emptied all the bullets from his automatic machine gun straight into the Vietnamese group dressed in black. We all hit the ground, but there was no return fire, so we advanced warily.

I soon came across seven dead bodies dressed mainly in black. They were all women and children and one of the women was heavily pregnant and had miscarried. I was shocked, appalled and completely horrified by this carnage of the innocent. Overcome by anger, I advanced towards the man responsible for this massacre.

But he, poor man, was staring at the bodies with a look of absolute horror and deep regret frozen on his face with his mouth hanging open. All my righteous anger dissipated immediately. His hands were shaking, and my heart went out to him so I said: "Mate, it was not your fault. They told us that these people were Vietcong and to shoot and to shoot them on sight. This is the fault of our armchair officers back in camp."

This man was dazed. He neither looked at me nor appeared to hear me. A short time later, safely back in camp, this poor man took not one, but two hand grenades, pulled both their pins, and held them above his head until they exploded. Now one grenade held above the head will kill a man 99.9 per cent of the time so two should really do the job. Not this time! It blew off both his hands and took large pieces out of his body, and especially out of his face.

I was to see him about nine months later in Australia where I was rushed to hospital when my body swelled up to twice its size from Agent Orange poisoning, thanks to the Americans' use of this deadly substance to destroy huge tracts of virgin forest throughout Vietnam while I guarded their low flying, spraying planes. That is another story for a later time, together with the story of this broken man.

We headed back to base camp, many of us still in shock, with no further encounters.

Chapter 4 - Saigon

A report came through from Saigon, the capital of South Vietnam, that the city was under heavy attack. This attack by the North Vietnamese Army was called the Tet Offensive. Ironically, Tet is an important sacred and celebratory Vietnamese holiday.

My platoon was lined up on parade and we were told that a volunteer was needed. I was ordered to step one pace forward.

"Thank you for volunteering to defend Saigon, private," my corporal from hell informed me. "You leave immediately."

No one else wanted to go, obviously, as no one volunteered. Too dangerous, I supposed?

I was used to packing and moving fast by now, and I knew exactly what to take. As dangerous as the Tet Offensive was, with thousands of Vietcong trying to overrun Saigon, I figured I had a better chance in Saigon with strangers than remaining here with the enemy both behind and in front of me. Strangely, I was not really angry with these men. Out of the six months' hurried training that the new recruits had been subjected to, I had missed nearly half and my earlier shortcomings could possibly have gotten these men or their mates killed. And even though this lack of training fast disappeared as I had learned quickly, I still realised that I could die at any moment. The odds were stacked against me as far as living was concerned.

Outside battalion headquarters, as I stood waiting for my plane to fly me to Saigon, a corporal came up and told me that another platoon had a casualty. "Who was it?" I inquired. He said he did not know his name, Cudo or something. My heart froze, as I suddenly knew who it was. Then I did something that no infantry soldier was ever allowed to do in a war zone or he would be "charged" then fined a large amount of money and given weeks of extra duties. I dropped my "precious" rifle in the dirt. Then I ran as fast as my legs would carry me, all the way to Cuddy's platoon about a kilometre away.

Yap was sitting in their tent and he was shaking like a leaf. Paul had already been sent back to Australia in deep shock, and Yap would probably follow. He told me that Cuddy had been on sentry duty and it had started to rain so he took his machine gun and moved under a tree for protection. Unknowingly, probably because of the heavy rain, he sat on a mine. When they got to him, he was still conscious and could not understand what had happened. They rolled him onto his stomach and his insides fell out. They rolled him back, and told him that he was going to be fine. This strong man was dead within minutes.

I think I would rather have known that I was dying so I could prepare myself to leave the body, hopefully with a smile on my face.

I fell down as a wave of utter despair overcame me. In Australia, when we were leaving for Vietnam, our fiancées had come to see us off. Cuddy's beautiful fiancée nervously pulled me aside and made me promise to bring him back alive. I told her he would always be safe with me and he would be fine. I assured her that I would bring him back.

The emotions that powered through me began with an amazingly deep sorrow and ended in a white-hot anger I would not have thought possible. I simply wanted to kill so I could revenge Cuddy's horrible death.

The ancient wisdom had no place in this hellish world: "The old law about an eye for an eye leaves everyone blind."

Or the Chinese proverb, "If you are patient in one moment of anger you will escape a hundred days of sorrow", was superseded by my blind rage.

Cuddy was the best of men. He usually had a smile and a helping hand for anyone. There is a saying in Australia that only the good die young. Those words were certainly written for this great young man.

I trudged back to my rifle and backpack where the officer was waiting. He was about to "charge" me for leaving my rifle abandoned in the dirt, until he saw my tears and my face. Luckily for him, seeing the blind anger I was experiencing, he gave me his condolences instead. I could not wait to get to the battle in Saigon and avenge Cuddy's death. The thirst for revenge had overcome me completely. A killer in waiting had finally emerged.

It has been written that a happy and contented man who has not killed, or seen or committed inhuman atrocities, has no past forever on his conscience. Those who have acted in ways less than human are unhappy because they have nothing else but a nightmarish past forever arising from their unconscious and their memories. If people were taught this then they would, for their own peace and safety, have morals, and think hard before they killed.

I did not realise this shocking truth until much later, after more sad experience in the war.

After landing at the busy Saigon airport, I was quickly driven to a luxurious hotel that housed officers, reporters and government officials. An American hotel around the corner had been bombed and attacked by the Cong the day before, and a corporal briefed me on my duties. If our hotel is attacked, he told me to kill the enemy immediately. "Good," I replied coldly and calmly, "let's hope they attack." The corporal gave me a strange look, but said nothing.

I was also instructed that if anyone approached the hotel with any sort of bundle that could be a bomb, I was required to fire two warning shots over their heads. Then, if they kept coming, I was ordered to shoot them, but only on my third shot.

As I sat behind the sandbags at the front of the hotel, I could hear the fighting in the streets around me, but unfortunately none of the attackers came near me. I desperately wanted to head towards the fighting, but that would be deserting my post.

On the second day I angrily volunteered to fight in the streets but was told that this was not possible as I was needed here.

This burning anger at Cuddy's shocking death was consuming me and pity help anyone who got in my way. I could not sleep at night due to this searing thirst for revenge keeping me awake. When I did manage to get to sleep, a nightmare of Cuddy with his insides hanging out would jerk me awake, sweating, from my dream.

On the third day while I was manning my sandbagged position at the front of the hotel, a man on a bike with a parcel strapped to the rear came towards me. I fired a warning shot over his head. He must have felt it whistle past as I aimed only an inch above him. He began yelling abuse at me and kept coming. I quickly fired the second shot, which must have almost parted his hair. This old man of about fifty kept yelling at me. I opened my mouth and said softly, "Here is my first one for you, Cuddy."

I aimed for his chest as a heart shot kills instantly. I hesitated for a second while I searched his angry advancing face. Is he brave or just plain stupid, I wondered? Yet that parcel strapped to his bike must be a bomb, so I took the slow, soft, first pressure on the hair trigger of my rifle.

Then a hand, out of nowhere, grabbed my rifle and jerked it upwards, so the killing shot flew harmlessly into the air. The tough looking corporal was behind me and ordered me inside and up on the roof, immediately.

"No," I yelled. "I am going to kill this man for my mate."

"No, you're not," he yelled back. "Get inside and up on the roof. Now!"

My training to follow orders blindly kicked in and I reluctantly ran with him to the roof where I then planned to immediately kill the man, and hopefully many others.

To my regret the old man was walking away.

"Why did you do that?" I yelled. "The fucking army has been trying to get me to kill and when I finally agree, you stop me?"

He took the time to tell me his story. This was now his second tour of Vietnam. On his first tour he found himself in exactly the same position as me, and he shot a man. The man was unarmed, and there were civilian witnesses to testify. This corporal was put in jail and his two stripes were taken from him.

He had just gotten them back after two years.

I cursed the army, but to this day I am eternally grateful for this corporal. In my mad rage I could have killed an innocent man, and anyone else. That would have been the end of my spiritual life as I would come to know it. Something was protecting and directing me, it seemed. Why did destiny want me here? For what purpose? I made a promise to myself that one day I would find out the answer to this question.

I did get to do some hand-to-hand fighting in the streets before I was flown back to Nui Dat. Some men in the street started attacking me because they were angry at foreign soldiers for invading their country. They were not Vietcong and were unarmed, so I did not kill them. This street fighting allowed me to release some of this anger.

Tet passed, the VC retreated and the fighting was over. I was flown back to Nui Dat where the company captain summoned me to his tent. He told me that he had heard that I was having a hard time in my platoon. Cuddy's captain must have spoken to him.

"Would you like to remain in camp, full time," he said, "and serve food and drinks in the officers' mess?"

This was a great way to get out of the war and I was tempted. I would be eating officers' food, fresh steaks, fish and salads. In our mess we ate mainly dehydrated food and I seldom saw fresh meat of any kind.

I declined this generous offer. I felt as though I would be deserting Cuddy and his memory, and so be a poor mate, if not a coward. I thought that it was my job to seek revenge for him and I could not do this back in camp with a cushy job. Plus, now that I was ready to kill, I foolishly thought that my platoon would accept me. I explained this to the captain. He looked at me with his mouth open and said, "I wish I had more men like you."

No, you don't, I thought. I am now a madman.

I was still crazy mad and that night in the canteen, as I sat alone drinking, I decided to burn a hole through a US dollar bill held against my arm. I had been told that this could not be done.

I simply kept puffing on my cigarette until the dollar had a hole in it. So, it can be done but only if you keep puffing on the cigarette to make it hotter. Of course, I ended up with a deep burn that soon became infected.

The acrid smell of my burning flesh attracted the ringleaders of my platoon. They began abusing me, calling me names, such as madman, and the usual "dickhead".

For a change, this made me very happy. I stood up and began cursing them for being idiots and "toy soldiers". One of the men stepped forward and took a swing at me. I ducked and hit him softly, then pointed at the largest man, a dark-skinned man who carried a heavy M60 machine gun and who looked like he was Middle Eastern. This was the man, together with the redhead, who most often threatened me with "fragging", the deliberate killing or attempted killing by a soldier of a fellow soldier.

The man I had softly hit was stunned and already beaten. I said, pointing to the machine gunner, "I do not wish to fight him, I want to fight you."

He confidently yelled, "Outside, dickhead!"

My time had finally come to teach these fools a lesson without guns, even though I was carrying my second skin, my rifle.

We pulled off our shirts. He was big and muscled. You have to be strong to carry the large M60, but I now weighed 90 kilos of pure muscle thanks to all the extra work that had been forced upon me.

He threw the first punch, a roundhouse haymaker, and I easily avoided it. "I am going to enjoy this," I said, as I looked at him with a smile on my face.

Unfortunately, it was closing time and a lieutenant suddenly appeared together with a corporal, who both came every night to see that the "boozer" was closed and there were no fights.

The officer took in the scene and yelled, "Attention, you men, there will be no fighting here! We are going on a three-week operation tomorrow. Save your fighting for the Vietcong."

On the way back to our tents, the redhead, who had previously held me down to have my genitals burnt, told me that I was going to die on this operation. They again told me the name for this when they kill their own. It is common knowledge in the ranks that this "fragging" is what they do, especially to national service second lieutenants who don't know what they are doing, which could get men killed. The death is then written down as death by "friendly fire". It has happened before, will happen again, and was talked about openly in the ranks of common soldiers.

I turned to Red, who was also physically bigger than me, and answered, "I know who will die, little man, when you come at me."

The officer, still behind us, barked another order and escorted us back to our tents.

The next day we were ordered to march to the far north to meet the enemy. We set out at a fast pace on the open road until one of the men again dropped from heat exhaustion. As usual, I was ordered to help carry a litter with his body on it. The other bearers were again given breaks, but I was not. My body again screamed out at me to stop, but my anger at Cuddy's death kept me going. I refused to drop or even ask for a spell, like everyone else. Anger gave me extra strength.

When we dug in for the night, I was made to dig the latrine trench first and then dig my bomb shelter, on my own as usual, in the rockiest area the corporal could find, even further away from the rest of the platoon.

The thought of impending action made me happy and if I was the first it came to – good! Anger, I noticed, nullifies fear.

The next day when we set off, we left the road and began cutting our way through thick jungle with the kids on the nearby roadway again screaming "Uc Dai Loi, number 10" – their usual catchcry that revealed our position to the Vietcong.

I was last man down, as usual, and it felt like the Cong would soon set up an ambush. They knew where we were and where we were headed. We had a New Zealand man, a Maori, attached to our platoon for this op. He was placed in front of me.

During one of our many stops I told him how stupid this all was. Because he agreed with me, I told him I often wandered off looking for the enemy and did he want to join me? These New Zealanders are tough soldiers who still think for themselves. He readily agreed to come hunting with me.

When the platoon began hacking its way through the thick jungle again, we slid away and quickly walked ahead of them along the open road. I suggested we get to the front of the platoon because that's where the Cong would be. After twenty minutes of walking, he got nervous and decided to head back. I did not want to go back just yet and felt to go on a little further. The platoon would take many hours to cut through this thick jungle behind us.

After another ten minutes I silently slid around a sharp bend and there stood two men in black with AK47s, the Russian automatic weapon. These are not very good weapons. They are not accurate over a distance, and tend to throw bullets way off course when fired on automatic.

I went down on one knee, raised my semi-automatic SLR and yelled, "Drop your weapons." They looked at me and then turned and took off running down the road. I aimed at their backs with plenty of time to pull the trigger. Here are two for you, Cuddy, I thought.

I found that I still could not shoot men in the back. I angrily wished that I had shot them when they were looking at me. Next time I hoped that I would give no warning and not be so hesitant. I cursed myself.

Dad's best mate had been in the Second World War and he told Dad that he had shot men in the back as they were running away. He then drank himself to death because the memory was too great for him to bear. This broke my father's heart. Dad tried, but he could not help him. His mate simply turned into an alcoholic, just to endure living, and whenever he could, he got into fights. He soon died a scarred and broken human.

I took my finger off the trigger and headed back to the platoon. I had been away longer than usual in my effort to kill but I picked up their trail and saw that after 500 metres it veered sharply off to the left. Strange, as I knew their destination was straight ahead. I followed the easy, and obvious, trail through the hacked jungle but it kept veering round to the left. I quickly caught up with them and the lieutenant, when the men saw me, called a halt and gathered all the men and me together in a clearing.

"Where have you been, dickhead?" the hero corporal asked in a loud voice. I felt like saying, "Up ahead clearing the track of enemy ambushes, fool." But I realised if I said that I would be charged with desertion, so I had to think quickly. I said that I had a bad case of dysentery and had stopped for a shit and then got lost.

Really? As if anyone, especially a tracker like me, could get lost following this platoon's obvious trail through the jungle.

The lieutenant and sergeant then both hurled abuse at me and threatened to charge me with desertion. They had already radioed base and named me as an MIA ("missing in action").

Then some of the men joined in with the officer and sergeant, yelling loudly and calling me names. When we finally set off again, three of the worst ones came over and said aggressively, "You are definitely dead this op, dickhead."

Bring it on, boys, I thought; I have had more than enough of your threats.

We finally reached our destination and I was taken even further away by my weird corporal. Far, far away from the rest of the platoon, up a small hill, and ordered to dig my bomb shelter in the rockiest place. It took me the rest of the day to dig a shelter and my hands were bleeding when I finished. But digging holes helped settle this anger that I carried. Anger at my mate's death and now anger at these men.

After the platoon had secured the area with patrols and booby traps, a brigadier landed in his little two-man, private helicopter. He stayed a very short while and then flew back to base. The following headlines in the Australian papers later read: "Leader on the frontline fighting with his troops". As if? This was such a political, bullshit, untruthful war.

Another helicopter landed with fresh supplies plus a nip of

rum for everyone, except me.

I was the last one ordered into the centre of camp to pick up my rations, or what was left of them. Everyone else had taken the most and the best. A pity, as I would probably need them very soon, when I was forced to leave this mob for good. Oh well, I thought, my Aboriginal and cattle station training would keep me alive. I had already taken note of edible weeds and flowers in Vietnam. Plus, I had four grenades instead of two and I carried extra ammunition hidden in my pack. There would be no helicopter to drop off more ammo when mine ran out.

When I approached the clearing to pick up my rations, my three favourite harassers were sitting behind their weapons in the middle of my path, while the lieutenant and the corporal stood about 30 metres off to their right. The rest of the platoon were spread around our perimeter but I could not see them.

"Here comes the useless dickhead," the large dark machine gunner said, while looking at his friends and smiling confidently. "Why don't we kill him now before he gets us killed?"

Don't look away from me if you are going to try and kill me, I thought.

"Yeah, why don't we?" the redhead chimed in.

That was it! I had had enough of their threats, of the extra duties, the digging in alone on the rockiest edges, the bullying, and their lack of decency and respect. All of it! I was also sick of waiting to die by a so-called "stray" bullet in the back, so I smiled in relief and calmly answered back.

"You know, dickheads," I said with the emphasis on dickheads, "I am now a better soldier, a better fighter, and a better man than you three combined, so shut up or it will not be me that ends up dead."

The faces of the two changed colour and they raised their SLRs and reached to cock them. At the same time the big gunner reached forward to cock the M60, the large machine gun that fires 650 rounds per minute. I wondered if they would really shoot me with the officer standing nearby?

But I felt a huge relief to be getting this over and done with. Death had, more and more of late, been looking like a better option than this life. Their threats did not scare me in the slightest. If I died here, at least I would be given the epitaph of "killed in action" and my family back home would not be shamed. I realised yet again how raging anger cancels fear.

Too bad I had not been given my rations, I thought, as now I would have to head into the jungle, alone with no food, after I shot them in self-defence if they cocked their weapons.

I really, even now, did not wish to shoot them, but I laughed and flicked my safety catch off.

I said, "Mine is already cocked, suckers." Aiming my weapon between the dark man's legs, I said, "Go ahead, cock it and I'll blow your little balls clean to hell where all of you yellow bellied pack-bullies belong."

They all froze and the colour drained from the brown face of the one behind the machine gun as my rifle was held steadily on him.

"At ease, men, and put down those weapons," the lieutenant yelled from the side.

"Yeah, right, sir," I sarcastically replied. "So you and your fucking war heroes can kill me. No thanks, sir."

My corporal standing next to the boss had also changed colour and stood frozen. I already knew that he was a man filled with fear and would not respond.

I began slowly backing away with my rifle now held firmly at my shoulder and aimed straight at the machine gunner.
He was still frozen in shock as were his two friends, while the officer and the corporal were not moving. It is amazing how my sight had focused with such clarity. They all looked like they were lit up by a shining light and I could see nothing else apart from them.

There was a bombed, deserted farmhouse to my right with the thick cement walls still standing. I had already scouted the area for escape routes, which I had become accustomed to doing since the death threats began. At the back of this house was a deep creek. I could run down the creek and make my escape, as soon as I picked up my gear. I had already packed and my gear was lying waiting for me on the bank of this ravine.

Sorry, Mum and Dad, I thought. Now these fools would call me a "deserter" and bring shame on you. One day I will try and tell you the truth, if I survive.

I slowly backed up to the bombed house, never once taking my rifle from my shoulder. The machine gunner's colour had not improved, I noticed, and I could not wipe the smile from my face.

"Wait right there, private," the boss shouted. He then picked up the platoon radio and spoke to headquarters at Nui Dat.

I began wondering what I would be charged with if I stayed to hear the outcome of this call. I did not expect anyone to take my side. If it was to be something stupid like "attempted murder", then I would take my chances in the jungle. Twenty years in an army jail would probably kill me anyway. I knew where my Montagnard Aboriginal friend lived and I was sure he would help me in exchange for my protection, until I learned the lay of the land. I wondered how long this war would last and how long I would be forced to live in the jungle?

The enemy within, sitting frozen in front of me, had not moved, and I still held my SLR aimed straight towards them.

Then a slight miracle happened. The boss approached my position wearing only a pistol, still holstered, and said that twelve men were needed from the battalions to join "Special Services" and work with the Vietnamese outside "the wire". Did I wish to volunteer and be one of these men?

I wondered if this was a way to get me back to base camp and throw me in jail but what choice did I really have if even a slight chance was being offered? I wanted to see my fiancée Jenny and my family again. I answered, "Yeah, what do you think, of course I do, or these useless fools are going to try and kill me again."

He turned away and called to the platoon's radio operator. The operator was quickly on the radio and asked for a helicopter to come in and pick me up immediately as I was volunteering for "Special Services". "Gather your gear," the boss said. He was obviously relieved to be getting rid of me.

Within half an hour I was flying free and looking down

from high on these men. I cannot explain the intensity of the relief and elation that was passing through me. Anything would be better than this pure hell I had found myself in. If they tell me I am charged with attempted murder, or some such lying story, I would still desert from BHQ. I kept my rifle cocked but with the safety catch on.

Back at camp I was picked up by a jeep with just one driver and no MPs.

The captain of our company, who had previously offered me the job in the officers' mess, told me that he had heard I was having a really bad time yet again and I should thank Tad, my sergeant mate from Australia. The captain said to my complete amazement, "Tad told me that you were the best soldier in his platoon going through the jungle training camp in Canungra. You were made honorary platoon commander while leading your platoon in and out of the bush."

Even though Canungra was recognised as the toughest jungle-training centre in the world, I was a bushman, taught by my family of whites, Chinese and Aborigines in the Australian outback. I could also read a map and follow a compass course if I needed to, so I easily led the platoon in and out of the jungle.

Back here in Vietnam in my mouldy tent I quickly packed and was then driven to army headquarters, a kilometre away in the centre of Nui Dat, where my new "dirty dozen" soldiers were to be based. That number of men, twelve, was soon reduced to ten.

My little friends who I grew to love. Some crippled by mines.

Chapter 5 - Special Services

The dirty dozen's first meeting and briefing took place soon after my hasty arrival, as I was the last man to arrive. We were told that we were to work outside the barbed wire. The barbed wire was the main wire that completely enclosed and protected our base camp at Nui Dat. We would be working with a platoon of ARVN soldiers (Army of the Republic of Vietnam). We were going to meet them on the other side of Wai Long village the very next day.

The bad news was that we had to drive through the village of Wai Long at dawn. Wai Long, like most Vietnamese towns, was controlled by the Vietcong at night. The colonel told us that he was sorry but he did not expect us all to live through this dangerous posting.

As I looked around me at the dirty dozen, I saw that there were some very tough-looking men among them. Heinrich was a German of massive proportions. I later saw him lift up the back side of a jeep. He was also a boxer of some note.

The lieutenant and the sergeant's faces were ashen and grave. It was obvious that they knew more than us about our coming dangers, and did not wish to be here. It was painfully obvious that their nerves were already shot.

Jack was the corporal, recently reduced in rank from sergeant for hitting an officer. He was an ex-underground coal miner from Ipswich and was also an Australian Golden Gloves boxing champion.

Ron, one of the other men, was a huge Australian and had recently won the heavy-weight championship of the army and navy combined. He weighed at least 230 pounds, probably more.

The rest of them were bigger than me because I now only weighed about 185 pounds. But I was fit and hard and carried absolutely no fat.

After the briefing, we were dismissed. We all gathered in the canteen where we discovered that, unlike the battalions, there was plenty of alcohol, not just two cans each, so we proceeded to get drunk.

Just on dusk, Jack and a few of the boys approached me, and Jack said, "I hear your nickname was 'dickhead' in your platoon. Are you here to get us all killed, dickhead?"

I couldn't believe I was hearing this again. This army with its bush telegraph had followed me, even to the new battleground.

I was still hurting from Cuddy's death and I had had quite enough of this baiting by people who did not even know me. An incredible white-hot anger, which lay just below the surface, boiled up in me again, this time fuelled further by alcohol.

"Why is it that you regular army men always hide behind your stripes when it comes to fighting?" I softly asked.

They all started talking at once, but Jack, the demoted sergeant, pushed them back. He loudly told me to get outside where he would take his stripes off and this fight would go no further than just us. Two men fighting with no rank. He was also angry. I wondered who, if any, of his mates he had lost in Vietnam.

I did not know at this point that he was an Australian Golden Gloves champ, so I was happy to oblige. If I had known, I may not have been so cocky.

We all spilled outside and Jack took off his shirt. I noticed that he was a very thickset, well-muscled man. I also noticed that he wore thick glasses. He whipped them off and gave them to the sergeant standing behind him.

We were about to fight in a cleared area in a rubber plantation. There were a lot of deep bomb shelters scattered about. These bomb shelters were about 8 feet deep and 10 feet long.

The sun was going down on the horizon and it was becoming a little difficult to see. Jack raised his left shoulder and took up the pose of a seasoned boxer. Hands up high to protect his face and elbows held in, protecting his stomach. I realised then that he was a trained boxer, but I did not give a damn. I would gladly have fought them all, one by one, if I had to!

We danced around and felt each other out, as good boxers do, landing harmless blows on each other's arms. After five minutes of this it had quickly become darker. I realised then that poor Jack must be almost blind because I was easily evading his punches. I gave him a good sharp left, with weight behind it, to his face, but he just shook his head and got angrier. This was one tough man as my punch would have stunned most men.

Now that I knew without a doubt that I could easily win because he could not see well in the dark, I suddenly remembered what happened to me the last time I knocked out an arrogant corporal. I really could not afford to be sent back to my platoon but I also realised that I could not pretend to let Jack beat me or my life may become hell, yet again.

So, I kept backing up while letting Jack throw his almost blind punches at me, which I easily avoided. Then, before their eyes, I suddenly disappeared.

The men were all stunned at my sudden disappearance until they realised that I had "fallen" down one of the bomb shelters.

This strategic move was my only way of avoiding knocking out another corporal, without letting these men think that I was a loser and easy to beat.

I was going to pretend that I had hurt myself falling down the hole if he wanted to keep fighting.

I had no idea a dangerous intruder was waiting in the pit. In Vietnam the deadly cobra snake is quite common, and I am deathly afraid of cobras. Standing in this black pit I suddenly felt these dreaded cobras striking at my legs through my long green army pants. My imaginary plans had transformed into a life-and-death situation.

Oh my God, I thought, there is more than one of these snakes trapped in this pit. Clawing at the walls of my prison, I screamed at the top of my lungs and then froze, quickly remembering that snakes only strike at someone who moves.

"Cobras," I screamed at the top of my lungs. "Get me out. Cobras are bitting me!"

Jack yelled, "Get a torch, quick."

The bloody blind and stupid snakes did not know the rule about not striking at a still object as they kept hitting me. It seemed like an eternity before one of the men finally rushed back with a torch and shone it down on me standing in shock. These deadly snakes were still biting my legs even though I had frozen. I was not game to look behind me at the snakes in case I fainted from shock. I thought that any movement of my head would be even deadlier as they may strike my face, though I felt I had already been bitten a number of times and was probably going to die anyway.

Suddenly the men began laughing so hard that some were falling down. Bastards, I thought, and glanced slowly behind me to see why they were laughing. A pit full of large trapped toads had been jumping against my legs.

They finally pulled me out and, after a moment's anger, followed by a sudden huge relief, I was rolling around on the ground laughing with them.

From that moment on, Jack became my new best mate. He also knew from our brief encounter that I was a fighter. He then warned the rest of the men not to get into a fight with me as they would lose. I had finally found respect in Vietnam.

This friendship would last into the future. In Australia, Jack would live in my family home while he was on leave from the army.

He always knew I could fight, and he knew I could beat him, and that was enough for him. No one ever called me "dickhead" again.

However, the thirst for revenge over Cuddy's death still engulfed me. I still looked forward to finding and killing the so-called enemy, who we would now be closer to. We had been assured that they would attack us.

The next day we all climbed aboard two jeeps, plus a gunship, and headed out through the town of Wai Long at dawn with the fog still lifting. The army orders to proceed at this hour was a textbook set-up, as vision was poor and conditions were perfect for an enemy ambush.

Of course, we were ambushed and we lost the lieutenant and the sergeant on that very first day. I noticed that I was excited about this opportunity to kill. I was not afraid in that noisy ambush, because now I had a chance to exact vengeance for my mate.

The rest of us jumped from the bullet-ridden vehicles and in the immediate silence I yelled at the men to split up and advance towards the shooters in staggered file.

That meant that half of us would advance several paces, quickly, while the rest of the men covered us with return gunfire directed at our ambushers' position. Then we would hit the ground, seeking cover in the rice paddy, and return fire while the other half advanced. This staggered file allowed us to gradually approach the shooters' positions, hopefully with none of us getting shot.

When we finally got to where the shooters had been, we found only empty bullet shells. Like nearly all contacts with the VC, they empty their weapons, hope to hit you, and then run into the jungle.

We drove straight back to base. The next day we were escorted through Wai Long by huge war tanks, guiding us to our meeting place where the Vietnamese soldiers were awaiting our arrival.

We met with the Vietnamese troops we were to train and fight with. They were heavily armed but some were even younger than me.

We set up checkpoints, which we heavily sandbagged. Then with the help of a "White Mouse" (a Vietnamese police captain) we proceeded to stop buses, bikes, cars, motorbikes and people on foot. We searched them for weapons and American dollars, which were illegal to carry. We had been told that they were used to arm the Vietcong and buy weapons. In this war zone, anything could be purchased on the black market with American dollars.

Also, whenever the enemy attacked in Bar Ria, a nearby

small city, we had been ordered to rush to defend the city against them.

We only worked in twos or threes each day. We had to man two checkpoints that were miles apart and we were already reduced to ten men. Every few days some of us were rested back at camp for a day, to relieve the stress of this nerve-racking work. It would be a miracle if we all survived, driving the same route every day at the same time. The army officers hard at work back in the safety of Nui Dat were making decisions, blindly getting foot soldiers killed through their obvious lack of frontline experience.

We manned the two checkpoints and were often shot at. Soon I was made temporary section commander because I took to this work with the added enthusiasm of revenge. I felt as if I was doing something for Cuddy, so I worked flat out and also scouted the jungle around the checkpoints, alone, to keep the enemy snipers away.

The weeks had slowly passed when one of the privates and I together with our section of ARVN troops were manning an isolated checkpoint to the north of Wai Long. Screaming bullets suddenly riddled our position. Our only protection was a tin shed barricaded with sandbags. Bullets penetrated through the tin wall and thumped into our sandbags.

I cocked my M16, a fine American rifle that fires a 223 round of ammunition. I had been practising every day since acquiring it. This rifle was accurate and deadly, reaching up to 200 metres. It fired a round at a higher speed than the Australian SLR, although the SLR fires a larger bullet even further.

The firing finally eased. I sighted my weapon over the sandbags. I could see people running through the bush. I picked out an older man of about forty-five. He was about 50 metres from me. I was amazed at how clearly I could see his features in this adrenaline-fuelled state of combat.

He was running as fast as he could, across my line of sight in front of me. I aimed my rifle a slight fraction in front of his head and gently squeezed the first pressure on the trigger.

At last, this one is for you, Cuddy, old mate, I thought. Time suddenly seemed to change pace and turn to slow motion. I noticed the terrible fear on the man's face and that he was not carrying a weapon. I wondered if he had fired at us and then hid the weapon. Yet I could not help thinking, what if he is just a poor farm worker caught up in this madness of war? Once again, even in my mad anger and deadly thirst for revenge, I automatically lifted my weapon slightly and fired a shot over the fleeing man's head, purposely missing my human target once again.

I screamed in frustration and jumped over the sandbags and ran after the shooters, hoping to find some who were still armed. After a kilometre I found no one, except some kids still running for their little lives.

Out in the thick forest, alone, I realised that I was once again facing death and was not afraid. I definitely should not have been out there by myself. I knew that this was foolhardy but really did not care.

I sat down under a tree and cried in frustration. Then I spoke to Cuddy and said in a soft voice, "I am sorry, mate, but I cannot kill someone who is running away from me. Not even for you."

I thought I heard Cuddy's beautiful voice say, "Mac, you do not need to kill anyone for me." A beautiful peace came over me while I shared a precious moment with the memory of my dead mate who seemed to be right there with me.

However, I still wanted to do something for him in this war effort, even without killing someone, and that opportunity soon presented itself. I trudged wearily back to my checkpoint and calmed everyone, assuring them that the danger had passed.

As days passed, the only man who was giving me any grief was big bad Ron, the heavy-weight boxing champion of the army and navy. And it was not just me that he was niggling at. Back in camp he would often slap the other men behind their head and knock off their hats. He thought this was a great joke. I noticed that he never touched Jack or the big German, and though he did not slap me, he began hounding me with snide remarks about my previous platoon. He was obviously sure he could beat me in a fight and seemed anxious to prove it. I did not wish to find out if he was right, so kept my mouth shut.

I was then issued with a 9mm pistol because I was still a temporary section commander. The army issue webbing holster is a useless piece of equipment if a pistol has to be drawn in a hurry as I often needed to. I rectified this problem by driving to the nearby city of Bar Ria. I paid a Vietnamese leather worker to make me a holster of my own design. It was made of black leather, and I could strap it down low on my right leg, like one of those cowboys in an American Western movie.

One day I suddenly realised, with a shock, that I looked like the men I first saw in my old platoon. I looked like a movie extra!

My personal holster was unique in design. I had the leather worker make a two-inch wide cut-out piece at the front of the holster with a clip at the top to hold it and the pistol in place. When I unclipped this strip of leather, it still sat at the front of the pistol, but I could draw the pistol, pushing this flap forward with the barrel, without having to first lift the pistol up and over the holster. Just a smooth forward motion and it was drawn, cocked, and ready to fire. Close to a second was taken off the time to draw and fire. A second is a long, long time in a gunfight.

I practised drawing and firing every day that I was "outside the wire". After a few weeks I could draw and hit a tin can 50 yards away, sometimes on the first shot, but by at least the second or third shot. This quick action saved my life on occasions.

The first time it saved me was on a bus. A man in black reached into his green coat and began to pull out a pistol. By the time his pistol was halfway out, I had mine cocked and pointed at his chest. He was taken away to the prison in Bar Ria by the Vietnamese police captain, ably escorted by three of our ARVN regular army troops.

I shall reveal the second time that it saved me later on in my story.

Chapter 6 - Terrorism in Vietnam

The command to continue driving at the same time each day, along the same route to our place of work, attracted gunfire. Fortunately, the Vietnamese soldiers were not good shots. Two events happened that once again increased my anger towards the Vietcong.

One day, while driving through the village of Wai Long in the foggy dawn light, I had to brake suddenly, as I could see a body lying on the road.

Jack and I immediately jumped out of the vehicle and into the deep drain on the side of the road, expecting an ambush. After five minutes of total silence, I cautiously approached the body lying in the middle of the road. Jack was covering me from the drain.

I was horrified to see the body of a young woman who must have been about eight months pregnant. Her stomach had been cut open in a crude cross. The dead baby, still attached to the umbilical cord, had been pulled from her open stomach and placed on the road next to her. Around her neck, which had also been cut open, hung a sign that read "Uc Dai Loi [Australian] Informer".

What blatant lies these heartless killers used to scare the locals into joining their sick army! This woman had been murdered to impress upon us, and the villagers, that these bloodthirsty, gutless human beings were in charge here and wanted no opposition from any of the villagers.

What sort of a person could inflict these sadistic and vicious depravities and then expect to continue living without horrendous nightmares?

What a waking nightmare this war was. These poor people wanted nothing to do with either side of the Vietnam armies, or any of these invading armies. They had to appear to serve both sides for their survival. They serve the ARVN (the South Vietnam Army) and us foreigners during the day and the Vietcong at night. They were forced to do this, as here was an example of what would happen to them and their families if they did not.

We could never win this stupid war by leaving these towns and cities every night and going back to the reasonable safety of our large, well-defended bases. Surely this was obvious, even to the armchair officers and their political, incompetent masters. This type of war cannot possibly be about winning. That's just another blatant lie to fool the Australian public.

I suspected that this war was about money, beliefs, political influence, the sale of armaments, and the riches of Vietnam.

Just a week later as I drove through Wai Long village at dawn, a woman was crying and screaming in the middle of the road. She waved frantically for us to stop.

We executed the same quick actions, diving into the drain and remaining still, while watching everything and everyone. We expected another ambush. The woman ran over to me in a terrible state of distress, speaking in Vietnamese. I let her take my hand and she led me to the back of her small home. Jack was behind me, waving our deadly M60 machine gun in all directions.

The backyard measured about 30 square feet and held a once well-tended garden with vegetables, plus some banana and fruit trees. This was the home of the recently deceased mayor of Wai Long. His body was lying in the garden, not whole, but cut into small pieces. He had obviously been dragged around and around his garden as he bled, because everything was covered in blood and knocked flat. A sign hung on his fence: "Uc Dai Loi Informer".

His dismembered head lay on the ground before me. I knew I had never spoken to him even though I had spent much time in this village. No other Australian soldiers came here and mixed with the poor villagers as I did. It was considered too dangerous. This poor man was just another innocent victim, murdered by warmongers to keep the community in fear ready to do their bidding. These cowardly and craven murders, done under the cover of night, always occurred when all the soldiers, Australian, American and South Vietnamese, had gone home to their reasonably safe fortresses.

The Vietnamese soldiers in my section had gradually become my friends. We would play games, plus wrestle on the checkpoints, and they had taken me to their homes and introduced me to their families. I had even bought a present for the bride and groom when I was formally invited, and attended, one of their weddings.

This poor mayor was not an informer. I wondered how we could ever win this war. With such terrible murder and fear visited on the suffering locals, how could they possibly side with us? They could not! We must eventually lose here. We should not even be here, invaders in someone else's land. Send the warmongering politicians and the generals out here with me every day and surely they would soon wake up to what is happening and put a halt to this expensive, both in money and innocent lives, madness.

What is it with people in government? Does the lust for power and greed for money overrule common sense, responsibility and decency?

Are all our religions and current beliefs failing us, I wondered in shock as I looked down at the mayor's head?

Surely it was high time for something new, both in governments, rulers and beliefs. But what is the new? I would find out one day, I decided right then and there. I had a long road ahead of me if I got out of here alive.

Meanwhile, we had been told at a meeting back at headquarters that the Vietcong were transporting money, donated by students at Monash University in Australia. We were ordered to search everyone we stopped even more thoroughly. We were also told that this money was being used to buy landmines that were killing our men.

This information was lies passed on to us by our superiors. Of course, I did not know this at the time and as a mine had killed Cuddy, I was determined to stop this transporting of American dollars.

As SS soldiers we now all hated the university students in Australia. We also hated the Cong who killed the locals, and us, if they got a chance. We suffered a lot of hatred.

One day I pulled one of the many large suitcases down from above the heads of the passengers sitting in a bus. When I opened it, I discovered that it was filled to bursting with American dollars. I could not even imagine how much money was in this case.

I called for my Vietnamese interpreter who questioned everyone on the bus and eventually found the woman who owned the suitcase.

I placed Walter, the only other Australian serving on the checkpoint with me that day, on the machine gun in our gunship and together with my Vietnamese interpreter we drove the woman who owned this suitcase, and her three children, to the old fort in nearby Wai Long.

A battalion of Vietnamese troops manned this great wooden fort and many of them sat in machine gun nests perched high on the walls. They watched us closely as we drove into their ugly hornet's nest.

The oldest child of this poorly dressed peasant woman was a girl aged twelve. She had another daughter aged eight and a small boy of six. The woman was forty years old. The children were just children but they seemed unafraid. That was about to change.

I still believed what I had been told by the military officers that the students at Monash University in Melbourne had sent this money and that this woman was a Vietcong. These American dollars would be used to buy the mines that had killed my friend Cuddy. I was still naive about the misinformation coming from higher up and I was still seeking some sort of watered-down revenge, without having to actually kill people.

The fort looked exactly like the forts you see in American cowboy movies. It was made out of large logs with lots of heavily armed soldiers patrolling back and forth on top of the walls. I again noticed that there were large machine gun nests, with huge and deadly 50 calibre machine guns manning every corner of the fort. The bullets from these weapons could turn concrete walls into dust in minutes.

I told Walter to stay manning our machine gun, and together with my interpreter I escorted the woman and children into a Vietnamese officer's private office. He was the head of the whole of Phuoc Tuy Province, and the boss at this fort.

He was sitting behind a desk with a pistol resting on its surface next to his right hand. Also on his desk was what looked like one of those ancient communication devices that send messages by Morse code as a finger is tapped on a small handle.

There were three of his heavily armed soldiers standing near the front of his desk, leaning up against the wall, weapons at the ready. The children were told to stand along the wall beside me, to the boss's right.

I handed him the suitcase full of money and he looked inside, his eyes widening at the amount.

He then screamed at the woman in Vietnamese.

What occurred next shocked me to my core and would remain with me for most of the rest of my life, often arising in nightmares.

I have read about soldiers being shocked and unconsciously changing their memories, totally believing their new stories. I never thought that could happen to me but it did. I had unconsciously changed my memory of this event and did not realise what had actually happened until many years later. Shock can do strange things to a man and his mind.

The commanding officer produced two long wires, which were thrown over a wooden rafter running across the ceiling of the small office. He attached one set of the wires to what I thought was the telegraph machine, and the other set he tied to the woman's thumbs. He then pulled the woman up by her thumbs until she was standing, screaming in pain, on her tiptoes. Then the boss proceeded to shout questions at her. When he did not like her answers, he attached a handle to his "telegraph machine" and wound it. This sent massive jolts of electricity through her thumbs, and her body jumped around in the throes of electric shock. Sweating profusely, when the winding stopped, she screamed even louder in what I could only imagine was unbearable pain, combined with the terror of extreme fear.

I noticed that her thumbs were bleeding where the wire was cutting into them.

The children were frantic and crying. A soldier stepped forward, raised his rifle and threatened to hit them with the raised butt. I took a step forward and said, "No, not the children." He gave me a stupid smile and backed off.

The sadistic "boss man" looked at me, glanced purposely at his pistol, smiled knowingly, then ignored me. Once again, he began loudly questioning the woman. Obviously not happy with her answers he again began winding the "telegraph" gadget vigorously.

I was standing against the wooden wall with my grenades, my M16 and my pistol, wishing more than anything that I could take the woman and children out of this office and out of this fort and free them. I was so distraught that for years I actually believed that I had somehow saved the children and set them free. It took a long, long time before my troubled mind would allow me to face the truth, and only after I wrote this story.

How could these people, from the same country, do this to each other? And these were women and children. I could not comprehend what type of man could do this to his own people.

I could not tolerate this cruelty any longer. I stomped outside and loudly ordered Walter to cock the machine gun and return any fire if it came. He stood, with mouth agape and eyes wide, clinging to the mounted machine gun after he cocked it.

In my mind I was going to return to the torture scene, and I was determined to shoot the boss man and his three stooges if they resisted. Then I was going to take the woman and her children to my gunship and transport them out of this evil fort.

As I walked back inside his little den, I was amazed at how quickly the mind works when the heart is pumping fast and loud. Even if I had gotten this family out of his office and into my gunship there were about a thousand Vietnamese soldiers, some armed with 50 calibre machine guns, others with mortars, watching me. They would quite easily kill Walter, myself, plus the woman and her children. Walter was not much good on a machine gun anyway, and I would have had to drive.

I thought about my fiancée and my family back in Australia and how much I wished to see them again. How I had escaped death by the skin of my teeth in my old platoon and did not want to be sent back to that platoon. How this animal in front of me was a high-ranking officer with at least a thousand men in his fort. If I acted, I would most probably die, as he or his henchmen would try to kill me and Walter without a second thought. I knew I could easily kill him and his three stooges, but I could not kill all the soldiers who would come rushing in.

I stood looking at him as he smiled up at me. I turned and rushed out of his office, again without asking for his permission. A big thing to do for a lowly "acting" section commander, according to my trained mind.

I was a real brave man – ha! No! Not at all. I should have died in that fort trying to save the woman and her children and I have hoped that I would have done just that if I knew what was to come.

I jumped quickly behind the wheel of my gunship and drove off as fast as my vehicle could go, expecting, in my paranoia, a burst of gunfire behind me at any second for deserting my post without permission.

It seemed like a very long drive through those massive gates and away from the fort. However, I had left all the money with the fort's boss, together with my interpreter, so he was obviously not too concerned that I had stomped out of his office without his permission. To my disturbed mind, the men on the walls, with their weapons directed at us, looked ominous as we passed under them.

After what seemed like an eternity, we were finally out of range and speeding back to our own base at Nui Dat where I hoped to persuade my officers to save the woman and her children.

I was sure that the Australian brass would not allow this torture of women and kids to proceed.

I still had some innocence and a little faith in authority

remaining, soon to be brutally destroyed.

I urgently called in on my two-way radio and was directed straight to the centre of operations. This was a large tent with a wooden floor, containing the big Australian boss, plus a bunch of high-ranking officers and a sergeant major. The boss was sitting behind a desk and the rest were standing to the front of him. I was marched into the centre of the room by an officer and ordered to stand stiffly to attention.

I gave my report, describing the electric thumb torture, the threats to the children, and stressing that the Vietnamese officer in command was going to kill this woman, and perhaps the children too, if we did not return immediately.

The general then opened his mouth and began to verbally abuse me. He interlaced his emotional tirade with dramatic threats of two years in a stockade in solitary confinement for questioning and walking out on an officer, my superior, who is our ally. A white-hot anger overcame me and I imagined, again, what it would be like in the jungle on my own.

Finally he asked me a question: "What the hell do you think you were doing walking out on your post, walking out on an officer, and one who just happens to be in charge of the whole of Phuoc Tuy Province? Are you trying to lose the war all on your own?"

Lose the war all on my own! These fools are doing a damn good job of that without my help even though this war is not about winning or losing.

I managed to reply in a calm voice: "Sir, they were torturing the woman with electric wires tied to her thumbs raising her to the ceiling. They were also terrorising and threatening the frightened children. I had to come here so you could stop this. It's not human."

He yelled at me, "Shut up, soldier. I'll tell you when to speak."

He had just asked me a question. I could not quite believe what was happening? Didn't he understand what they were doing to this woman and her kids? That's all that counts. Aren't we here to save women and innocent children?

This was just another wedge in my faith in humanity. My own kind were condoning this brutal and shocking torture by turning a blind eye towards it.

By this time, I was staring open-mouthed at this man, with my body gone a little slack, when an arse-licking captain near my right front yelled at the top of his voice for me to remain at strict attention. My nerves were already damaged by this time and I jumped in fright.

The big boss then yelled at me again. "You are going to spend the next two years in the stockade and then you are going to be dishonourably discharged. Do you understand, soldier?"

What was this? Was he playing some sort of a game? Were these officers only worried about politics and their careers? This was a nightmare, a crazy, relentless nightmare. It had now driven me over the edge.

The same cold hard anger I had lately often felt, boiled up inside me again. Two years in prison for trying to do what I knew was right? Fuck them all. I had finally had enough of this disgusting, uncaring, ignorant madness that was now openly supporting terrorism for political reasons.

It was amazing how much passes through the mind in this precarious, highly adrenaline-fuelled state. I remembered my childhood, my parents, my friends, my horse, and my dog, and lastly my fiancée. I bade them all farewell in what seemed like a second.

These arrogant idiots with their 9mm pistols strapped to their sides, and clipped safely in their useless webbing holsters, were in for a rude awakening. I was standing there with an M16 sub-machine gun, a pistol and two grenades, and my rifle was still cocked after what I had just come through. It was a habit I kept, just to stay alive.

And my M60 machine gun was also cocked and ready, waiting for me just outside this tent. I knew these office-bound fools had not a hope in hell of stopping me when I chose to leave.

A funny smile came over my face and gave me a feeling of deadly power, relief and calmness. I was suddenly transported beyond this sick world and their heartless orders. The shackles of obedience fell off. Their every slight movement took on a slow-motion appearance and their abusive voices faded away. I decided that I'd rather live alone in the jungle and take my chances, as it appeared that these officers were no better than the Vietnamese sadists I had just left behind.

I relaxed my stance. I had been at attention with my rifle against my shoulder, pointing upwards. I slowly let the barrel wander down from the ceiling.

Nobody yelled at me this time. Tension immediately filled the room and they probably were wondering if I had gone mad, which maybe I had. I could not control this smile on my face. Maybe I was crazy at that awful moment!

Make your move, boys, I thought. There was no possible turning back from here. Scream at me again, fat man behind your desk, I thought. Let's see how brave you and these officers really are. Which of you thinks that he can actually arrest me?

I did not wish to kill anyone, and I would not if they did not pull their little pistols and try to kill me. I would simply walk to my gunship and leave this sick army forever.

In retrospect I suppose that their hands were tied because we could not afford to make an enemy of the South Vietnam Army. In my shocked state, back then, I did not have the patience to consider this.

Anyway, I now had Vietnamese friends in Wai Long and Bar Ria, so it would be easy for me to get supplies and survive, hidden under the canopy of the thick jungles. My Vietnamese friends and my mate Jack would keep me in supplies and ammunition when I needed them.

So come on, I thought, try and arrest me, try and shame me and my family, and then drive me insane during two years of solitary confinement in your shocking prisons.

An experienced infantry major who actually fought in this war (I found out later), standing to my left, could obviously see what was about to occur and wisely broke the deadlock, addressing "fatso" behind the desk.

"I think he is correct, sir, and he is trying to do the right thing. We cannot leave the woman and her young children back there."

And then from another totally unexpected source I also received support. The tough sergeant major spoke. "We could take them into Bar Ria, sir. There is a just officer there. I've witnessed the torture in Wai Long, sir, and would not want to see it again."

These were two combat-hardened men, not "desk jockeys". I supposed it was pretty obvious to them what I was ready to do to defend myself. No other officer had dared to yell at me for not remaining at strict attention.

The general looked down at his desk while drumming his pencil on the map in front of him, and after what seemed like a long time, he looked at the major and addressed him by his first name. He said, "All right, Tom, if you think that is best. I suppose this man was brave enough to do what he stupidly thought was right. Dismiss him, Sergeant Major."

It was sergeant majors who really ran this military.

The calmness and feeling of power dissolved immediately and was replaced by a sick feeling of shock at what I had just been prepared to do.

A voice inside me said that the poor woman and her children would be treated better in Bar Ria, but deep down I really did not believe it. A hundred and fifty years of war had made many of the Vietnamese soldiers quite heartless. The wages of war.

The sergeant major ordered me to about turn, in a softer voice, and directed me outside.

I would never forget the woman and children and what I had not done. And I may always distrust blind authority. But I wanted to go home one day. I was truly sorry.

Outside, the SM spoke to me with a civil tongue, man to man. He said, "I'm attached to that fort in Wai Long as an adviser. I know what you saw and I want to shake your hand." I give him a limp handshake and he grinned. "You only did one thing wrong, you know. You should have let her and the children go and brought the money to me. We could have split it. Next time eh? I'll drive into Wai Long now and have them removed to Bar Ria. The man in charge there is much fairer."

I just nodded, as I was too sick at heart to answer and I was still in shock.

Walter and I headed back to the checkpoint closest to the fort. As acting section commander, I decided that we should test fire all our weapons and then clean them. I needed to shoot, to dispel some of this awful energy that still lingered inside me.

Our personal weapons worked well. I wasted a lot of ammunition, releasing my energy by shooting at tin cans. Then Walter tried to fire the M60 machine gun that we had mounted on our gunship. It fired one round then stopped. I asked him to cock it again, fire, and the same thing happened. Now this weapon has belts containing about one hundred rounds each, all attached to and extending out of the breech. I was surprised because this weapon is capable of firing 650 rounds per minute. I began to take it apart and inspect it.

Walter had just cleaned and checked it the day before, so it should have been working perfectly. When I pulled out the gas plug it became obvious that it had been put in back-to-front and this was why the gun was not working on automatic fire. I realised then that we had had no firepower in the foreign fort and little protection if we were ambushed while driving.

A hot anger came over me at this incompetent idiocy, and I advanced with fists closed, ready to beat some sense into him. His lack of care could have gotten any of us killed. But as he cringed away from me, I realised that I had suddenly become like the men in my old platoon. I could easily have killed Walter if I had hit him hard enough. This shocked me into sad awareness at what I had become.

Poor Walter did not know any better. I breathed deeply, calmed down, then gave him a long lesson on how to reassemble an M60. Walter should not be an infantry soldier, just like I should not have been posted to that platoon without proper training. I wondered how many men had died in battle because the army did not adequately train them in advance.

Two weeks later I met the friendly sergeant major while on patrol in Wai Long village and he regretfully told me that the woman was tortured to death in Bar Ria, but the children had been released. He sadly apologised to me and a tear came to his eye. I doubled over and vomited.

Bar Ria was attacked by a large enemy force a few days after that, and I sped to help confront the attackers. More despair confronted me in this horrific useless war. I saw the decent man, who had spoken up for me back in Nui Dat, get killed. His head was sticking out of the turret of a rolling tank, slowly moving through the dangerous terrain. His head was blown clean off by an enemy rocket. I wondered again why the good guys seem to die here?

My death wish became stronger in my conscious mind, but I observed how my mind and body vacillated between fearless rage and the sad longing for survival. My body won the battle for now.

Anger, revenge, madness and deep hatred are the rulers here. Wars exist only under a mantle of these things and loving, caring, soldier saviours on our side are a thing of the imagination fed to people back home.

It suddenly dawned on me again, during these depressing recollections, that when I think that I have been talking to God, I have simply been talking to myself.

After the Vietnamese woman had been brutally murdered, I made it a point to never capture another prisoner, even though I would often discover illicit goods and money in the months that followed. I enjoyed seeing the look of surprise on the face of the people when I let them go. Jack supported my choice after I told him about the brutal murder of the woman I had captured and the shock and despair of her innocent children forced to look on in the military fort. We refused to take the responsibility of getting any more of these most likely innocent couriers, who had been forced to carry money, tortured and killed by their own government.

I never declared the money to the Vietnamese soldiers and I never took it as this would also get the female couriers killed if they arrived at their destination without the cash.

I never again found entire suitcases filled with money and I simply did not believe that the smaller amounts of American money carried by these frightened people was only used to buy mines. This understanding would be confirmed at a later date.

Boredom can quickly replace horror in a meaningless military existence, and Jack and I were bored with all this bullshit. So we decided to find a more pleasurable adventure. We folded the G on our armbands so they only showed the letters MP (Military Police) instead of Garrison Military Police. We then drove to a large brothel in Bar Ria that was a known hangout for US soldiers. Two of their vehicles were parked outside on the road so I got Jack to creep up to an open window and be ready to back me up with the M60 if need dictated, to scare them into submission. When he was in place, I confidently walked through the front door to confront the American soldiers.

Inside, there was a huge bar with seven working women, plus the old mama-san who was in charge of the brothel. There were four GIs, all officers, with a battle-hardened captain in charge. I casually walked up to him and in a friendly manner told him that this particular brothel was soon to be attacked by the Vietcong and we had been sent to clear it of all Australian and American personnel.

The captain laughed and said that he would take care of any Cong and that he and his buddies were not under my jurisdiction. Then he turned his back on me.

I liked him. He was tough, but Jack and I had already made up our minds that this place was ours for the rest of the day and anyway he was an officer and at that moment I had no time for them. I needed some rum and Jack wanted a woman.

"I'm sorry, sir, but I'm under direct orders from the Australian and American command to clear this place of all American and Australian personnel. You will have to leave now." I loudly emphasised the word "now", so that he would turn around.

The captain stood up, holding his Armalite rifle, and turned and faced me. "And who is going to make us leave, buddy?" he asked, stressing the word loudly.

He was a real tough guy, standing up to what he thought was an MP. I was not very happy with most officers right now so I smiled and got tough with him in return.

"Listen, buddy, you either walk out of here and fuck off or I will arrest you and escort you back to our stockade on the base. And, believe me, it's fucking hot in there, buddy."

With this announcement his friends also got to their feet with their M16 Armalites held ready. The captain met my threat with a sneer. "And how do you propose to do that, buddy?" he sarcastically replied.

Enough fun, I thought, and pointed up at Jack in the window who obligingly cocked the M60 he held and smiled down at them like a big kid.

"With the help of my friends, arsehole. The sergeant you're looking at with the M60 and the others you cannot see. You want to do it the hard way, by attacking me, or the easy way, by walking out of here? Either way I am not going to be held responsible for the death of you and your friends. I am under direct orders – sir."

They looked at Jack. It's pretty awesome looking down the barrel of an M60 machine gun.

He looked around for the rest of my men but obviously could not see them because they did not exist. He knew he was out-manoeuvred.

To help him along, I said, "They can see you, buddy."

I again put the emphasis on the word "buddy", as he had done to me. Without another word he walked out and his men followed. I made sure that they drove away.

The brothel was ours!

I turned to the mama-san who was watching me with wide eyes and open mouth and informed her that this brothel was now being closed, forever. Jack came in and played the role of the "good cop" and talked me out of closing the poor woman's business. After a while I reluctantly agreed and Jack was the immediate hero. Drinks and women were on the house, the mama-san happily announced, and as I sat down and started to get drunk, Jack proceeded to do something quite extraordinary.

There was a fancy wooden stairway leading to rooms upstairs, and as everyone here now loved Jack because he was the saviour of their profitable business, he began to love them in return, one at a time.

He would take a willing, laughing girl upstairs and ten minutes later they would both walk back down the stairs.

The rest of us were downstairs having a party, complete with free drinks. I had told the mama-san that my men outside would stay and protect them from the coming Vietcong attacks, should that eventuate.

Meanwhile, after Jack's third girl everyone started to clap as she came out smiling and nodding. By the sixth girl the mama-san and I were both drunk and also suitably amazed at Jack's staying power.

She would question the current girl, and I would question Jack, and sure enough he had had an orgasm with every one of them. There were seven working girls here and like a prize fighter he grabbed the hand of the last one and led her upstairs. The rest of us were laughing fit to cry and I told the mama-san that she was next. The grey-haired old woman fell off her chair in hysterics, or perhaps in hope.

This time we had to wait half an hour until the last woman finally appeared at the top of the staircase. We all went totally silent as she looked down on us, savouring the expectation of the moment.

I could not believe that a man can have an orgasm seven times in an afternoon, and so I thought he had surely failed this time.

Slowly, the girl raised her hand above her head and screamed in delight as she punched her fist several times into the air and confirmed his amazing feat. As Jack came to the top of the stairs, tucking in his shirt, the place erupted and we welcomed him like a hero. The girls hugged and kissed him and all of them wanted to touch him.

I suppose some people would think this a bad thing to do but it was simply the best day I had had in nearly a year in this hellhole and it was better than constant war, which is all we had had so far.

However, I still could not get the killing and the torture of the poor Vietnamese fort woman out of my mind, no matter how much I drank.

And sex was far from my mind, or body, even though the willing girls kept grabbing me by the hand, trying to drag me upstairs into the bedrooms. I wanted to remain true to my faithful fiancée back in Australia.

That night back at base camp I was feeling particularly troubled so I approached our Catholic padre and asked him if it was okay to kill people.

He looked me in the eye and said, "Yes, it is. These people are Communists and if you do not kill them, they will come to Australia, kill your father, and rape your mother and sister."

I trudged back to my tent and thought about this.

I was still confused and in despair the next day after my evening talk with the padre, and I went into the jungle to try to commune with this elusive God and get some answers. I asked him why his representative here would speak such obvious lies. I told God that it was impossible that all Communists could be rapists and killers, so how could I kill a man just because he called himself a Communist? This name, surely, was only forced upon them by the current rulers and believers of Communism. And these Communist peasants probably didn't even know what that word meant or if it's good or bad to be a Communist.

I wondered what a future world would be like with no belief in Communism or capitalism or Catholicism. Would all wars stop?

Obviously from our study of history, all beliefs pass away and change, that's for sure. Yet these days, the human mind seems easily persuaded by priests, imams and leaders.

In the silence that followed I screamed out, "Answer me, God."

No answer came so I began abusing and threatening God to force him to react.

"When I die, I am going to make you pay for all this," I screamed up at him, "just you wait and see. Come on, you idiot, kill me now."

The silence continued and I gave up, thinking once and for

all that the God I had been taught about did not, and could not, exist. It would simply be impossible for an all-powerful God not to act in this hellhole where innocent victims, including young children, mothers and farmers, were being tortured, burnt and slaughtered every day. And not just a few innocent deaths. Two million of them to be exact, by the time we left this sad land.

Suddenly Jack came bursting through the jungle thinking that I was in trouble. I told him again what had happened to the woman in the Wai Long Vietnamese fort and it was this that was still troubling me.

He, being in a better state than me, said, "You did not know what they were going to do to her and the children so come on, mate, let's get out of here fast. Your rage revealed your position to all the Cong within two miles."

The terrible shocks were not going to let up, as a week later I was in position on this same bloody checkpoint when a huge explosion occurred about half a kilometre away on the road to Bar Ria. We had been trained to take immediate action with any enemy attack. I thought I may be able to help someone so I jumped in the gunship and sped towards the rising cloud of dust and smoke.

A short time before, two young brothers from Wai Long had passed through our checkpoint. I now knew most of the kids in this village. I would give them lollies and sometimes a little money, and occasionally play with them if I had time.

I had no idea that these two boys, aged about six and ten, had found a type of grenade that we, the Australians, use in our hand-held grenade launchers. These grenades are made in America but one would think that they were made in Australia because the shell is a beautifully coloured green and the head is a pretty, bright gold. The Australian Olympic colours of green and gold.

Of course, it would be obvious to everyone but the greedy people who make and sell them that small boys anywhere in the world could not resist something so pretty.

These two little innocent brothers were attracted to the colours and tried to see what was in the deadly grenade by banging it with a rock. This very powerful and deadly grenade had quite simply blown them both into little pieces. When I arrived at the explosion, I fell to my knees and vomited.

The father of the boys came running from the field where he was tending his buffalo. When he saw what had become of his sons, he did not vomit or cry. He began cursing over and over, non-stop. It sounded like "Du Mau, Du Mau, Du Mau." This is a Vietnamese swear word which means something like motherfucker.

I had two small sugar bags made of hessian in the gunship, so I handed him one and we proceeded to put the pieces of his two sons in these bags.

A recurring vision that I also carry with me came from when I picked a small thumb off a nearby bush. It was neatly speared on a thorn. The remaining pieces of the small bodies hardly filled the two small sacks.

The father then trudged off down the road with the two bags over his shoulder, still repeating the words, non-stop, "Du Mau. Du Mau."

My heart went out to that poor man.

I fell down onto my knees and cried for Vietnam and its children.

That night back at camp I called some of the men together and told them about the blown to little pieces boys and to double-check other little kids to see if they were carrying pretty little bombs.

I then stressed to the men once again that I was no longer going to capture any Vietnamese and nor was I going to fight again in this useless, sick and tragic war.

Jack and the boys agreed with me and we placed our hands together and said, "Fuck the war."

The next day I told my section of Vietnamese soldiers exactly the same thing. I had become closer to them over the year. They were mostly young, like me, and we still wrestled in fun and I also taught them how to box when I had time. They understood my position and told me they had never wanted to fight, for either side, in the first place.

A week later as we drove to our checkpoints, we were astounded to discover the road littered with A4 pieces of paper with a very strange message. These papers stated that the Vietcong did not want to fight Australians, only "Imperialist Americans".

From that time onward the Australian soldiers, who were forced to drive during the dark of the predawn along the same route day after day, were not attacked.

Guarding hotels - TET Offensive, Saigon, 1967.

Chapter 7 - The Minefield

One of my favourite Vietnamese soldiers, together with a young man from Bar Ria whom Jack and I had taken under our wing because he was such a lot of fun, came to me one day on the checkpoint and told me about the minefield that the Australians had laid nearby the Long Hais. They told me that kids had strayed into the unguarded field and some were killed while others lost their limbs or were badly wounded. They said this field of death had existed for months and many dogs and buffalo had also been lost.

They also told me that the Vietcong send young girls into the minefield at night and that these thin-wristed girls would steal the mines. I asked them how this was possible as we not only laid the mines but put anti-lifting devices under them.

The boys told me that the Vietcong had been stealing mines for fifty years, from the French and the Americans. They were even taking the mines with anti-lifting devices attached.

Not quite believing this to be possible, I asked them to take me to this minefield near Dat Do. It was quite a long drive so I took the gunship. Halfway there I felt that I could die at any moment if we ran into the VC. I was a single white soldier travelling on this isolated dangerous road without military backup. I could not allow myself to turn back. I needed to find out if this story was true. I had lost my best mate Cuddy to what I had been told was a stolen Vietnamese landmine.

The minefield was originally protected by hundreds of tons of razor wire. I could see where the wire had big gaps, probably cut open by the Cong at night. All through the minefield there were freshly dug holes where the mines had been lifted each night by the young girls forced to perform this dangerous work for the Vietcong. Sometimes, the boys told me, the mines go off and the girls die, but they have no choice in the matter as the

Vietcong, or at least their secret police division from the north, would kill them and their family if they refused their orders.

I was angry and concerned about what was going on here and later made inquiries back at camp, through a mate of mine in the Intelligence section. He told me that most of the mines were M16 jumping-jack mines that were made in America.

We had laid 20,292 of these mines and only 12,000 odd had anti-lifting devices because that was all that were available. He told me that we did not have anywhere near enough troops to guard the minefield and that the officers in camp were aware that thousands of our mines had been stolen and were being used to kill our own troops.

I realised that this must have embarrassed both Brigadier Graham, who was the commanding officer, and the Australian government leaders overseeing these war games.

An obvious conspiracy was perpetuated when they had tried to pass the blame onto the students at Monash University in Australia. They told us that these mines had been purchased by the enemy using funds sent from these Australian uni students. I then realised that Cuddy had been killed by an American made M16 mine and obviously Brigadier Graham was directly responsible. Another deskbound soldier hunting glory and recognition and perhaps a coveted VC medal. How absurdly ironic as we fight against the VC, the Vietcong.

I was stunned. Who was the real enemy in this war, killing the Australian boys forced to go to this war? Or rather, what is the real enemy? Reputation, status, complicity, ignorance? Conspiracy has many faces, and I did not respect the face overseeing this war.

My friend told me that the brigadier had been warned in advance that his intention to create a minefield to win the war was madness and would not work. Experienced officers, plus engineers and the soldiers who were forced to lay the mines had warned him. Some of these Australian soldiers had been killed while laying the dangerous mines, but this killing field remained, despite the future deaths of innocent children and Australian soldiers.

I checked the numbers, later in Australia, and found that between May 1967 and February 1971, fifty-five Australian soldiers had been killed by our own stolen mines while 250 were badly wounded, some with arms and legs dismembered. The number of men killed by this less than strategic Australian minefield represented 10 per cent of all Australian casualties in the Vietnam War.

Emotion consumed me once again as I heard the true story about the minefield. I now knew who was responsible for the mine that killed my mate and many others. I was determined to do something about this.

From this madness, Brigadier Graham was eventually promoted to second in charge of all the combined Australian armed forces back in Australia.

The lieutenant who was eventually promoted to major general and then promoted to the head of all Australian armed forces did at least lead a platoon of infantry soldiers in Vietnam. However, the story in the ranks about him was that when an enemy machine gun nest pinned down his platoon, he ordered an immediate attack, jumped up out of his concealed position and charged. Halfway there he realised that not only was he running alone, but he had also left his webbing belt with all his ammunition and grenades back in his foxhole, so he had to turn and run back as fast as his legs would carry him. Luckily for him, the Cong had done what they usually did and opened fire, then retreated into one of their numerous booby-trapped tunnels.

My one intention now, during this war, was to expose this terrible folly to Brigadier Bloody Graham. There had to be a way.

I took photos of the empty holes where the mines should have been. I was also lucky enough to take a shot of a dead buffalo that had wandered into the minefield.

Then I asked my Vietnamese soldier friend to gather together as many of the wounded kids as he could. There were some with only one leg, one with none, others with missing arms. Kids on crutches, kids bandaged, and kids limping from their minefield wounds. These young children still gathered food for their families, if they could, and tended their buffalo herds, even in their wounded condition. They could not read signs designating the minefield location. They had been following the same tracks for centuries, mines or no mines.

I placed one of these children on my shoulder and gathered the rest around me for photos. I then took these photos into the fort where my sergeant major friend was now stationed. He smiled when he saw me and asked if I had another suitcase full of money for us to share.

"Unfortunately, no," I answered, "but I have something else and I would like you to give it directly to Brigadier Graham. I would love to give the account and hand him the photos personally, but there is no way that I will ever be allowed to see him."

I told him the story of the minefield and showed him the photos. A tear came to this decent man's eye and I realised once again that all officers were not the same. Some actually cared about the consequences of our invasion. He told me that he was aware of all this and he had confronted Graham about it himself, before the mines were laid, and he was told to mind his own business and get on with the job.

"Please," I said, "just show him these photos. The man must have some humanity about him and even a little common sense."

"It is not possible to show these to Graham," he said, "as he has already returned to Australia. But I will show them to the new commander, Brigadier Wilton."

"You are kidding! Graham has only served here for six months!"

"That is the amount of time brigadiers serve here" he said with a sigh.

I couldn't believe it. Foot soldiers serve a year, high-ranking officers only six months. How the hell could they understand anything about this war in six months, sitting behind a desk, in the safety of the headquarters, making decisions that would kill and wound so many?

Eventually, this photo of me with the children was seen. It was published later in an Australian paper, but not with the story of the Australian laid mines. No one was publicly made accountable for this travesty.

The minefield was left in place for another long year. Then, what was left of it was removed by our suffering engineers, some of whom were killed in this duty. Tragically, this strategic killing field would continue into the future because thousands of these mines were now in the hands of the Vietcong and would be used against us with devastating effect.

Years later something happened that I felt showed the utter madness and heartlessness of the earth's warlike cultures. The Queen of Hearts, Princess Diana, had been killed in a car accident. I had just emerged from the great red desert in the heart of Australia where I often spent time alone. Her death touched me deeply. She had been working tirelessly to get landmines banned worldwide in order to save the children killed and wounded by these mines. After her death her effort gained momentum and the world stood united and ready to ban these terrible weapons.

Behind the scenes in the White House, the patriarch won the battle against the Queen of Hearts. President Bill Clinton reversed the coming world decision with this statement: "We cannot ban these mines because they protect our boys in the field." The mines were never outlawed.

Chapter 8 - Napalm

Back in Vietnam the battles continued, and more shock and horror followed.

A combined land assault was ordered. The entire population of Wai Long was rounded up and placed inside a barbed-wire enclosure outside the town. Everyone who didn't leave the town was killed. Again I was shocked, not only at the mass killing, but at the method that was used.

We bombarded the forested area – where the so-called enemy was holed up – with napalm. I remembered reading somewhere that according to the Geneva Convention, the use of napalm was forbidden in war. It was forbidden because once napalm hits any part of your body it slowly and painfully burns right through it. Napalm was considered to be a brutal way to die and a terrible torture of not only humans, but also unlucky domestic animals and wildlife.

I was ordered to go to the isolated northern part on the outskirts of Wai Long with one Vietnamese trooper and my gunship to block the attempted escape of any Vietcong during the extensive bombardment.

At the end of the day, I was summoned to the headquarters that had been set up at our checkpoint. I was informed that six bodies had been collected and were lined up in the main street of the town. However, there was still one body in the long grass and I was sent to find it and bring it back. Only my Vietnamese interpreter, who knew nothing about combat or how to use a weapon, was allowed to go with me.

I asked why I was being sent in alone with no backup. An officer told me it was because I knew the town and its people.

I finally found the body of the man in the long grass, transported it to the centre of town and laid it down next to the other six bodies.

As I stood there, I was saddened to watch the Vietnamese troops order all of the townspeople, many of whom I knew, to walk past in single file and look at the dead bodies of their friends and family. What a lovely war!

The bodies of the so-called enemy were burning slowly from the effects of the napalm and the smell of human flesh was nauseating. Two of the bodies displayed a freakish phenomenon. One had a distinct round, burning hole over his chest exposing his still twitching human heart. The other had a hole in the forehead exposing most of the brain and this brain was still twitching and moving as well. The dead heart and dysfunctional brain of wars, paraded before me, and I was once again helpless to do anything.

As the population filed past these very young, very dead, very mutilated men, an old grandmother could take it no longer and screamed out, running forward and hugging the young man with the hole over his heart. Her grandson, my interpreter later informed me.

Some Vietnamese soldiers rushed forward, as this was what they were waiting for, and roughly seized the grandmother, dragging her away screaming. I suspected that she was to be tortured and probably killed, like the many others before her.

My initial response was to rescue this tragic, loving woman. Anger boiled up at this disgraceful, brutal display of war. I took two paces forward, yet how could I rescue her, outnumbered as I was by hundreds of foreign soldiers, and outranked. How could I? My legs soon turned to jelly and I felt nauseated by this display of sadism and horror, combined with the stinking wounds of the dead young men lying before me. The terrible sickness of this war and, I suspect, any war of the past or future.

I stumbled like a drunk man to my gunship, vomited, and then sped away.

And still this war would not give me any peace.
That evening, when I finally arrived back at our base, I had to walk past the military police compound at Nui Dat to get to my tent. Sitting on the side of a cot was a Vietnamese man who had

been captured that day. He was dressed in peasant's rags.

Standing in front of him was an American CIA man who had been with us in camp for a week. I had been amazed at this man's seemingly dead eyes, wondering what he had done to look like that. He always carried an automatic shotgun. Too bad if he was caught out in the open, I thought, as a shotgun is a short-range weapon. Standing next to him were a Vietnamese policeman (a White Mouse) and an Australian officer.

They did not notice me in the dark as I watched them questioning this man. Their form of questioning was ruthless. The Vietnamese policeman would shout a question at the man. Obviously, the answers were not suitable, so the policeman would hit the man hard on his bare shins with a bamboo cane. The cane had razor blades embedded in it and the man was forced to spring to attention at each strike. Each time he would scream. This must have been going on for a long time because his shins, and the ground around him, were covered in blood.

I wanted to stop this cruel torture but I meekly walked away with my head hung down. It could be worse, I thought. The colonel in Wai Long could have the poor man instead.

There was always an excuse, in my mind, for me not to act, as I clung to my own freedom and my troubled life. I desperately hoped that I would one day see my family back in Australia.

I went to bed but got no sleep that night or the next. I felt like a coward because I wanted to live. The army had trained me well – above all, to be subservient to officers and obey orders, without thinking.

Chapter 9 - The Medal

A few weeks later, while I was working to the north of Wai Long with my mate Jack, I was ordered to a meeting with Dat, the Vietnamese officer who was the head of this province.

Together with our Vietnamese interpreter we were led to a large, well-built compound in Wai Long, situated to the south of the massive fort. Inside the iron gates I was quite surprised to see a large building, raised about 6 feet off the ground, with no front walls. On the veranda were about forty Vietnamese sitting on high-backed chairs around a huge table. A feast of magnificent proportions adorned the table. Mekong whiskey, beer and wine were plentiful, together with an abundance of exotic Vietnamese foods.

Many of the people sitting at the table were already a little drunk, including the captain of the White Mice. He was the one who often came to our checkpoints. There were about fifteen beautiful young women sitting around the table with them. They must be the wives of these high-ranking officers, I innocently thought.

Much to my surprise, Jack and I were invited to sit down and partake of this magnificent food and alcohol, which we devoured with much appreciation.

We had been eating mainly tinned or dehydrated food for almost a year. Only the officers' mess back in camp served fresh steak, salads and really good food. This was truly a wonderful feast and we washed it all down with Mekong whiskey.

At the end of the meal the colonel stood up and made a speech in Vietnamese, which my interpreter translated. He thanked me for the brave work I had done in capturing the enemy. What he really meant, I thought, was thank you for making me a rich man with that case of money. So he had not been insulted by my stomping out of his torture chamber without asking for his permission or saluting him.

He then asked me to come forward and he pinned a Vietnamese medal on my shirt. As he was also a little drunk, he first stuck the pin in my chest and I jumped back. He laughed and motioned me forward again. This time I held my shirt away from my chest so he could pin it without making me bleed again. It was the highest Vietnamese medal he could give me. The Vietnam VC? How ironic. I had unwillingly paid for this medal in more ways than one.

Then he pointed at all the lovely young women sitting around clapping, and said I could have any one of them. I was surprised, to say the least. I told my interpreter to tell him that I did not want any of them. He whispered to me that it would be a grave insult to refuse the offer of a beautiful Vietnamese woman. These women were the best and most beautiful concubines that money could buy.

I felt like telling him that I would much rather shoot Dat, and was almost drunk enough to do this.

Jack interrupted and said, "Don't be a fucking idiot, mate."

So I took a long hard look at all the beautiful women. I smiled as I thought that this looked like an ancient scene out of a movie.

I picked an exceptionally beautiful girl about the same age as me.

As soon as I picked her the captain of the White Mice police force, dressed in his immaculate white police uniform, and drunk as a fool, jumped up and began shouting at me.

Now all our military men did not like this man with his big white teeth and his pearl-handled 38 calibre pistol which was always strapped low to his side. He would intentionally hold his hand on it to frighten people as he screamed at them.

Whenever he worked with us, we observed his cruelty and his sadistic behaviour towards his own people and he was especially arrogant towards women. Jack and I had previously confronted him about this cruelty to his own people and he had merely laughed at us.

"What is he screaming about?" I asked my interpreter.

"He is saying that you cannot have that girl," he answered. Suits me, I thought.

"Why not?" Jack angrily inquired.

"Because he thinks that she is his girl and he does not want a dirty foreigner like you touching her," our interpreter answered.

If it had been anyone else other than this ugly old sadistic copper, I would have been happy to back off, but I felt it was wrong that he should even be touching this young woman. I had finally backed off once too often.

The little prick kept screaming at me and threatening me with a closed fist. I told the interpreter to ask the now red-faced boss man, who was watching all this with interest, if his offer still stood. Smiling in anticipation, he nodded yes.

The White Mouse then ran up to me and made a challenge, which I did not understand. I soon found out that he was challenging me to a duel with pistols.

The all-powerful Dat thought this was wonderful and clapped his hands, then made another loud speech in Vietnamese. My interpreter told me that he said we were to stand ten paces apart and then the interpreter was to count to three. Only on the count of three were we allowed to draw and fire. If either of us drew our pistol before the count of three, Dat would shoot that person himself.

This was a primitive place I found myself in and this officer obviously held the power of life and death over anyone he chose, even Australians.

Jack yelled, "Come on, mate, you can take this prick."

I realised, on hearing Jack's voice, that I was just drunk enough for this to suit me. I had my 9mm pistol, which weighs about half as much as his 38. I kept my pistol tied down low on my right leg in a leather holster which had the strip of leather on the front. I quickly unclipped the leather strap to allow for an easy draw. This allowed me to grab the butt of my pistol and push it straight forward, rather than first have to lift it straight up from the holster and then push it forward. I could draw and fire my pistol in under a second.

Standing face to face with him I realised I was about to die if I did not kill this man. A slight shudder ran through my body and my heart began to beat at a furious rate. This man was a sadist, and I would probably be helping a lot of people by killing him. Yet, I also realised, he had been carrying his pistol for many years and I had seen him use it.

These thoughts passed through my mind in seconds but I dismissed them one by one, finally allowing a deadly calmness to come over me as I faced the now inevitable outcome.

At least if I did die here my parents would think of me as a hero as it would be reported that I died in action. Some action!

I could not help but smile as I looked at him, standing like a little cowboy, with a snarl on his face and his hand hovering over his heavy pistol.

In the years to come I would always find myself unconsciously smiling before a fight. I could not help it and did not know why it spontaneously occurred. Perhaps this arrogance of booze and state of "no mind" had become addictive.

After what seemed like an eternity the colonel yelled out, and everyone became quiet. The interpreter began his slow count as we, the combatants, faced each other from ten paces away.

"One" … a long pause … "two", and then what seemed like a pause that lasted forever – "three." And we both moved together.

I swung my hand forward, gripping my pistol as it passed, whipping it up and forward. I had it pointing straight at his chest by the time he had lifted his 38 not more than halfway out of its holster.

His imminent death was suddenly reflected in his face when he realised what had just happened. My pistol had almost magically appeared in my hand and was pointing straight at him. He realised that if he kept pulling his pistol out of his holster, he would be a dead man. The colour drained from his brown gnarly face.

My pistol had seven rounds in the magazine and it was a semi-automatic. That meant I could squeeze the trigger seven times without cocking or reloading. As I could hit a tin can from 50 yards, and he had seen me do this, he knew he would most likely die after my first shot, as my pistol was aimed at the middle of his chest. I softly squeezed the trigger.

Fortunately for him he froze solid and his pistol dropped back into his holster.

I held my pistol steady for what must have seemed like an eternity to him and then I said to my interpreter, "Tell him that next time he challenges me I will kill him." Then I very slowly lowered my gun in case he tried to pull his again.

Everybody stood up and cheered. The beautiful young woman ran up and hugged me. She then led me by the hand around the back of the big hall into a courtyard and then into a well-appointed, free-standing cabin with a large bedroom.

She was very happy that I had beaten this man as, obviously, the only feelings she had for him were those of contempt and fear. To her he was an old, powerful man who could do whatever he pleased and she had to obey him.

Smiling, she began to undress, but I took her hand and sat her down on the bed. I explained to her in broken Vietnamese and English that I did not want to have sex with her. I just wanted to sit and hold her hand for a while. I explained to her that I was engaged to a woman back in Australia. Also, I did not feel it was right to take her as a gift.

I felt sorry for her and offered her what money I had. She laughed at this gesture and pulled out a large rolled wad of US dollars, which she then tried to give me. Obviously, the Vietnamese officers were sharing the money I had found and were rewarding their concubines. I refused her generous offer and laughed at the thought that I could suddenly become her expensive concubine. This world could be crazy at times, especially here.

We sat for a while holding hands and shared some human kindness in this ruthless place. She spoke a little English. How delicately beautiful she was, I thought, but I missed my Jenny back in Australia.

Eventually we went back to the party and Dat asked her

about our encounter. She smiled and held her hand high, and they all cheered again.

We had to get back to our checkpoint and our Vietnamese men, so we took our leave, with Jack doing much silly, drunken saluting on our departure.

Back at the checkpoint I walked off into the jungle, unpinned my medal and shot it with both my pistol and my M16. I did not deserve any medal for what I had done. Jack told me later that Australians were not allowed to wear Vietnamese medals anyway. I would never have worn this medal. There was no honour in the memory of that poor woman and her children being tortured by this man. This experience and accompanying wound eventually melt away into my unconscious mind, only to be revived forty years later by a strange encounter with another woman.

Yet I am sure that the poor Vietnamese woman's fate is one of the main reasons I wrote this story.

Chapter 10 - The Long Hai Mountains

One morning, while working at our checkpoint, we had been warned to stay low, behind our sandbagged position, as American B52 bombers were going to be flying over us and we probably would not hear them as they would be too high. The huge bombers would then bomb the Long Hai Mountains close by to our north.

The Long Hai are a large range of massive, cavernous mountains, covering perhaps 50 square kilometres or more. It is common knowledge among the Vietnamese, who have shared this knowledge with me, that this mountain range is also a haven for the Vietcong and North Vietnam army regulars who live in the caves and hide in its huge, deep, man-made tunnels. They then come into the villages at night or set up quick ambushes during the day.

We lay behind our sandbagged positions because the Americans had already killed and wounded some of our men in the past. Their deadly bombs are not always accurate.

We heard a faint drone and then all hell broke loose. The ground shook and it looked as if they were bombing the entire mountain range out of existence.

I wondered how all this earth-shattering bombing eventually affects this poor planet we live on?

I asked in a loud voice how anyone or anything could survive this holocaust? My young Vietnamese army friend answered: "Their caves are so deep they will hardly feel these bombs. They have been bombed before but they are always there and when your army goes in, after the bombing, they will step on the many mines that were stolen from your minefields."

"Surely this bombing will detonate all the landmines," I said.

The soldier answered: "No, the Cong will quickly lay mines on your path when they see your army advancing up the mountain. This is the way they operate, and they have many thousands of your mines."

This man had grown to trust me and knew what he was talking about. These Vietnamese had been at war for 150 years, with many different countries, and they had learned the hard way how to survive and how to kill.

After the bombing ceased, Jack and I went back to base camp at Nui Dat. We went straight to the sergeants' mess where I passed on this information to a sergeant and a staff sergeant. I then asked them to pass it on to the current brigadier so that he would not send anyone else on suicidal missions into these mountains.

Meanwhile Graham, back in the safety of Australia (not that he was ever in a dangerous frontline position in Vietnam) was promoted to major general and Deputy Chief of the General Staff. This promotion was done at the height of these mine deaths and the shocking maiming of so many of our men, as well as Vietnamese children. His lofty, undeserved position after such a monumental and callous blunder ensured that silence surrounded the minefield and aided in the deaths of Australian soldiers and their continued disgraceful treatment in the field of fire.

Only this dinosaur of an army could make such a monumental stuff-up yet for me, and most "frontline soldiers", this had now come to be expected as a matter of course. Ah! The terrible cowardly politics of the military high lords! It seemed that to some high-ranking, politically minded officers the chance of a medal and promotion, plus the extra cash that comes with it, is a carrot too hard to resist so the lives of expendable soldiers are not considered with the utmost care, as they deserve to be.

As the officers so often reminded us: "Many of you will not be coming home." And many of us never would come home, even when our zombie-like bodies did, but I will explain that phenomenon later.

And Graham's continued silence, to his shame, cost so

many more lives and limbs than it otherwise could have. And, of course, the "officers' club" stick together and dare not criticise each other and especially dare not criticise higher ranking officers. In 1969, back in Australia, General Wilton even commented at a ministerial inquiry on this issue, and I quote him: "I would not criticise any army task-force commander."

Their mafia-like code of silence will continue to cost lives in any war that we so quickly partake in, usually at the bidding of "John Wayne's America".

What madness, apart from our misplaced fears, made us the ever-faithful deputy of the US of A? America, the self-nominated sheriff of the world.

I bet Wilton would "criticise" if he were forced to fight on the frontlines and could witness the carnage and the terrible loss of life and limb which was the direct result of their monumental blunders as they played their war games from the safety of Nui Dat.

Thereafter, Wilton continued to assert that Graham's decision on the minefield was "sound in all circumstances".

In our newspapers these brigadiers and generals, together with the Prime Minister of the day, are all praised and lauded as great "hawks of war". What utter nonsense! Send all these brave "hawks", who call for war, straight to the frontlines of these wars that they start so easily and willingly! Invasions of other countries, and wars, would then be over in about three weeks. We seem able to begin wars in three weeks but take years to cease from them.

While we are at it, please also send to the frontlines all the priests, the mullahs and the holy men who all call for sacred wars in the name of one god or another. Give them all a weapon and a hundred-pound pack to carry and let them practise what they preach, amongst the pain, blood and guts of the dead.

Back in Australia I went to one Anzac Day parade to honour my dead mates. A priest, a politician and a general spoke, and praised us the lowly soldiers. They then praised, and offered their support for, our seemingly endless wars where we send troops to foreign countries to "save" the local people. I looked at the children watching that parade and saw the longing in their faces to become a soldier when they grew up, so that they too could be a hero. Don't do it, kids, I thought, as it is nothing like they are telling you!

I walked off this first parade in disgust before they could finish their bullshit. I could listen no longer to their inexperienced words and their support for the politically motivated perpetuation of wars. I have not marched since but go quietly to the dawn services to honour and remember my dead friends who gave their lives.

We continue to use soldiers by praising them to high heaven, giving them medals, and holding up these same soldiers almost as objects of worship, so that governments may fill their outdated cumbersome army's savage ranks with the innocent young of our nation.

And if you doubt that armies are trained to be savage killers then pay a visit to the 2nd Battalion in Townsville. There, carved in bold letters on a large wooden sign facing the road for all the world to see, is printed the mortar platoon's motto: "OUR JOB IS TO KILL". Perhaps it should read "OUR JOB IS TO SAVE", even if this is not the truth.

As always, soldiers are still trained to kill without thinking or being made aware of the consequences to themselves of killing. I recall just how many times during recruit training the words "You are not paid to think, soldier" were screamed at us. Well, a man should think before he quickly kills another human.

And this perpetuation of the thirst for these endless wars is supported by the forced silence of an entire nation, as any criticism of war is cunningly misconstrued as a criticism of soldiers fighting for peace. How can you fight for peace? The simple truth is that you cannot "fight for peace", and if some soldiers need to be criticised to save the lives of future generations, then so be it. This is what I am attempting to do, and of course I expect that I shall be one who is criticised severely.

A war zone is a dark place where frontline soldiers quickly learn to hate, and then thirst for revenge killing, just like me, after we see our best mates or brothers blown to pieces or horribly wounded. It is not the fault of the young soldiers. It is the fault of our foolish and ignorant masters.

And whom do the politicians ask whether or not we should go to war? Why, they ask the generals and brigadiers. How stupid is this? Generals, plus brigadiers and officers, spend their entire lives preparing for war, and often long for a war to further enhance their careers, their promotions, and their already fat pay packets. This is an obvious and sad fact. Definitely do not ask them if we should go to wars in foreign countries because we know that their answer will usually be yes.

And certainly do not ask the priests and mullahs because the Koran and the Bible are filled with precise instructions on killing and going to war. Men wrote these books, not gods. Show me a god with a pen.

At this point back in Vietnam I thought about my old platoon. I really was a raw recruit when I joined them as I had missed months of training. The men in that platoon had seen their mates shot or blown to pieces, and the unthinking, uncaring army had sent a soldier not fully trained to replace their friends. It is little wonder that they treated me as they did and would not take the time to help me out.

I realised now that I cannot blame them or hate them. The blame rested squarely on the shoulders of these desk jockeys who write the orders, and they were about to show themselves in all their heartless stupidity yet again.

I was still very angry with Brigadier Graham for his colossal blunder and now held him directly responsible for the shocking death of my mate Cuddy who was obviously blown up by one of our own stolen mines.

And now more men were going to die. This was vividly obvious because of the information I had received and from what I had seen of the number of stolen mines in the Australian minefield. Sadly, it quickly came to pass.

They called it Operation Pinnaroo, in the Long Hai Mountains. Seventy Australians and New Zealanders were wounded, many losing limbs, all by stepping on our stolen mines. Twenty-six men were killed. Only one of the twenty-six killed was shot, not blown up by a mine.

What I later found out, which was terribly ironic, was that an American brigadier named Sandy Pearson, together with the CIA's Principal Adviser, had visited Nui Dat and advised (ordered) the Australian forces back into "the box". The box is the mined area in Phuoc Tuy Province that had been left unpatrolled while the Australian forces had been supporting the Americans in Bien Hoa Province.

During Operation Reynella (May 15 to June 15 1969) forty-one men were wounded and six killed, yet again by our own stolen mines.

The silence of the commanders about these mines continued as they disgracefully protected each other and this, of course, allowed the horror to continue.

Please do not tell me that the heads of armies care about their common foot soldiers. "Kill one for the good of the many" may be their mantra to erase their guilt.

Meanwhile, in the mess one night as I spoke to Jack about these dumb commanders, two other men, both staff sergeants, had been listening to our conversation and they addressed me.

"Who do you think you are, Corporal, talking about our commanding officer like that, and what would you fools know? You are not paid to think. You are paid to kill. I should have you thrown in the stockade. We did not expect you 'Checkpoint Charlies' to live this long anyway."

They were obviously drunk but I realised yet again the truth of what they said about us living this long. We had been ambushed and shot at many times. The only reason we had not been mined was because at least half the ARVN soldiers we worked with also worked for the Vietcong. We would not find that out for certain for a while yet, and then all hell would break loose.

Jack's face changed colour and he stood up. I grabbed his arm. As much as I would have liked to fight these fools, I had learned my lesson the hard way. If we fought with staff sergeants we would be jailed and then sent back to the battalion.

That would be my death warrant. I gripped Jack harder and walked him outside.

I smiled and said, "Remember the fire crackers we bought in Bar Ria, mate? Let's teach these fat fools a lesson in a different way."

Jack laughed, thank goodness, as he was a wild-tempered man and had been busted from sergeant for fighting with officers of a higher rank.

We ran back to our tent and grabbed the crackers then went back to the mess hall and waited.

The two desk-bound staffies eventually came out and we followed them to a tent where a movie was being shown. A war movie of all things, being shown in a war zone. The only war they would ever see, I thought.

We let them get comfortable and waited until a particularly loud war scene appeared on the screen. We then lit the crackers and threw them under their seat. Jack yelled "Contact" (which means enemy contact) and after the first explosion, the speed at which those men launched themselves from that tent was awesome to behold. Jack and I, lying in the dark, could not move for laughing.

Chapter 11 - The Fight

The days then passed ever so slowly on our checkpoints, surprisingly without any more attacks.

Big Ron, perhaps through boredom, became more of his usual bullying self. Very often, he would come up behind the other guys and slap them behind the head, just for fun. He had never done this to me but must have thought I was ripe and ready in my by now depressed, gaunt and withdrawn state. On Checkpoint Charlie this particular day were Ron, Jack and I.

He could not help himself and finally slapped me behind the head and knocked my hat off.

Never knock a cowboy's hat off unless you want a fight as it is a deep insult. Our hats protect us from the sun, allow us to drink water out of rivers, and we use them for a pillow at night.

I spun around, grasping my pistol. I thought, you idiot, I could easily beat you over the head with my pistol. Are you stupid?

Luckily Jack was there and yelled out, "Hold it, you can settle this back in camp, in the ring."

"Suits me," big Ron said with a confident smile.

It did not suit me as he weighed at least 240 pounds and I was only about 175 pounds by now as I had been off my food. He was a big man who had won a heavyweight boxing contest between the army and navy. And he was about 65 pounds heavier than me. That's a lot of weight to give away in the ring, wearing gloves, against a proven heavyweight champion.

Apart from shooting him in the foot or hitting him with the butt of my Armalite rifle I had little choice, so I agreed to fight him.

Back in camp the word was passed around and over the next few days Jack was betting large amounts of money on me.

He had more confidence than me and I reminded him who Ron was, what he had won, and his size. Jack just laughed and said, "You can beat him, mate."

Now I had to wait for three more days for the fight and as my anger had been replaced by common sense, my mind told me again just how huge this man was. I also found it difficult for my busy mind to get over the fact that he had won a heavyweight boxing championship.

I was laughingly informed by some men, who had bet on him, that he was also never beaten in a fight.

Now I had my own inner battle to deal with. I could lose, beaten by my own mind, right there and then, before we even stepped into the ring.

An MP, who had bet on him, came into my tent and told me that Ron had held heavyweight titles in Australia as a civilian before he joined the army.

I had to get my mind on top of this looming fight so I thought in the positive. He obviously could fight but could he box? There is a world of difference. But my monkey mind arose again and I thought that if he hit me a few times, with his weight and strength behind the punches, I could easily end up broken and badly beaten. Especially as the gloves we were to use were lighter than the gloves used in normal boxing.

Once again, I wiped these negative thoughts from my mind as I felt them slowly draining me. I reminded myself that most bullies are cowards when it comes to the really tough stuff.

I also remembered that three Australian champions, plus my tough father and uncles, had trained me since I was four years old. And my Uncle Boof is the best street fighter in Northern Australia. He famously beat seven men on his own, in a hotel in Cairns, after they knocked his large Akubra Stetson off his head. Unfortunately for them they then tried to throw him out because he looked different to them. This was how he got the nickname Boof.

Also, I had quite a successful boxing history of my own. I had fought in many rings – in Jimmy Sharman's tent and in the famous Festival Hall in Brisbane. I had fought on the streets, and in other show tents, against all comers, so my family could place bets on me and win money.

Remembering the cowardly punch, his bullying ways, the blown-up kids, the decapitated major, who had become my friend, and my growing hatred of any form of authority, my anger surged back. Fear and doubt in my mind were gradually defeated.

On the night of the fight the large tent was packed to overflowing with men. News had travelled fast.

The military police corporal, who won a light heavyweight boxing title in Malaysia, was there to act as referee.

We stripped down to our shorts and took our shirts off. My "second", good old Jack, tied my gloves. I whispered to him not to tie them too tight as I may wish to get them off quickly.

Looking at the gloves I realised that I was probably lucky that they were so small and light. They would give my punches more speed and hitting power. His too, of course, but I felt that I would be quicker than him, and hopefully much more experienced.

As we stood in our respective corners, I avoided looking at my opponent. I only wanted to look, unblinking, into his eyes when we finally faced each other in the middle of the ring. I had been taught to do this by King Keddy, one of my past trainers.

When I am speaking to people in a normal conversation I hardly ever look directly into their eyes. I look at the middle of the face, or at the mouth, as these are neutral positions. Yet when I am fighting, or angry, I stare only into the eyes because here any coming movement can first be detected plus the eyes can win a battle before a punch is thrown. The eyes can be deadly.

I noticed that the tent had a pole in the middle of the ring and this I could use to my advantage. Also, the ground was slightly sloping and I made up my mind to always be on the high side as he was at least 6 inches taller than me. This ground height would somewhat nullify his height advantage.

We were called to the centre of the ring. There were no ropes, but a ring had been formed by the many watching, yelling men. It was here that I used my unblinking stare. I was not actually doing anything, just watching and waiting. He grinned and broke eye contact first, turning and waving confidently to the crowd and his corner men.

I had won the first encounter whether he realised it or not.

The MP referee told us the rules and asked us to shake hands and then come out fighting. We both declined to shake hands and returned to our opposite corners to await the bell.

As I studied his actions, it was obvious that he was definitely overconfident and I realised that I could also use this against him.

This overconfidence had revealed his second weakness and it was a major one.

When the bell rang, he walked straight up to me so I stepped to the side of the post to slow him down. He sure was confident, walking straight up to me like that.

I let him throw the first few punches and took them on my arms and gloves as I sized him up and worked out his style and any further weaknesses he may have had. I had always been taught to use my head and carefully study an opponent in the first round.

He went for a knockout with every punch and the power of his massive hits hurt even my forearms and pushed my gloves, with unnatural force, into my face with every connection. This was unusual.

But I needed to let his punches land on my gloves and arms. This strategy allowed me to find out a lot about his boxing style. He did not throw true, straight punches with his left hand although his right, which he led with, was, unfortunately, almost a straight punch. I could see he had had a fair trainer in his past. Yet even when he was about to throw this punch, his eyes dropped just a fraction and his elbow moved slightly outwards, so I could see the punch coming long before his large fist actually began to move. This is aptly called a "telegraphed punch".

Any boxer worth his salt should know to at least respect any unknown opponent and observe him carefully in the first round. He did not do this, the arrogant fool!

I stalked him and when he threw the next punch, I let it miss me and flicked my left into the centre of his face. Not a hard punch as I wanted him to grow even more in confidence and uncontrolled anger, while forcing him to think that my weak punches could not possibly hurt him.

This worked and he confidently dropped his gloves a little and, growling, threw more huge haymakers with all his weight behind them.

I knew that I must eventually use his uncontrolled weight against him to beat him. That means I must hit him when his weight is coming at me, propelled by one of his huge punches, which I must make miss.

For the rest of round 1, I let him throw a lot of punches and softly boxed him in return. None of his punches hit my face. But mine hardly stung him and he believed, even more so, that I could not possibly hurt him with my weak punches.

Three minutes fighting in a ring is a hell of a long time when you are not in training and at the bell he was breathing heavily and I was not, as I hadn't used anywhere near the energy that he had been forced to use.

However, my forearms were hurting where his punches had landed. I realised that I would have to avoid any more of that or my arms would grow weak.

I was now feeling calm. I had his style worked out.

I decided to carry him until round 3 and then I planned to slowly take him apart and humiliate him in front of everyone, as well as hurt him. This, I hoped, would teach him a valuable lesson.

The crowd had grown and as I looked about, I noticed that even officers were now here.

Everyone was yelling excitedly and Jack was taking more bets. Ron was the firm favourite by this time and only Jack was still backing me and taking more money at even better odds. Good on you, mate!

I had whispered my plan to him in the break, between rounds, so he could see what the others could not.

Ron was not sitting down between rounds, as I was, and he busily guzzled cans of beer. He was still smiling and even more sure of himself. Very good! My plan was working well.

I concentrated on completely relaxing all of my body while I breathed in through my nose and then forcibly pushed the poisonous carbon dioxide out of my lungs with a strong whoosh through my mouth. Another trick I had been taught.

The bell rang, the crowd roared, and after a minute and a half of this second round I slipped in under his punches and put a short sharp right into his stomach to remind him of his beer drinking mistake. The smile left his face and he picked me up bodily and threw me into the crowd. This was not in the rulebook of boxing and he had caught me unawares. My breath was knocked clean out of me and the men just picked me up and threw me straight back at him. Struggling to gain my feet while stumbling forward, Ray grabbed the opportunity and let fly with a flurry of punches. I managed to avoid most of them but a haymaker of a left hand, the one he swings, caught me flush on the side of my head, over my right ear.

Lights and stars suddenly became my entire vision as I was lifted off my feet by the force of the punch and thrown back into the yelling men. Once more they picked me up and threw me back at him and this time I landed at his feet, on my face, still stunned and seeing stars.

He ungraciously kicked me in the ribs and screamed at me to get up, while the many men who had backed him cheered loudly.

The referee pushed him back and warned him about kicking, then began the ten count on me. I quickly realised I desperately needed as long as I could possibly get to recover and would have to use all the tricks I knew. One more punch like that and I would be out of that fight for good.

I slowly came up onto one knee and waited until the count of nine before I stood up. Then I asked the ref to do up the lace on my glove, which was not even undone. He grabbed the lace and I fell into him and jerked the glove away, causing the lace to come undone. "Take your time doing it up," I urgently whispered, and he took pity on me and did just that.

Ron was jumping around like a mad bull. He now sensed an easy victory. I was still shocked at the sledgehammer-like weight of his punch. Normally a punch to that part of my head does not even slow me down in the slightest, and yet the side of my head would be swollen and very tender for the next two months. He had very nearly knocked me clean out for the count of ten with just one punch. I was just lucky the punch landed on the side of my head and not the front of my face.

By the time my glove was done up and the ref slowly asked me if I could continue, the bell rang. I felt that I was, virtually, saved by the bell.

Jack had a bucket of iced water which he towelled onto my head and neck. Ron was still standing while drinking yet another beer. Gradually I began coming to my senses and feeling okay. The stars began to fade away. Jack was holding up one finger not two, I told him when he asked.

"Get into it," he yelled in my ear. "No more fucking around."

He was right – that was too close, and this man was dangerous if he got any opening at all. I was going to give it all I had in the next round. I had to. He was full of beer and breathing heavily.

I decided to let him think, for one moment, that I was already beaten. Encourage him to make a deadly mistake.

The bell rang and I advanced slowly towards him with a slight stagger to encourage him. But now I was deadly serious, although very wary, because I had to let him throw the first punches. I made sure that his telegraphed haymakers missed me and my sore arms altogether. He jeered at me in a loud voice to stand and fight. The overconfident bully, breathing heavily now, threw two weaker punches as he shuffled towards me and I easily avoided them.

In blustering frustration, at again being forced to miss, he finally gave me the opening that I had been waiting for. He let his left foot cross over his right foot, and let fly with a big left-hand haymaker. As his feet crossed, he was totally off balance and all his huge weight was finally being projected towards me.

Never let your feet cross if you find yourself in a fight because you are off balance for a split second and all your weight is coming forward at your opponent. And never throw a punch, like he did, from there!

I easily made him miss and hit him with the best true-straight left-hand punch that I have probably ever thrown. It was directed straight from my shoulder with no movement of my elbow or eyes whatsoever to warn him of what was coming. Three inches before the punch landed squarely on his exposed face, I twisted my fist.

The punch felt like someone had rung me like a bell and the vibration ran all the way down to my toes and back up to my head. He was stunned, knocked back into the crowd and then thrown back to me.

And now it was my turn as he stumbled forward. I hit him with another straight left followed by a straight right and left hook combination. All in a split second. My hardest combination of punches in the fight and one I had practised a thousand times.

I was expecting him to at least crash to the canvas before he got up. But no! He tottered backwards, dropped his gloves, held up his hands in a stop sign, and said, "That's it. I've had enough. I am beaten."

This gladdened me, relieved me, amazed me, and then made me angry.

The bastard! You never give up while you are still standing!

I rushed over to him and said, "If you ever hit me or anyone else again, we are going to fight with bare knuckles and I am going to put you in hospital.

Have you got that or shall I do it now?" With this last statement I ripped off one of my gloves.

He nodded his head and held up both his gloves defensively and it was all I could do not to lay into him again.

The fight was over and men rushed forward to congratulate me, as Jack, whooping in ecstasy, tried to lift me onto his shoulders.

The MP corporal who had refereed the fight approached me and said, with his mouth agape, "Can you teach me how to do that?"

I smiled and thought, it takes a lifetime and a father like mine.

I imagine this all sounds a little farfetched and hard to

believe, yet there were many men present in the tent for that fight, and that is how it happened. I am sure that some of them who witnessed it are still alive.

Another day in this madhouse of a war-consciousness zone had gone by. One day closer to going home.

By now I was actually counting the very minutes till I actually left Vietnam.

Chapter 12 - Death of Another Friend

The days melted one into another ever so slowly. My little teenage Vietnamese mate, named Wan, who called me Matt, often came to greet us on the checkpoint. Jack and I took him for rides in the gunship and tried to teach him to shoot, though he soon told us that he did not like guns.

Wan was a lot of fun. He had a quick mind and a good sense of humour. One day he took us into the nearby large city of Bar Ria and proudly introduced us to his family. His mother and father were a lovely couple and he also had two younger sisters. They lived in a two-storey, somewhat run-down house, so we insisted on buying them some furniture and food.

When Wan could come to our checkpoint, we would often wrestle playfully. I also taught him a little boxing so he could at least defend himself if he needed to.

He warned us about certain things that we should avoid, such as Coca-Cola filled with crushed glass or cobra poison hidden inside the bottle by the Vietcong. What a lovely war.

One day, the White Mouse copper captain who I had nearly shot pulled up in his vehicle with three of his policemen. Ignoring us, he began abusing Wan in loud Vietnamese. This made me angry and I immediately stepped in and faced this excuse for a human being. Wan put his hand on my shoulder and said, "No, Matt, it is okay." Wan called me Matt as he could not pronounce Mac.

Then I did something that very nearly cost me my life. The captain challenged me to a game of Russian roulette with his .38 pistol. I silently exploded inside with anger at how he was treating my little friend. I immediately agreed as I thought he may give Wan grief later. My body stiffened and any common sense still left in me was replaced by irrational revenge.

Perhaps it was time he died, and by his own hand, I thought.

The game of Russian roulette required me to put one bullet in the six-bullet chamber and then spin the chamber. I then had to hold the gun to my head and pull the trigger, hoping I wouldn't blow a big hole in my head.

I was confident that I could outsmart this sadistic copper because when I spun the chamber I could see, out of the corner of one eye, that the bullet was not to the immediate left of the barrel.

I went first. I placed the barrel of the gun to the side of my head and pulled the trigger. Click!

The captain then took the pistol and did the same. We passed it back and forth. Click, click, click. It continued to fall on empty chambers.

It was again my turn and I noticed that I was sweating a little. I did not know if that was because I thought I was going to see his brains eventually blown out or if I was nervous about making a mistake myself.

That insane irrational experience was a critical point of destiny as I faced death due to my out of control inner rage. I experienced a powerful realisation, one which had saved my life. This experience showed me that there is a feeling intelligence that surpasses my analytical mind which thinks it is always in control. Some call this spontaneous intelligence, intuition. But this momentary event was deeper than intuition because I took no conscious action to follow or respond to an intuition.

That day I realised that I must listen to my intuition, which can be strong. In the future I would be taught, by women, more about this seemingly psychic sixth sense.

Once again, I spun the chambers then cunningly glanced down and saw that there was no bullet to the left. Luckily for me this pistol had a hair trigger and as I raised it to the side of my head, I squeezed.

Bang! I was deafened by the terribly loud blast and suddenly found myself sitting on the ground. I calmly wondered if I was dead or perhaps dying. I felt for the blood. But none was pouring out of me. For the first time in this game, my finger had pulled the hair trigger a fraction of a second before the barrel reached my head. The bullet came close enough to singe the hair on my head and also deafen me. And the blast of this large bullet had thrown me from my seat.

This cunning captain had somehow reversed the track of his pistol. Instead of it spinning to the right, it spun to the left. I was super-lucky to be alive, because every other time I had pulled the trigger with the barrel jammed against my head.

I sat in shock from the nearness of my close call while the captain laughed. He then jumped into his vehicle with his men and sped off.

The very next day Wan arrived at the checkpoint, called me aside and began to cry. He was just a young innocent boy. He told me that the secret police of the VC had come to his home last night. After threatening him, they told him in front of his family that they would kill him and all his family unless he joined them and carried supplies at night to their regular army based in the nearby Long Hai Mountains.

I was outraged and I wondered if the White Mouse captain also worked with the VC.

I told Wan to wait there on the checkpoint and I rushed back to our base at Nui Dat in the gunship. I managed to quickly arrange a meeting with a colonel and told him what my little mate had told me. I then volunteered to sneak into his house, that night and every night, to wait on the top floor for these animals to return.

I had to ask permission in advance. We were always checked back in to camp at the end of the day unless we were ordered to stay out in an ambush position for the night.

The colonel was sitting behind a desk and a couple of lower ranked officers were standing.

One of these officers laughed and asked me if I was still trying to win the war on my own. He had obviously heard about me. He told me that I should know better than to trust a Vietnamese or to make friends with them. "They will get you killed," he said. He sneered, "No, you cannot go into Bar Ria at night."

He dismissed me with a wave of his hand and a shake of his head.

It was a terrible thing for me to tell poor little Wan that neither Jack nor I could help him. I felt sick in my stomach.

He graciously said, "Don't worry, Matt," and trudged off with tears in his eyes.

Two weeks later I was having my allotted day off in camp. It was about 8 am in the morning when a Land Rover, towing a trailer, pulled into our camp. An officer called me over and ordered me to bury the two Vietcong that they had killed in an ambush last night. He pointed to the trailer and told me that the Cong were in there.

I grimaced at the thought that I was about to become a grave digger.

I walked down to the trailer and was stunned when I saw little Wan, staring at nothing with his eyes wide open. Lying next to him was one of the quieter Vietnamese soldiers from my section, also dead and staring open-eyed at nothing.

My mind snapped and I ran back to the Rover and jerked the door open.

"You fucking idiots," I screamed at the officer and the soldier sitting next to him. "These are my friends. Wan is a young boy, not a fucking Vietcong, and I work with the other young man on my checkpoint. What do you fucking hero clowns think you were doing killing them?"

I was not carrying my rifle but my fists were closed and I was about to hit a major. He was stunned. To his eternal credit he stared at me open-mouthed and then a tear ran down his face as he answered, "It is this dirty war. They walked into an ambush last night. I'm sorry, I did not know."

I sat down and cried. He looked down at me and then drove off with the trailer.

I did not even get to bury poor Wan, my lost little mate.

The next day I angrily confronted Ho, the Vietnamese soldier in my section who always gave me inside information. He told me in a matter-of-fact, sad voice that this is why the VC would win this war.

At night when the Australians and the Americans went back to their safe bases, the towns and cities belonged to the Vietcong. The local people had no choice but to support them, even though they may not wish to. If they did refuse, obviously they would die a horrible death.

Their brutal death would be a terrorist warning to the villagers – the vision of the mayor of Wai Long hacked into little pieces or the pregnant woman found lying on the road with her belly cut open and her dead baby lying next to her, still connected to the umbilical cord. These visions would remain as a terrifying reminder in the hearts of these people, who never wanted a dumb war.

I just couldn't comprehend what in hell were we doing here? We could not win this useless war like this, yet the politicians were always telling us, and the Australian public, that we were winning. Why do they lie to the world? What is in it for them? I know that the American economy relies heavily on the sale of bombs, mines, planes and armaments of all types, which they supply to the rest of the world. One of their late presidents, I think it was Eisenhower, warned America of the direction it was taking, and that their economy would eventually rely on wars to survive.

And what about Australia? Were we here because we are afraid to stand alone without America, our "big daddy, saviour and protector", who we believe will come running to our aid if we are attacked? Don't the politicians realise that America is a self-centred nation and if they are threatened, they will not be thinking of Australia. They will take care of themselves. Damn this war, and our fear to stand alone!

Years later we would obediently join America in the wars in Iraq and Afghanistan. In these countries the unthinking American army created their own enemy, the Taliban, by supplying them with weapons and men to fight the Russians.

What did we learn from the sacrifices in Vietnam? What have the millions of lost lives taught us? Nothing, it seems, because the Taliban, I have been told by returned soldiers from Afghanistan, use the same terrorist tactics at night, just like the

Vietcong did in Vietnam when we were relatively safe back in our fortified bases. So, the populations of these countries we have invaded are forced at night to support the over-zealous, religious Muslim Taliban, even though they do not wish to. Just like poor little Wan in Vietnam.

And still the politicians appear on national television and say we are winning the wars in Afghanistan and Iraq. What a lie, what nonsense. It is impossible to win this type of war. Soon the Americans will sell out the people of Afghanistan and make peace with the Taliban. The retreat will be political.

How much more of our resources will we, Australia, be coerced into spending, as we purchase from America (motivated by our fear) submarines, second-hand tanks and planes that do not even work? America's throwaways. It is no wonder they think of and refer to Australia as a "Banana Republic".

Perhaps it is time for us to face at least one stark truth. The American and Australian war machines are called the ADF which stands for the American DEFENCE Force and the Australian DEFENCE Force. Let us call them what they really are as we are the ones forever invading other countries. They should be called, in truth, not the Defence Force but the ATTACK Force. The AAFs not the ADFs.

America, before its eventual defeat in Vietnam, would drop at least a million mines in nearby Cambodia and Laos, simply because the Vietcong may have been using the jungles of these innocent countries as a supply route.

America's covert soldiers, who were not even supposed to be in these two neighbouring countries, would also lay thousands of deadly mines.

Thousands of locals, including a lot of children, have unknowingly stepped on these mines over the past forty-five years since the war ended, with devastating results.

Even today, ex-soldiers with a conscience search these countries and blow up the many mines, at least the ones they can find, that still remain.

Yet, in our past history, most great ATTACKING armies have been defeated eventually. From the ancient Romans, to

Hitler's army, to us in Vietnam, to name but a few. Why don't we learn from our history?

We could spend a very small fraction of what we now waste on armies and their toys by following the example of the Swiss. Dig huge tunnels in the forests and outback of Australia. Put food and arms in these underground tunnels that a competent guerrilla army would need in case of attack.

Obviously, from the lessons of history and wars, we would not be defeated either. A large attacking army has always been defeated by a guerrilla army that is properly prepared and trained.

And it would cost us a small fraction of what we now waste on these outdated war machines and the trillions, yes trillions, of dollars the world is forced to spend every year on arms. Huge amounts are also spent on returned, wounded soldiers.

Learn from the mistakes of our past wars! Stop spending at least 70 per cent of all our resources and energy on wars, on war machines, massive armies and police forces. Spend this money on peace, health, education, invention, art and creativity to name a few things. In so doing we could make life much richer, healthier and rewarding for everyone.

Anyway, this is my war in Vietnam, so back to it and how it shaped my life and my search for answers.

Chapter 13 - Ending

When I began writing this book, I thought I would write a small chapter on Vietnam, but the pen had a mind of its own and the story has grown to this. But I shall wrap it up as quickly as possible and get on with the rest of this life I have been led to lead.

I was wounded twice in Vietnam and flown by the "dust off" helicopter to the army hospital in Vung Tau. One of my injuries changed the shape of my face because my nose was smashed.

A shocking and recurring nightmare then haunted me. My mother, father, sister and I (my other self) would be standing waiting to greet me, the returned soldier. But when I approached and told them who I was they would say that the man standing next to them was their son and they did not know me. This nightmare would continue to haunt me. What had I become in this terrible war? Would I even recognise myself? I guess it meant that I was not the innocent young man who had left them.

Back on the checkpoints, life went on. There was a great deal of distrust, through misunderstanding, between some of our men and the ARVN troops, since one of them had been killed in that ambush.

Our men also hated the sadistic White Mice, the Vietnamese police, we were often forced to work with.

I tried to explain to our men that it was not the fault of our Vietnamese troops. They had to protect their innocent people and their loved ones. These men also had to spend some of their nights in the Vietnamese fort while the heartless secret police of the VC threatened or killed their families if they did not do their bidding. I could not see how any of these people had much of a choice. Protection of family was more important than their own personal life.

Fear, however, has its own agenda. During one of my days off, Jack and some of the boys had a full out war with a section of South Vietnam troops and some White Mice. Thousands of rounds were exchanged. The dead were never acknowledged, as this, of course, would have created a political uproar.

Our "Special Services" group, however, was disbanded immediately and we were all sent back to the battalions. We were then purposely split up and sent to different platoons, and different battalions, alone.

Reluctantly, I found myself back at 2nd Battalion where I was brought before the friendly captain who was still the company commander of my old platoon. He told me to report back to my old corporal and my old platoon.

I told him that if I did that some men were definitely going to die and I might well be one of the dead. I told him that I could no longer submit to that corporal after what I had been through.

The captain looked at me and revealed that a lot of the men in that platoon had already been sent home while others had been wounded and also sent home.

I told him that the corporal would quickly turn the new men against me as I was now a lowly private once again. My stripes had not been officially given to me as a Special Service soldier.

The captain, to his credit, finally saw the dilemma I was in and agreed. He attached me to the 2nd Battalion headquarters. As we only had a month left in Vietnam, he said they would use me on overnight ambushes and other services such as driving vehicles into the city of Vung Tau.

That was fine by me, even though when men are getting close to being sent home, they do not like going "outside the wire". Yet I knew that I had a better chance of surviving alone, outside the wire, as that is what I had become accustomed to.

Two weeks later, I was sitting alone one night in my tent when Jack came through the door. He had a black eye, a broken nose, and one lens of his glasses was smashed.

"What happened, mate?" I asked.

He told me that another corporal had beaten him up the night before in the boozer. Jack had been busted back to corporal yet again because of what happened on the checkpoint.

"Who did this to you?" I asked.

"A Golden Gloves champion from Western Australia named Taipan," Jack said.

"Why did he do this, mate?" I asked.

"Because he said that I was a copper so he king hit me," Jack said.

"You know you are blind at night without your glasses. Why don't you fight him now, during the daytime? Maybe you can beat him because you are also a Golden Gloves champion. I will come with you and act as referee."

I smiled and said, "Maybe we can make some more money."

"I'm sorry but I cannot, mate, because he has also broken my jaw," Jack replied.

Red-hot anger once again boiled up inside me. This soldier had treated Jack just like I had been treated in my old platoon.

"Okay, mate. When will this prick be in the boozer again?"

"Tonight," Jack answered, "but you do not have to fight my fights for me."

"Well, Jack, if I had a broken jaw and a busted nose, I would expect you to fight for me, my friend. Tonight I am coming over to your company. Do not act as if you know me. I am going to pretend that I am very drunk. Just point this bloke out to me and stand well away."

That night I stumbled into Jack's bar. Jack nodded towards three men drinking at the end of the bar. "The one this side," he whispered as I slowly walked past. I sat myself down on a stool next to them and ordered a drink in a slurred voice. They glanced at me, noted my presence, and then ignored me.

Over my drink I began talking to myself, cursing war and war heroes. Jack had warned me that this man was called Taipan because he had such fast hands. He turned to me and said, "Fuck off and shut your stupid mouth, private."

I ignored him and kept raving, while swaying over my drink, trying to look like an easy mark.

Taipan was about 6 feet tall with a rangy, solid build and I could see why he could possibly be so quick. He had light brown hair and what looked like a permanent snarl on his face that said "annoy me at your peril".

As my cursing of war heroes became progressively louder, Taipan finally stepped towards me and cocked his fist in my face. "Fuck off or you will spend the night in sickbay," he snarled.

With those words I erupted off my stool and hit him three times before he could move. I didn't hit too hard though. He was dazed, but still standing.

Then I proceeded to lecture him, as I was hitting him, about never beating up any of my mates again.

He is the one who spent the night in sickbay, and more than one night, Jack told me later.

I had quickly disappeared back to HQ after finally knocking him down, and no one from Jack's company could identify me.

I had to help my mate because his senseless and shocking beating was unjustifiable simply because he had been judged to be a copper.

There is only one more story that stands out in my memory about Vietnam. Yet again it has to do with how the military establishment has forever had little regard for their frontline soldiers, sending them into places they should never be sent. We, the fighters, are just little pieces on a map that they, the big brass, move around.

I had a week left "in country" before being shipped out on an aircraft carrier named the Sydney, when I was ordered to report, fully armed, to the helicopter pad. There were four other men waiting in the helicopter. I had never seen any of them. We were then flown a long way over thick jungle to guard two American bulldozers which were about to totally destroy a pristine jungle by flattening it completely. These huge machines were lined up and waiting for us.

Our helicopter could not land so we had to jump out while it hovered over a small clearing.

I wondered why we were in American territory, guarding them, instead of their own soldiers doing this job?

The five of us fanned out around the machines. We were all

privates and we spread out and took up positions around the edges of the thick jungle.

The machines were started and I was expecting the Vietcong to arrive in numbers at any minute, attracted by the great racket.

Near the end of the day a large part of the jungle was flattened. It was completely gone, together with all the birds and wildlife.

The bulldozers left and we were ordered, over the radio, to wait for an American light aircraft and to guard it as it flew over us.

One of the other infantry soldiers came over to me to have a cigarette while we waited for the plane to arrive. We had been told that it would be flying low over this cleared jungle. We were not told why it would be flying over this huge clearing.

We were both sweating profusely in this humid heat and had rolled up our shirt sleeves. We talked about going home next week to our families and girlfriends and how relieved we both felt about escaping this hell.

The bulldozers were long gone when the small plane finally arrived. We watched, alert, with rifles cocked and ready to fire as it swooped in low over the cleared area and unleashed its coloured payload.

As we concentrated on guarding it from enemy fire, an orange mist, which the plane spewed out, settled over our bodies, arms and faces. I held up my arm and asked this man next to me, "What is it, this orange stuff?"

The jungle was a very hot place at that time of year, and he answered, "I don't know but it is very cool."

"Yes," I said, "it sure beats this heat."

Then the burning began. "Run," he screamed. "It's poison!"

And we ran for the protecting cover of the thick jungle behind us but it was too late, as I shall explain a little later.

I wonder where those other four soldiers are now? I would not be surprised to find out that they are dead after what happened to me. I found out, back in camp, that this spray was called "Agent Orange", the poison used so excessively by the Americans to stop the jungle from growing back. This deadly poison kills everything in its path. Nothing will grow again where it has been dropped.

I realised that this was also the liquid, in an oily form, that I had been ordered to spray around our Nui Dat base when I was back in my old platoon.

If this burning orange spray from that plane kills tough trees and animals, why would they forget to tell us that we could be killed by it as well? Because a plane is more valuable than a few Australian soldiers. And, as always, lives are easily sacrificed for "the greater good". Especially the lives of lowly infantry privates.

How many times had I heard that statement about "the greater good" which made wasting lives okay?

Chapter 14 - Home

Finally, after over a year in Vietnam, I was loaded onto the huge aircraft carrier, the Sydney, and we set sail for home.

Everyone was ecstatic but I was beginning to sicken and my body was hurting badly so I flatly refused to wash dishes when ordered to by a naval officer as I lay in my bed. I was sick of taking orders from officers, yet instead of charging me for disobeying a direct order, he let me be after looking at my now pale face. He said I looked sick and should see a doctor.

When we finally landed in Brisbane, I was feeling a little better after the long rest. We were ordered to line up on the wharf where an officer told us that we could go on leave after we marched through the main street of this city. He warned us that there were people who may throw paint, and also spit on us and want to start fights, but if we did retaliate, we would be the ones thrown in jail, not them. Army justice?

After we marched, and had abuse, spit and paint hurled at us, Jack and I met up with our other mates in a hotel just off Queen Street, the main street of Brisbane.

The boys were very upset so I explained to them that our war had been uncalled for and these angry people were probably right in protesting. Maybe they would help to stop the next war.

The boys proceeded to get very drunk but I was still not feeling very well and could not drink much at all, so I soon said goodnight. The bar we were in was in a basement so I had to climb a flight of stairs to get back onto the street.

On the footpath I was approached by two young men and a woman. The men asked me if I was one of the soldiers who had just returned from Vietnam. I innocently answered yes, then the men began abusing me in loud voices.

They said that I should have been killed along with all the other soldiers and innocent civilians that were killed in Vietnam. I smiled and thought that they may well be correct.

They asked me if I knew that millions of innocent civilians were being killed in Vietnam.

I answered no, I did not know that, but told them, stupidly, that my best mate was killed over there. They laughed and said that was good.

I was not feeling very well but something snapped in me because of my memory of Cuddy, so I returned their abusive aggression and told them, in a harsh voice, to apologise for that remark and fuck off.

They just laughed and then the two foolish men attacked me. I may have been slowly sickening but within a few minutes they were both lying on the road near the gutter. The police then pulled up, saw the fallen men, and told me that I was under arrest

I told the police what had happened and they turned to the girl and asked her if this was true. Thank goodness for an honest woman! She answered yes, it was true, and so they arrested the recovering men instead and told me to go home.

The next day we were granted a week of leave and so I took the train to Sydney to pick up the new Holden car that I had ordered before I left for Vietnam. It cost just over $2000 and the salesman had traded my old Fiat car.

I was very excited because I was also picking up my fiancée, Jenny. Then we were going to drive back to Brisbane where I was to serve another two weeks in the army before being honourably discharged. Finally, we would really be going home!

Jenny and I met when we were only sixteen. She was playing vigoro, a game similar to cricket, and I was training for rugby league on the oval next to hers. Jenny hit a ball high in the air and I caught it. That is when I noticed her. She had light brown hair, beautiful legs, a great figure and a lovely, kind face that was very pretty.

As I walked towards her and the vigoro team, with the ball in my hand, she was smiling at me. I could have thrown the ball back but looking at her I was almost in a trance. Handing her the ball I said, "May I see you after the game?"

Jenny laughed and said, "Sure."

Some of the other girls whistled and cheered.

From that day on we spent as much time as we could together. When I was at Teachers' College, we wrote letters and often rang each other.

When I returned from college, Mum gave Jenny a job in our hotel and taught her how to be a barmaid. Being in love with Jenny made me very happy and we went everywhere together. She had a beautiful heart and was kind to everybody.

I was suddenly conscripted into the army and bound for Vietnam, so we decided to become engaged. My family gave a hearty approval. Everyone loved Jenny.

We drove to Sydney together. The trip was like an early honeymoon as we took our time, stayed in motels, swam at pristine beaches and made love at night.

We were just nineteen and Jenny said she would wait in Sydney for my return and would think of me every day. We eventually wrote often.

While in Vietnam I had been paid $51 per week and had saved this money, about $2500, along with the little extra I had made from the earlier gambling. I had ordered a white Holden sedan with four gears on the floor. It was a beauty, and in this I proudly drove with much anticipation to a hotel in Western Sydney where Jenny was working.

I innocently wore my army uniform to pick up my beloved Jenny as I had no civilian clothes. This was very naive of me. One did not venture into a hotel, dressed in an army uniform, in Western Sydney in those days. It could be a bit like venturing into old Harlem in New York City dressed in a Ku Klux Klan outfit. The guys in Western Sydney were tough and they did not like uniforms or authority of any type. I did not know this, and I am sure that dear Jenny knew nothing about Western Sydney when she took the job there.

I knew I looked different, with the war wounds on my face, from the way I looked when I left, but seeing Jenny after thirteen months was a shock indeed. She looked so different. Her hair was dyed red, straightened, bunched up and pulled back. I was shocked that she did not run over to hug and kiss me like I imagined she would. Did I look that bad?

She did look up and smile at me. I also noticed that she had a tooth missing in the top of her mouth, plus an ugly scar on her pretty face.

Who did this to her, I wondered? She did not look anything like the young country girl whom I had left behind. Never mind, I thought, as my looks must also be a shock to her. We would be all right as we were engaged to be married. Jenny wanted to have children and so did I.

I wanted to be normal again, if that was possible.

I was sweating heavily, and feeling nauseated from this sickness that was once again growing worse each day. I quickly sat down and ordered a beer. Jenny handed it to me and then went to the other end of the bar and served the men there.

There must have been about fifty of them and they had suddenly grown silent when I had walked in. It was obvious that something was wrong here.

Jenny came back over. She was smiling again and did look pleased to see me, though she seemed very nervous. She told me that her shift ended in five hours at 5 pm and we could leave immediately for Brisbane. This made me happy.

Then four men, ranging in age from twenty to forty, slowly approached and formed a semi-circle around the stool I was sitting on. I turned to face them. They did not look friendly.

Without any introduction the older one spoke. The bar had gone quiet again. He simply said, "You are not taking Jenny anywhere."

"Why is that?" I asked him.

I had thought that I had been given enough shocks in Vietnam to weather anything thrown at me back here but his answer sickened and shocked me beyond anything I could have imagined.

He answered loudly and clearly, "Because we have all fucked her and she belongs to us."

I also thought that I was beyond ever wanting to kill anyone again. I thought I had dealt with that in Vietnam after Cuddy's death but as I looked at these poor excuses for men, with their dirty mouths, I suddenly saw myself hitting them with killing punches.

I quickly stood up and moved to their right to escape their enclosed semi-circle. My sick legs nearly gave way but they did not notice this.

I asked them who gave Jenny the scar and knocked out her front tooth? If one of them answered that it was him, I was going to attack. I would be badly beaten because I felt like death warmed up and obviously could not beat them all in my weakened state.

The older one said that it was not them that had hit her and they tried to protect her. Then he threatened me by saying that they were going to belt me and send me packing in my nice army uniform.

During this confrontation Jenny stayed down the other end of the bar and offered me no support, which hurt me deeply.

The only way out of here was to use my head so I said, "Listen, boys, I have just returned from Vietnam with a platoon of Special Service infantry soldiers who are as hard as nails. You may well try to beat me up, though I doubt that you would be able to. However, remember this, I will return with my platoon and it does not matter where you hide or how long it takes, we will find you. Jenny is a big girl so why don't we act like men and let her decide for herself. This was a free country when I left."

They looked at each other and then, luckily for me, the older man turned and walked away and the rest quickly followed.

Jenny then came back down to my end of the bar. She did not deny anything, though she must have heard what the loud-mouthed leader had said. She spoke with her head hanging down, "I will be ready at five."

I got in my car and drove away. Tears came and I was sick to my stomach. I pulled over and vomited until only black bile was being thrown up from the pit of my empty stomach.

The shock of their words, plus this aching, painful body

sickness, had really done me in.

I drove to a shop in the city and bought some civilian clothes. Then I drove to a gun shop to purchase a rifle and a bayonet. The shopkeeper told me that I could only buy the bayonet as I did not have a permit for a gun.

I purchased a stone and steel, then spent the next hour sharpening the bayonet until I could shave the hairs off my arm with it. I also made the point needle sharp. Then I tied a piece of leather through the handle, which I could wrap around my wrist. I did not wish to lose the bayonet in the coming combat, if it eventuated, when Jenny tried to leave with me.

So now the odds against me were more in my favour and the thought of dying had no hold on me after facing death so many times in Vietnam.

I parked at the back of the hotel in case I needed to make a fast getaway. I stepped out of my new car but vomited again. This sweat-induced sickness was steadily growing worse.

Thoughts of concern and confusion raced through my mind. Why did Jenny have her hair dyed that awful red colour and fluffed up on her head like that? Why didn't she get her tooth fixed when she knew I was coming home? Most of all, why on earth would she let me walk into something like this? Where was the Jenny I left behind? Where was the person I was? All gone! Everything gone! Do all beliefs need to be taken from me before the new may arise?

I walked up to the door of the large hotel at 5 pm and saw that the bar was packed. I had the bayonet out of sight, hidden up the sleeve of my shirt. I was ready to use it if these men attacked.

At the door to the bar, before I could enter, Jenny ran out to meet me and said, "Let's go."

None of the men came to stop us so we walked to my nice, white, shiny new car.

Then I did something else that I have always regretted. I let her down. I had no right to judge Jenny. She had come to Sydney to see me off and was waiting for me to return. She was a young and innocent country girl who had no chance in a place like

Western Sydney in those days. I did not know what I know now. I unexpectedly turned to her and said, "I cannot take you with me."

I will never forget the look on her face as I sped off. Being so young myself, I was confused. Either I was a coward, or a man who judged women for enjoying sexual freedom.

It seems obvious now that there was one rule for men and another rule for women in my male, religious-based society, and it had been firmly planted in my judgemental unconscious. On the long drive to Brisbane, I vowed to explore why women are condemned but men held in high esteem for their sexual adventures.

My ensuing years of guilt for judging and abandoning Jenny was finally alleviated when I learned from my mother, who had stayed in touch with her, what actually happened after I drove off, leaving her standing on the street alone with her bags packed. Jenny left Sydney immediately. She eventually got married and had two lovely children. Jenny was a big-hearted woman. I guess that is why she loved all those men.

I swore never to desert a woman in trouble again and I have put my life at risk in New York, Melbourne and Queensland to keep this promise whenever I saw women being attacked.

A case in mind stands out. While visiting New York for the first time, I was walking down the main street, lost in a virtual sea of moving people. I saw an older woman walk out of a type of hotel that rented apartments or rooms. A tent-like structure extended out from the hotel, reaching all the way to the edge of the footpath. As she emerged from the door two young white guys attacked her. They tried to drag her bag from her shoulder, which she was fiercely clinging too. They began punching her. She was screaming for help.

My mother and father had both warned me not to interfere if I saw anyone being attacked or robbed in New York. Robbers in the USA carry guns and would shoot without hesitation. Apparently, no one interferes because of this, they told me.

I was still really amazed and shocked watching the huge sea of people continue to walk down the sidewalk ignoring the unfolding situation. They had formed a moving U shape in order to avoid this robbery and vicious attack on an elderly woman. No one even looked at her as she screamed for help.

I already had far too much on my conscience because of the woman in Vietnam, and Jenny, so I ran straight in to help this woman and stood very close to these boys in case they did have guns or knives. I then roughly swung one of them around and told him to leave her alone. He took a swing at me so I tapped him lightly and he fell back into his friend's arms with blood pouring from his nose.

Still crowding them, I yelled loudly, "Leave her alone or I will really hurt you both!" They ran off and the woman raced back into the hotel in a panic, saying nothing to me.

During my short visit I could see that New Yorkers seemed to be a different breed of people and are forced to fend for themselves. I wondered if this is because of the fear that everyone is armed.

I quickly stepped back into the moving sea of people in case the boys did have a gun.

After leaving Jenny behind I drove all night from Sydney to Brisbane. When I got back to my army base, I told Jack what had happened. Being Jack, he wanted to drive straight back to Sydney and find these fellows, but I told him I was too sick and I went to bed. In the middle of the night, I awoke to a pain I had never experienced before or since. I staggered to the mirror and looked at the "Elephant Man" in the mirror. My face and entire body were swollen to twice their normal size and every joint was racked with pain. I collapsed on the floor.

Jack found me on the floor in the morning and immediately called the army ambulance. They rushed me to the military hospital where I spent the better part of a month, in isolation, as I wrestled with death, extreme pain, and recurring nightmares.

My body continued to swell. I spent a lot of time in a delirious, raving state and the pain was almost unbearable. I am allergic to morphine so that could not be given to me for the pain. They gave me other painkilling injections but these did not defeat it or seem to ease it.

The doctors tested my blood every day. When I could finally speak, I told them that I had recently been sprayed with a lot of Agent Orange during the war in Nam.

I was given more injections, including large needles into my chest, supposedly to keep me alive and my heart beating.

I had a plastic wall enclosing me and a large sign warning everyone to keep away.

Jack had been promoted back to sergeant. He was on leave, and seemed to be always near, outside the plastic wall.

He often said, "Come on, mate, you can beat this too."

The operation to repair my face was cancelled because I was too sick. The terrible nightmares kept on coming. There were two in particular. One had me back in Vietnam and Vietcong soldiers were coming at me from behind, while my old platoon was coming for me from the front. I was never armed and I knew they were, friend and foe alike, about to try and kill me. I guess I was close to death. I would wake up in shock, sweating, just as they all opened fire.

The other nightmare saw me coming home to my family who were standing on the veranda with someone who looked like me. I would say, "I am home," and again they would say, "Who are you? This is our son standing next to us."

This nightmare also left me with a terrible feeling and I would again awaken in a hot sweat. Of course, the truth was that I could never be the same person that left his family, both in looks and psyche – how could I be?

After a few weeks I began to improve, though my body looked like a skeleton. All my muscles were gone and I could barely stand up.

A few days later I was cleared of being able to infect anyone and could finally stagger along with the aid of Jack and a walking stick.

One day as I slowly grew stronger, I was stumbling along outside to get some welcome sun when this seemingly never-ending nightmare continued to haunt my life.

I ran into the forward scout who had unintentionally killed the seven women and children and then held the two grenades above his head. His left hand had been rebuilt using calf bones, while his right hand was now a steel hook. His face and head were terribly scarred.

I stopped him on the path and spoke to him. He did not appear to remember me and did not speak.

I shall never forget his eyes and the dull look on his face. Those lost eyes were vacant and stared straight past me at nothing. This was the awful thousand-yard stare that afflicted many returned soldiers who had seen and done too much.

I told him that I was there with him on that awful patrol and that the officers back in the safety of the camp had repeatedly told us, by radio, to cock our weapons with safety catches off and to shoot these so-called Vietcong, all dressed in black, on sight. I stressed that it was not his fault, but their fault, which caused him to open fire when he did. I told him that I or anyone else on that mist-covered patrol would have done the same if we had been the first forward scout to reach them, as he was.

This poor man said nothing and just kept staring at nothing. He then became slightly agitated and, pushing past me, continued on in his shuffling gait. I never saw him again.

I was discharged from the hospital, and the army, soon after that encounter. My hospital file contained the fallacy that an "unknown substance" had poisoned me. When I asked the doctor why it did not state the simple truth that it was Agent Orange, he answered that he could not put that in my file. More army and politically motivated cover-ups?

Jack was still on leave from his army posting so we drove towards my home in the far north, 1800 kilometres away.

I was very nervous about meeting my family in my skeletal condition. Now I looked even worse than just having a busted face with a hooked nose wandering all over it.

Because I was still very weak and had no muscle on my body, Jack drove. One day we stopped at a motel in Mackay which is a city situated on the road about halfway from my home.

We had a shower and then Jack took me to a hotel for a counter meal. He had a few large beers but I have never really been able to drink beer again since being so sick.

The bar he took me to was not very big and had about ten men inside. At one end of the bar were four bikers, dressed in their intimidating leathers with skulls on the back of their vests.

After we had eaten, in my weakened state, I dropped my plate and it smashed. Everyone in the bar cheered but one of the bikers called me a useless bastard and told me that if I couldn't hold my drink then I should piss off.

Brave bully of a man, I thought, abusing a skinny, sickly looking weakling, but he did not count on Jack looking after me.

Jack turned to these four men and asked the big, very dirty, bearded man who had spoken to me, what he had said.

The man answered, "And you can fuck off too!"

Jack started towards them and I grabbed his arm and whispered in his ear, "Jack, I cannot help you, I am as weak as a kitten, and there are four of those big bastards. Let's just go, mate."

"Fuck them," he said, then shoved his glasses into my hand, shaped up to the big bearded biker, and began to fight.

I soon felt like vomiting. My nervous system was already badly damaged and my legs were close to collapsing as sweat broke out all over my body.

I did not realise just how tough Jack was until I saw him fighting in this bar. He was a coal miner from Ipswich before joining the army and he had also won an Australian Golden Gloves title. In this bar, the lights were bright so he could see with his glasses off this time.

When he hit this man, who was bigger than him, Jack would make a noise out of his mouth and nose that sounded like a steam train. I also noticed that he tucked his head into his shoulder and always led with his left hand. He was a typical well-trained boxer.

I stepped forward and said to his mates, "Stay out of it."

Thank God they did because I was useless, but they did not know that, maybe because I wore a long-sleeved shirt and they could not see my skinny arms.

Jack was easily winning this fight and then he did

something quite strange that shocked me. He dropped his hands and invited the big biker to hit him. The man obliged and hit him in the face with a swinging right hand. The sound of the punch was very loud and blood flew off Jack's busted face. He shook his head and yelled at the fellow to hit him again, which the man quickly did. More blood.

"Jack," I yelled, "what are you doing?"

He shook his head and let out a roar and flattened the large man with a combination left, right punch. The fight was over.

When the man came to, Jack picked him up and bought him a beer. Everyone laughed and said it was a good fight.

Finally, and thankfully, we left.

I asked Jack why he let the bloke hit him like that. He just said, "I needed it."

I would not find out for a long time what he meant by that.

We finally arrived home and of course my family welcomed me, though Mum asked me, through her tears, what had happened to my face and body. I just shook my head and said, "I shall tell you one day, not just now, Mum."

Jack stayed for three weeks and got on very well with my family, but we did not talk about our time in Vietnam to them, to each other, or to anyone else. Best that we tried to forget.

I decided to get my strength back so I soon began training and eating healthy meals. It took me many months to regain my fitness and my lost muscle but I would never be the same again. I would never play football for Queensland even though this had been my ambition when I left at the age of nineteen. I was once a good footballer who represented my city, and North Queensland, and had been told by coaches that I could possibly, one day, play for Queensland.

Jack wanted to visit all the bars in town while he was there. I do not particularly like hotels as some drunken men like to fight.

Jack got into another fight and scared the hell out of me again when he kept letting the man hit him in the face. His face was becoming scarred. Did he want to look like me? Lucky for me, he won, as I still would have been of no use to him whatsoever.

He finally left, much to my relief, to report back to his battalion. I went back to work in the large office where I had been studying accountancy before the war.

At least once every couple of weeks my sight would fail and I would get a severe migraine headache which would last for a long time until I vomited. These headaches were debilitating but I refused to give up my training. I was running, though very slowly, plus boxing and going to the gym. I even began studying karate but I found that my boxing, plus a couple of street-fighting skills, served me better. There are far too many moves to remember in karate and fractions of seconds are important in any type of fighting. Karate is a fighting style that I felt needed to be studied and practised for at least ten years to fully master it.

Chapter 15 - The Hotel

My mother and father owned a working man's hotel near the wharf and after work each day their hotel was packed with men, many of them tough wharf workers. In those wild days, nearly every night there would be a fight and on Fridays and Saturdays it got so bad that my dad built a boxing ring out the back of the hotel. A lot of the fights would not get to the ring, erupting in the hotel instead. I stayed out of the fights while trying to help Dad stop them. I was still much too weak to become caught up in fights because my muscles were taking their time to rebuild.

I hated seeing my dad end up in fights. He was too old to be doing this and he did not have the ability to fight like my Uncle Boof and me.

After about eight months of serious training, combined with much pain, I had finally regained a lot of my lost muscle and fitness.

Finally, one day after I had finished work at the accountant's office, I arrived home to find my mum on the floor, my older sister screaming, and Dad being beaten up by a large man. My sister screamed at me that he was the one who had also hit my mother.

I was still dressed in my work clothing wearing a white shirt, a tie, and shorts, with long socks pulled up to my knees. I looked like anything but a fighter.

I dropped my briefcase and tapped the man on the shoulder. He was still punching my dad who was thoroughly beaten and covered in blood.

When he turned around and looked at me, I told him in a quiet but deadly voice that he should not have done what he did to my family and that a real man would never hit a woman.

He took one look at my clerk's dress then laughed loudly. Then he tried to hit me.

I easily avoided his telegraphed punch and began to teach him a lesson, with Dad, Mum and my sister yelling at me to flatten him. But I did not want to knock him out.

I began hitting him, not too hard as I wanted to lecture him as I slowly punched him down the corridor and out of our hotel, step by step, until he was on the footpath. I then warned him that he should never again hit women or old men, and he must never come back to our hotel.

Finally, I did what my family was yelling at me to do and knocked him out. When he fell, his head made a loud noise as it hit the concrete footpath. I wondered if he was dead and hoped that he was not. I felt his pulse. He wasn't dead so we rang the ambulance and they took him to the hospital. He never came back to our hotel again.

Finally, I had regained my confidence and my boxing ability. I let it be known that if anyone touched any member of my family, or started a fight in our hotel, then they would have to answer to me.

Over the next few months some of the men who were always fighting, decided that they had to try me out. I tried not to hurt them as they were often drunk when they attacked. Often, I only had to slap them to bring them to their senses. I did not enjoy causing anyone to bleed or hurting people.

There was one gang of tough guys who drank at our pub. Their best fighter finally challenged me. After a few minutes of fighting, he laughed and shook my hand. That was a big turning point for our hotel.

Finally, my parents were running a peaceful hotel as my reputation quickly spread. The young man who had not been able to fight was now a fighter again after eight months of hard training.

I eventually passed my accountancy exams but the headaches grew worse and the pain became a nine out of ten.

Even more serious than the pain was the depression that overcame me at least once a month. A depression episode felt a little like a volcano inside of me trying to erupt. On those awful nights I followed Jack's example and went to other hotels where I definitely did not fit in nor was welcome. To make matters worse I purposely had grown long hair and carried a shoulder bag. This was unheard of in the 70s but became common practice for men when they followed the European fashion in later years. I did not want to fit in. I didn't have the ability to fit into the common culture.

My excuse for going to these hotels was my belief that this was a free country and people had a right to go wherever they wanted to, whenever they wanted to. I also had a grudge against gangs, or bullies who preyed on the weak. These people suited my crazy moods.

One of the hotels I went to was for black people only and the other was for bikies or tough prawn fishermen. I would go and plonk myself down in the middle of one of these public hotel bars and order a rum and Coke to sip. Usually, within a short time I would have three or four men surrounding me trying to pick a fight by calling me a white cunt or a poofter.

I would not speak or move from my stool until someone swung a punch at me or grabbed my long hair. Then I would let the hidden volcano erupt.

I was a boxer, and now a street fighter, so the fights never lasted long even though I would not hit the perpetrators too hard. These awful fights made me quite sick after they were over and I would go home and lock myself in my room and try to understand why I was doing this to others and myself. I know Jack did it but I had no answer to why I did it, or even why he did it, so I went to see a doctor who referred me to a psychiatrist.

4X4 crossing river. My escape from the world heading bush.

Chapter 16 - The Psychiatrist

When I walked into the psychiatrist's office for my first appointment, I noticed two things. One, he had a large photograph on the back wall of himself on his game fishing boat, standing proudly next to a thousand-pound dead Marlin, hanging by its tail from a weighing station. This was a magnificent creature that would have been shamelessly dumped back into the ocean after he took his egotistical hunter's photos.

Secondly, he had a strange recurring twitch to his face. Every so often one eye would shut and his head would jerk spasmodically to the left. Strange indeed, I thought, for a man who, I was told, fixes people like me. Why couldn't he fix himself, I wondered?

I had been writing stories and poetry while trying to figure out this strange world and my uncontrollable moods. I innocently gave the doctor these writings, hoping that he could figure out what was wrong with me and why I went to dangerous places where I knew people would pick fights with me. I suspected that it had something to do with my feelings of guilt and shame.

The writings were about Vietnam with a lot of poetry mixed in. I felt that I had to finally trust somebody so he was the first person I had shown my stories to.

He spent an hour with me, then prescribed some tablets called Valium which I obediently took for two weeks. He also assured me that he would read my writings first, then see me again.

These tablets calmed me when I took them but next day, when they were wearing off, my nervous system would suffer and I would get shaky, nervous, and easily angered. I noticed that this debilitating condition worsened, the longer I took Valium.

I told the doctor what they did to me, the next time I saw

him, so he gave me some other tablets called Tuinal, which were a strong barbiturate. They had somewhat the same effect on me as Valium, and I did not like taking them, so he experimented with other tablets. They all had the same, eventual, negative effect on my nervous system when they wore off.

Finally, he prescribed some tablets that worked well called Mandrax. These did not seem to leave me with a hangover, or anger, or a damaged nervous system.

After finishing work one Friday afternoon, I went home to the hotel just wishing to sleep as I was very tired. The good doctor assured me that if I took one of these Mandrax I would sleep all night. I took one at 5.30 pm, expecting a long night of sleep.

The enormous relief I suddenly felt when the tablet kicked in after an hour was quite amazing. I mowed the back lawn, bottled a keg of beer for Dad, then served in the bar, which I usually hated doing. During this time, when I should have been asleep, I was energised but relaxed as I laughed and joked with our customers.

I felt like my old self, happily human once again. Then, when our hotel shut, I went dancing, joining in with a crowd of happy people. I had been avoiding crowds since Vietnam.

I woke up next day feeling refreshed and again ready for work in the family hotel. What an incredible relief this tablet had given me.

Now some people are just natural born addicts and if they discover something good like Mandrax they will take them both day and night. These people are in the minority and I have never been like that. I would take only one of these tablets just once a week on a Friday night, so I could relax and feel somewhat like I used to before Vietnam, not someone who was nervous in a room full of people and who reacted to loud noises.

I even had a date with my first girlfriend (since my split from Jenny), meeting her on a Friday night after taking one of these Mandrax.

The tablets continued to help me until the government banned Mandrax because the drug addicts were abusing them.

The minority rules. This ban didn't matter much to the addicts. They just switched to more alcohol, cigarettes and street drugs, as an addict is an addict unless he or she alone decides to change, perhaps after enough bad experiences.

The doctor experimented with other drugs but after a while I realised that there was always a price to pay for taking tablets, probably even Mandrax, so I stopped taking any of his experimental prescription drugs.

I finally realised that this psychiatrist could not help me so the next time I went back to his office I asked the receptionist for my writings and told her that I did not wish to see him anymore. She was a nice girl with whom I had become friendly, but her face changed colour and she became embarrassed. She softly replied that my writings were gone.

"Gone where?" I asked.

She glanced at the doctor's closed door, turned to me and whispered, "He burnt them."

I felt like bursting into his office, lifting him up and shaking some sense into this ignorant man, an expensive pill pusher. My writings were important to me. They helped me cope in times of despair. They are something I had to do. He had no right to burn them and I wanted him to know that.

Then I thought of something much worse that I could do to him to wake him up so I scheduled an appointment for the very next day.

Still quite angry at him for being so heartless by burning my stories and poems, I walked into his office and sat down. I immediately pointed to the picture of him in his large boat with the dead Marlin.

"That is a beautiful boat, doctor. What is its displacement and what size motors do you have in it?" I asked in order to pique his interest.

For the next fifteen minutes this man told me about his boat, his fishing, and the number of Marlin he had caught, never, of course, using the word killed.

During the entire time that he was raving on I would exactly mimic his spasm and twitch every time it occurred. I did this for fifteen straight minutes and not even once did he give any indication that he noticed my mirroring of him.

Gradually my heart softened as I realised that this man spent a large part of his life creating stories by looking into the sewers of people's minds and this job had finally driven him more than a little mad. I felt genuinely sorry for him, and, bowing my head, took my leave without even mentioning my lost poetry or stories. What use would that have been with this man who was obviously so damaged?

I never went back to see him and he died soon after.

If someone is looking for the negative in other people for their entire adult life, this probably acts like a slow poison that eventually kills the therapist.

Meanwhile, my debilitating and painful headaches were not improving and I had to leave my job in accountancy, as my stomach troubles were also becoming worse. There were so many things that I could not eat or drink or the deadly headaches would follow.

Over the next few years, I worked in many jobs such as brickies labourer, waterside worker, truck driver and carpenter. My parents thankfully sold their hotel so I tried being a car and then an insurance salesman. I was even a cowboy on a vast cattle station but for some reason I could not stay in a job for very long.

Unsettled, and worried about my inability to stay at a job, I had begun going back to the "out of bounds" hotels when the depressions overcame me. But the men in these hotels now knew me and left me alone and I would never, ever, pick the fights with anyone.

Eventually I found a solution which allowed me a constructive, instead of destructive, coping mechanism. I took a job as a bouncer (crowd controller) in the largest hotel in town. The owner of this hotel, a lovely old lady whose husband had recently died, told me that she was having a lot of trouble with fights and it was affecting her business.

On the weekends the hotel would get a crowd of rough customers as this was the place known as the Barbary Coast. There were fights every week because the customers ranged from army and navy men, to mine workers on leave, prawn fishermen, cowboys, footballers, wharf-side workers, bikies, and other likely lads. Alcohol encourages some men to think that they are better than they really are and so they fight.

After months working as a bouncer, I took over the job of managing all the entertainment and the security. This meant that I hired the band, plus different individual entertainers, while also taking the entrance fee at the door.

I also brought the crowds under control by hiring my own bouncers.

I realised that if I wanted to run a successful establishment, I had to stop the fights. I gathered together nine of the best men I could find in the city. I sat them down and told them that I did not want men working for me who, because of nerves or ego, would punch or attack anyone. If a fight broke out, I stressed that we must all go in together and two men were to pick up each fighter and carry them outside without hurting them.

This, of course, did not always work, as the really tough fighters could not be stopped in that way, but most could. My business became very successful with large local crowds. In a short while I had four establishments and I was running all the entertainment in the large city.

I began to fly in the best bands and biggest stars of that time. Stars such as Johnny Farnham (a really nice guy), Kamal, the suave Indian crooner, Dinah Lee from New Zealand, Freddy Paris from the US, Jeff Phillips, Frankie Davidson, Tony Worsley, and many others.

When they came to town, I would be their personal bodyguard as well as promote them. I also became the MC on stage and would introduce them to the audience.

Because the crowds were so large and diverse there was still occasionally trouble. For instance, one night I arranged for Johnny Farnham to appear in two consecutive venues because my hotel was not big enough to hold all the people who wished to see him perform.

My hotel was the first venue and the event went off well with no trouble, but his second gig did not. I had booked him into the largest hall in the city as he was super popular and I needed the huge area to hold the already booked-out audience.

To get into this hall we had to go through a side entrance that was packed with the people who had not been able to get tickets but wanted to catch a glimpse of Johnny.

I could only afford to take one of my men with me as the rest had to keep the peace in the other crowded hotel. I took one of my best men, a big Maori guy named Eddie – a truly big man with huge muscles and a good heart. Maoris have a reputation as wild men but the ones I knew, and hired, would much rather sing than fight.

I had Eddie walk in front of Johnny and clear a path through the crowd while I stayed close behind and watched his back. A young strong-looking man, whose girlfriend was screaming madly for Johnny, suddenly erupted in a fit of raging jealousy. He was drinking from a large bottle of beer, which he smashed, then advanced towards Johnny's back with the raised broken bottle. I stepped in front of the jealous, drunken fool and quickly disarmed him then hit him with a short sharp punch to the solar plexus, which hardly anyone saw. He collapsed on the ground and without missing a stride I followed Johnny into the hall and introduced him to the sold-out crowd. The rest of the night went off well probably because there was no drinking allowed inside this unlicensed venue.

Dealing with the public allowed me to meet a lot of people and I became even more disillusioned with the society I was living in.

I had decided to run a weekly women's only night in one of my smaller hotels. Strangely enough when women gather for a "ladies only" night on the town they are wilder and louder than men in the same situation.

At the end of a night of male strippers and other organised frolics, the strippers would stand up on the bar and the ladies would bid to see who got to take one home. Married women, many of whom I personally knew, would usually bid the highest.

One night I saw a beautiful young blonde woman, who had earlier that night told me she was engaged to be married next week in Sydney, win the bid for a stripper who had a bad reputation. I did try to warn her away from him as he was an indiscrete and less than fussy womaniser. Unfortunately, she was raging drunk and informed me that she was determined to have sex with him before being "tied down". I told her to at least definitely use condoms.

The men were no better when it came to easy sex, though they were much quieter than the women on the "men's only" nights. A lot of these married men had other lovers. I knew some of their lovers as well as their wives.

Christianity and all religions may one day realise that they cannot control humans sexually. The more they try to suppress sex, the stronger the sexual urge becomes. Look at the poor priests who are ordered to be celibate for life. They end up having sex with young boys, girls or women in their congregations. Sexual suppression creates its opposite.

Of course, if either of the couples found out about the hidden trysts, then all hell would break loose. On a couple of occasions, I had to take a shotgun, then a rifle, from two different men, hell bent on murder.

I took other assorted weapons from enraged and jealous women. One lovely lady had a small, black, hand-held contraption that she said delivered 50,000 volts of electricity and she was intent on attacking her husband with that.

This made me think, yet again, that there was something awfully wrong with our religious teachings that make sex a taboo unless it conforms with Christianity and Islam's laws of marriage. This is ingrained in us all as we grow up in our society where religion is taught in schools, kindergartens, mosques, temples and churches.

Religions cannot control nature by making futile laws about sex and marriage. This just creates insane jealousy and judgements that too often lead to murderous rage, death or beatings.

I realised that love and freedom can never co-exist when jealousy controls the ego.

How can anyone make love with judgement in his or her

heart? That is not possible. These men were more like frustrated, quick "fuckers". The women certainly did not seem satisfied and sometimes laughingly referred to their vibrators as better lovers than their men.

Oh, my goodness! This crazy world I again found myself in. I sometimes wondered, as I did in Vietnam, if I had come from another planet?

I noticed that the women who the men spoke the most disgustingly and harshly about were usually the most loving and kind-hearted.

Why were they treated differently to men? Because in society, women are treated as "lesser beings" and I found this, harsh, ridiculous and quite primitive.

Sexuality is a natural thing and some women enjoy sex more than others, perhaps to share their love. Why are they stoned, burnt at the stake (as they were in the past by Christians) or ostracised and criticised, even beaten or murdered?

Then a memory hit me like a hammer and drove the wind out of me. I remembered my poor darling Jenny. How could I have left her in that awful hotel? She had the best heart, and just because sex was not shameful to her, I judged her. I realised that I was as judgemental as these men I was watching. I sat down and did something that I very seldom did now. I cried. I was one of these very men who were so offensive to me. Ah! The honest mirror of life. I realised that life was this strange mirror that was teaching me, or trying to, albeit ever so slowly.

Chapter 17 - Mary

I was making bucketloads of money in this entertainment work at these popular nightspots, but, as usual, I could not stay very long at anything I did, no matter how successful. When this something once again surfaced it felt like a force was stuck in my throat and my stomach contracted in spasms of pain. Gradually I would become sick and weak and my body would begin to break down. It seemed like an invisible force was controlling me against my will or was it an uncontrollable force of destiny?

In order to rest my body and mind, I bought a 47 foot, steel-hulled dive boat, and dived for crayfish and trout, while sailing in the waters from Australia to New Guinea. I took a crew of three with me – two divers plus my new girlfriend, Mary, as the cook.

Mary was a blonde lady with striking features. When she walked into a room, she became the centre of attention because of her good looks. Because she came from Melbourne, Mary also had a sense of fashion as city girls often did in those days, so she was always well dressed.

She also had quite a fiery temper and I had seen her use it on people who became too familiar with her, including me.

Yet we got on well together and enjoyed a happy love affair on the sometimes wild ocean. I noticed that I held part of myself back and never committed to a future together. I could not say the words "I love you". Even when I wanted to express these feelings of love, fear immediately arose to shut me down. Perhaps I was protecting myself as I did not want to be badly hurt again or maybe I couldn't bear to be normal and perhaps lose my freedom.

After three months on the sea Mary wanted to go back to Melbourne where she had a job waiting as a popular hair stylist.

She said she missed her large family but was also probably fed up with my casual way of relating.

We kissed goodbye, knowing we would always be friends.

Six months later, after the boat was almost shipwrecked in a savage cyclone, I sold it. I returned to living back in the city and one day Mary rang for the first time since our bitter sweet departure. She called me asking for help. Mary was pregnant and her parents were giving her a hard time. She needed the support of a friend and asked if she could come up and stay at my place where she would have the baby and then adopt it out.

As Mary was my friend, and past lover, of course I agreed as I had a large apartment in town and two well-paid jobs.

Unfortunately, I was working extremely long hours during her pregnancy both for the government during the day and as a bouncer again at night. I was not by her side, even though I had planned to be, when she was unexpectedly rushed to hospital and had the tiny baby. When I was contacted, I immediately hurried to be with her and found her extended local family of uncles, aunties and cousins crowded in her hospital room, but the baby was gone. I asked her what she had done with the baby and she tearfully said that she had already signed the adoption papers. Her visiting relatives told me that they had encouraged her to do so.

I demanded that her family leave the room immediately. I wanted some time alone with Mary. These people had talked her into adopting out her baby girl while she was in a confused, weakened state after childbirth. They were callous and uncaring of her personal desires. Giving up a baby was a huge decision which she could regret for the rest of her life.

I sat down on the bed beside her, held her hand, and asked her if that was what she really wanted. She said "yes". Then I asked her what the father of the little girl would think about this. She looked at me with her mouth agape, eyes wide, and said, "You are the father, stupid."

I nearly fell off the bed in shock. When I could speak, I stammered, "Why didn't you tell me?"

"I thought you would obviously know. I would not have

asked you if I could have the baby up here with you if it was not yours."

I was stunned and speechless. I eventually stumbled out of the hospital room and walked down the corridor to a desk where a matron and a doctor were sitting. I told them I was the father and I wished to see my daughter. They pointed to a large glass window a little further down the corridor.

Behind this window was a small cot with a plastic cover enclosing it and a scary looking tube protruding from it. Lying on her little stomach with her eyes closed, and looking so very vulnerable, was my daughter. I stood and stared at her for a long time then I walked back to the desk and said to the doctor that I would like to adopt my daughter back, because I was the father.

He said, "I am sorry, but in Queensland the father has no rights of adoption." Then he looked back at his papers and ignored me.

I stood there in turmoil for five long minutes, never moving, hardly blinking, thinking of the cruelty of this world. I then said something that surprised me and could easily have landed me in jail.

"That child is my daughter and you will be in serious trouble if you imagine that you can take my daughter from me."

Because I still always carried a large knife on me at all times at this stage of my life, this was not an idle threat. I was prepared to die to save my daughter from the moment I laid eyes on her and was ready to take on the whole world if I had to.

The doctor and the matron both looked up with a start and I noticed the doctor's face change colour when he looked into my eyes. He could see immediately from the look on my face that I was deadly serious.

Without another word the matron handed me some adoption papers and said, "Fill these out and get your girlfriend to sign them, as well as you, and the baby is yours."

I filled out the papers. Then, with the matron's help, I took this little bundle of wonderful innocent joy into Mary's room where I again angrily chased her uncle and aunt, cousins and friends out of the room. I placed my darling daughter on Mary's stomach and said, "She is our baby. Sign this."

Mary held her little daughter and cried. She has loved her daughter dearly ever since, even though I was obviously not the ideal partner. However, I promised to be forever in my daughter's life and always support Mary.

Thanks to the papers I filled in, my daughter has kept my surname throughout her life. We have always been very close, often travelling alone together to many wonderful places in the world.

This child changed my life. She gave me a reason to open my heart again and to love unconditionally. I am totally convinced that if it were not for my love for her, I would not have been able to remain alive as long as I have. Love gave me a reason to live.

After the adoption I bought a home, using my war service loan, and the three of us moved in together.

I again took over the running of the hotel where I had been working as a bouncer. I left my government job, which was boring beyond belief.

Life was good for a time. I was making lots of money, but this unknown force in my solar plexus and head still drove me into despair on bad days and seemed to cause my body to turn to stone. I wondered again why I did not seem to be in charge of my own life. I just wanted to be a "normal" father and loving partner, living a happy life like people in the movies appeared to do.

Settling into family life was a challenge when things would get out of control in my hotel on the esplanade. On one particular night a large Australian navy ship docked and about 120 of the crew came to my hotel. I was concerned about a particular sailor who weighed at least 18 stone, was all muscle, and who behaved like an animal. During the course of the night, he flattened several of his own men with just one punch. A large Irishman backed him up and no one dared to stop them.

I did not wish to start an all-in brawl with 120 sailors in my hotel but I realised that I had to stop this blood bath. I asked Maori Eddie, who weighed about 16 stone 7 pounds of all muscle, and Big Tony, a huge, 22 stone professional footballer, to tell the big sailor and the Irishman to leave.

The two sailors loudly demanded that Big Tony and Eddie go outside with them. They both did, together with me, the rest of my security staff and, unfortunately, the rest of the sailors.

I then witnessed that big sailor do something that I would not have thought possible. He was arguing with Big Tony, who was doing his best to placate him, as Tony was a gentle man, for all his massive size, not a fighter. They were standing next to a Holden sedan car when the sailor picked Tony up under both arms and sat him on the roof of the car.

We could not believe this feat of strength and were all taken aback with no small amount of fear. How could one man be this strong, and able to lift a 22 stone man so effortlessly? I was now more concerned because all the sailors had watched him do this. I wondered if these men would back him up after the way he bullied them. Probably they would, because men in uniform usually stick together against all comers.

I quickly raced around to the front of the hotel and told the bar staff to make sure that the doors of the hotel were locked. The night was definitely over. Luckily it was a Monday night so none of my regular clients were there.

The big man and the sailors had followed me around to the front of the hotel, together with my security staff of nine good men.

Let me tell you that none of us, or almost none of us, wanted to fight with these sailors. How could we possibly win if a fight broke out?

I stepped to the front of the large crowd and addressed the big man and the sailors standing behind him. I surpassed myself and quickly had them all calmed down and had even gotten some of them to laugh, so they appeared ready to go home.

Then Alfie, my smallest, redhaired bouncer who was standing behind me, did something that froze our hearts. The big Irishman (who weighed about 55 pounds more than Alf) stepped forward and said something derogatory to him while I was still talking to the rest of the sailors. Alf immediately punched him and they began to fight.

Alf was coming off second best and I knew I had to help him. Before I could reach him, the big sailor stepped forward, addressing me with his huge fists clenched preparing to strike, and said, "You are dead meat, little man."

When someone talks to me that way, I immediately know they have a weakness, but what was his? I could not see it. He was as strong as an ox and had a deadly knockout punch which I had seen him use time and again over the course of the night on his own men.

I quickly observed the way he was holding his fists. I could see that he was not a boxer so probably did not have much technique, not that he had probably ever needed it with his size.

He was standing about 2 feet in front of a large four-cornered cement post that supported the front of the hotel, while he was still roaring threats at me.

In moments like this the first thing I feel is fear. This is a wild energy that courses through my body and makes my heart beat at an alarming rate. But then a deadly calm comes over me and I smile. I can't help it. In this fighting mode I know I am much stronger and faster than I usually am. It's probably caused by the huge amount of adrenaline pumping through my body.

I quickly worked out that I must hit this raging monster so that his head bounced off the concrete post. This may be the only way I could win, and win quickly. To arrange this tactic, I had to get him to throw the first punch with all his weight behind it. That should be easy, I thought, as he was already angry and red faced, and continued to spew threats at me.
He would definitely go for the knock-out punch.

Now if, on the other hand, he got a hold of me with his ape-like arms, then the fight would be over and he could easily kill me. He was simply too strong.

All this information flashed through my mind in a seeming second, while I dealt with any doubts that could slow me down, and I remembered to quickly reject any negative thoughts that would drain my energy.

I took my fighting stance and called him a fat, useless puss.

I knew that would make him retaliate because "puss" is the derogatory term that army men call sailors. He immediately took the bait and swung a huge right-hand punch with all his weight behind it. I leaned back quickly, making the punch narrowly miss my head. Then using his forward momentum, I hit him with the usual left-right-left combination. The first two punches knocked him back the two feet I required, and his head was then inches from the corner of the cement post. My third punch was a beautifully timed straight left. It was my best punch and caused his head to bounce off the corner of the concrete post.

I stepped back and time seemed to stand still while this huge man fell forward as straight as the post, unconscious. When his head hit the concrete, it sounded like a watermelon exploding. I hoped he was not dead with all that body weight hitting the cement.

There was a stunned silence and then a few of the sailors cheered. I was so relieved to see that they were on my side. They had witnessed a fair fight and the ship's bully was finally beaten, probably for the first time, and probably much to their relief.

I turned back to assist Alfie. The Irishman had him up against the wall of the hotel and was laying into him with punches. Luckily little Alfie was as tough as nails.

I quickly turned to my friend Peter, whose mum owned the hotel, and quietly told him to grab one of the Irish sailor's arms and I grabbed the other. My back was facing the sailors and I threw one sweet punch. Because I could not afford to let the other sailors see me hit the Irishman as that would not seem like fair play, I used my left fist holding it in front of my body out of view so the sailors could not see. I threw this punch a distance of only three inches, holding my fist and arm as used in an uppercut. I aimed the punch for his solar plexus, a blow which will stop anyone if landed in exactly the right spot. A fraction of a second before it connected, I twisted my fist into his stomach. The Irishman collapsed in our arms, doubled up and gasping and no one knew why.

I spun him around and handed him to the sailors, and said, "Take them home, fellas, they have had enough." The sailors picked them up and carried them off. They sailed next morning and that was the last we ever heard of them, thank goodness, as I did not want a rematch.

Thinking about my life and the number of fights that I was getting into, it made me feel sick again and I knew I had to find Jack and ask him why we kept putting ourselves in harm's way.

The very next day I took my leave, jumped on my Honda Four motorcycle, and rode the 400 kilometres to the nearby city of Townsville. I had recently found out that Jack was still in the army and was still with the 2nd Battalion which was now based there.

At the Army Headquarters after I informed the officer in charge that we had served together in Nam I was told, with a knowing smile, that he had been promoted back to sergeant, yet again, and that I would find him at a certain hotel in Townsville. As I turned to leave, the officer warned me to be careful, because that hotel was off limits to most men. I wondered why, but was not surprised that Jack would choose to be there.

On finding the hotel I remembered the officer's words of warning and quickly took note of where the exits were if I had to leave quickly. I cautiously went around to the back of the hotel and entered from the exit door.

It was obvious when I stood observing the scene that this was a hotel for black and mixed race people. Prejudice against whites was the norm here in their nominated hotel. These men had probably suffered the same prejudice against them in a predominantly white society and were thoroughly sick of it. This was their hotel, and as they were not allowed into most other white hotels in this city, so they did not allow white men to come in here. Fair enough, I thought!

Sitting around the bar were quite a few harmless drunks but also quite a lot of very tough and dangerous looking men.

In the middle of this "black only bar" sat a sole white man with his thick unexposed arms resting on the bar. He had his back towards me. Behind him stood a black woman holding a bottle by the neck, raised dangerously above her head. She was yelling threats at the white man sitting at the bar who was completely ignoring her and her deadly looking bottle.

I could see now that it was Jack, so I walked up behind the woman and took the bottle out of her hand and told her to leave my mate alone. She turned and began cursing me so four of the men stood up from the bar and came towards me. One of them said, "Hey, what the fuck you want in here, you white cunt."

These men were not the real black men of Australia. One was nearly white while the other three could be called half or quarter castes.

I experienced racism coming from both whites and blacks and I could never tolerate such ignorant and violent human prejudice. Any type of prejudice is all the same to me. I do not like it whether it is coming from red, yellow, white or black men as, really, we are all the same. I noticed that these four men were the more dangerous types because they were only slightly coloured. They did not really fit in anywhere in those days. I had noticed in the past that mixed race people from any country are usually quite big and strong because of the healthy cross-breeding that produces them.

I answered, "I have come here to see my mate and you would be real smart to leave us in peace, brothers."

After saying this I placed the large bottle on the bar. The men looked at me, probably wondering why I put down a perfectly good weapon. Then one of the men ran at me and swung a haymaker of a telegraphed punch which I easily avoided. They did not realise that I grew up in a coloured family. Some are my family. My great grandfather had married an Aboriginal woman and one of my grandfathers, on my father's side, had married a Chinese woman. How in hell could I be racially prejudiced?

So I simply slapped this man's face as he was coming at me with his misplaced weight. This was enough, however, to send him skidding across the floor and crashing into the far wall. The other three looked shocked and one of them said, "He knows karate," then they all turned and walked back to their seats.

I did not know karate but this was the second time someone had said this, which saved me, or them, from further harm.

I turned back to Jack who still had not moved. I was amazed

that his back was still towards me. This man, my mate, had always had my back in the past, with no questions asked, yet he had not moved when these men threatened me.

I slowly walked over to the bar, placed myself on his left side and said, "G'day, Jack. It's been a while."

What turned towards me would be indelibly imprinted on my mind forever. My mate was drunk and spitting white froth. He slowly stood up and yelled, "Fuck off, cunt!"

What shocked me much more than his anger was his face. The lens on the left side of his glasses was completely shattered, just like it had been after his fight in Vietnam. He could not possibly see anything through this lens. And his face! Good God, his face! It was just an ugly mass of scars covering scars. His nose looked as if it had been broken many times and was spread over his poor, almost unrecognisable face. He stood and faced me, with fists clenched.

I said, "Jack, it's me, Mac."

Recognition slowly dawned and his twisted, angry face relaxed.

"Hey, mate, what are you doing here?" he asked in a normal voice.

"I came here looking for you," I answered.

"Why?" he asked.

"Because you are my best mate and I have a problem."

"Hey, give my mate a rum and Coke," he told the barman. Then he said, "Now, what's your problem, mate?"

I told him that like him I was going to the most dangerous hotels in my hometown, when I got in a certain mood, and I would wait for people to attack me as I knew they would because I stood out like a sore thumb with my long hair and the shoulder bag I carried. I told him I had been to doctors and psychiatrists and they could not help me to explain why I did this.

"What is wrong with us, mate? We come into these places where we know we are going to get into fights. We are not supposed to be here. Do we want to die?" I asked him.

He slowly turned fully towards me and told me his

shocking story, in his usual caring effort to help me.

Looking at his mashed, scarred and battered face, I felt like crying. I had a few scars as well, but nothing like his. I had scars from where I had been hit with a pool cue, another where I had been bitten over my right eye, and a couple of scars on my lips from sometimes unavoidable punches. However, Jack's mouth looked as if it had been put through a grinder. His nose was beaten to a pulp, and deep scars criss-crossed his entire, once handsome face. He could not possibly see out of his left eye. I hoped I would not end up like this but I couldn't figure out how to stop being like Jack.

Jack was calm and acting like my old mate when he asked me, "Do you remember Operation Coral in Vietnam?"

We were outnumbered by a force that just kept coming, especially at night. One of our platoons was totally overrun. All the men in this unfortunate platoon were killed or wounded. One of them, an old school mate of mine who had survived, had both his legs blown off and spent the remainder of his life in a wheelchair. I eventually attended his funeral.

They say that there are no atheists in the foxholes in war, and I could understand why, but I have always refused to become a beggar by asking some god to save me just because I was scared.

Jack continued. "My section had become isolated [a section is usually seven men] and we had been fighting on and off for days and lost some good mates. We were angry and bloodthirsty and had finally pinned down a section of the Cong and had fought until there was only one of them left alive, a North Vietnam officer. We were behind cover on the side of a hill and he was in a gully, also behind cover, about 80 yards to our front. We could not get a shot at him."

Jack took a deep breath and continued: "Using our interpreter, we spent a long time trying to talk him into surrendering but he kept replying that we would torture and kill him if he surrendered. We told him that he would be treated well, have hot meals and a safe bed, until the war ended, and then he could go home.

"He kept holding something up in his right hand trying to show us what it was, but we could not see it clearly from this distance. Finally, after hours of negotiation he agreed to surrender. He slowly stood up with his arms in the air, still holding this piece of paper in his right hand. I, and every one of my men, opened fire. We had an M60, plus automatic weapons. You could see pieces flying off him as he died on his feet. When he fell, I told the men to cover me, and walked over to what was left of his body. In his right hand he was still tightly holding the paper that he had been waving at us."

Jack reached into the left-hand pocket of his shirt, the one covering his heart, and handed me what this man had been holding. It was a photo of him sitting on a chair with his wife next to him. On his knee was a little girl of about three, and standing next to him on either side were two slightly older children, a boy and a girl. They were all smiling at the camera and looked like a lovely, happy family.

Jack gently took the photo from my hand and placed it, almost lovingly, back over his heart with the picture facing inwards.

Then, with spittle flying out of his mouth once again, he reverted to this tortured, strange self. He screamed at me, "Now fuck off!"

I slowly stood up and addressed him before I walked out.

I said, "Jack, it was not our fault. Those thoughtless bastards trained us to kill at the drop of a hat. They never warned us about the emotional consequences of our actions or how to handle our hatred after our mates were killed and maimed. Come with me out of here, mate. You have suffered enough."

Jack stood up and for a moment I thought he was coming with me. But instead, he threw a weak punch that I let hit me, and then yelled at the top of his voice, "I said fuck off, cunt!"

The rest of the people in the bar were totally quiet and you could hear a pin drop as they watched two white blokes, who looked like they were about to fight.

I stepped back with a tear in my eye and said, "Jack, it was not your fault." I then slowly backed out of the bar.

I never saw Jack again. He was sent back to fight for the second time in Vietnam and that was where he was killed. I am sure that this is what he wanted. He would have died by purposely putting himself in the line of fire to try and atone for what he felt he had done to the lone soldier and his family back on our first tour.

Does anyone really believe that countries and governments have the right to kill people? Would Jesus approve of this? Of course not.

Does "Thou shalt not kill" apply to everybody or did God make an exemption for governments, countries and religious zealots?

Anyway, Jack's story made me wonder what I had done in Vietnam that I was not seeing and that was causing me to act like my mate Jack. Was it the woman in the fort? Or one of the many other nightmares of war I experienced? Did I wish to be like my mate Jack? I still did not know but at least now I had a clue. It was because we felt guilty.

I expect that what helped save me from following Jack's fate, to some extent, was that I never directly killed a woman or a child or an innocent or unarmed man even when I was directly ordered to. I chose to play the fool.

Most of my mates are dead now. Gordon, the young ringer who raced me across the flooded, crocodile infested Laura River when we were only sixteen. We were encouraged by my father and all the older men who were making bets on which of us would win. He had killed himself three weeks after his return from that needless, accursed war. What happened in Vietnam is his business and best left unspoken. But the reasons are the same for all of us.

We were all faced with "moral suicide" after the war and young soldiers in training were not taught, nor told, anything about it. Can't have soldiers thinking before they kill, can we? We were told once, and only once, not to kill civilians, yet warned in the same breath that even women and children would kill us if they got close enough. We were taught to trust no one in this war.

I have learned that turning the other cheek, or showing mercy to the weary, has been strictly left out of all army manuals and trainings.

We were purposely trained killers with no understanding of morals, and were not taught about a conscience and what killing can do to a man. Killing unmercifully has its own deadly revenge and no god can take that away by some holy man forgiving you after confession. And we come from a Christian society. What had religion done to our morals as individuals and a country? I would do my best to find out.

As I rode my motorbike home in despair, I recalled a quote by a famous man named Arthur C. Clarke. He said:

"The greatest tragedy of mankind may be the hijacking of morality by religion."

Hughie, another very close mate of mine from the 1st Battalion, drank and smoked himself to death. That is the long difficult road to death and how most of these tough, haunted men choose to die. Slowly but surely. Hughie said to me, three weeks before he died, when I told him that if he did not stop smoking and drinking so heavily, he was going to die: "Mate, I do not want to linger longer."

He always had a way with words, my old mate Hughie.

Another of my funniest and seemingly happiest friends, who was always cracking jokes, took a full bottle of Rohypnol sleeping tablets one night. I found him next morning, cold and blue, dead and gone. No longer telling jokes and wearing a false smile to fool us all.

We have lost our understanding of moral fortitude. Our religions no longer work for us. There must be another way to heal ourselves and transform into different more conscious selves on this, our earth – to see how we continually separate ourselves, in our minds, from each other, from the earth, using one of the thousands of gods worldwide to whom we kneel and beg. And we call this begging, worship.

Prayer is not begging to a god. Prayer may well be listening to the silence, as a great spiritual teacher, J Zee, said when she lay dying.

Anyway, I chose not to die. I chose instead to die to myself – a damaged self of guilt and judgement; a self, created by others who believed in war and killing to gain control.

I remember that Jesus in his Gnostic Mystery teachings answered the question, "What should we do to be perfect?" His reply was, "Be ready in every circumstance. Blessings on those who have found the strife and have seen the struggle with their eyes. They have not killed nor have they been killed, but they have emerged victorious."

Judas asked, "Tell me what is the beginning of the way?"

Jesus replied, "Love and goodness. If one of these had existed among the rulers, wickedness would never have come to be."

So, this is why I would be forced to continually leave home and seek answers and understanding in other countries and other societies, just as soon as I could – to discover an understanding about true spirituality and morals. I had to try and find out for the sake of my child, the people I love, and a better world to come.

My road would eventually lead me to become a bodyguard and confidant to many "New Age" spiritual leaders. I would also meet and guard some of the greatest "minds" in existence on this planet.

Destiny has many faces. I was being driven by my past to find answers and a new way to exist in this world that did not create minds of separation and judgement.

As two of my future gurus would tell me, and which I now understood: "Everything is perfect."

I would add this to their words: Everything is perfect for what it achieves in the ever, ongoing evolution of humankind.

In leaving this first part of my story, here is an apt quote by someone unknown:

"One of the most important things a person can do in their life is to take in new information and change the way they think."

It is certainly the right time, in our quickening evolution, that we do just this and put a stop to wars, false beliefs, and the ever-increasing proliferation of armaments.

The world recently spent three trillion dollars on wars and armaments, apart from the cost of growing police forces and security.

It is, of course, never an easy transition or task, as history shows us, to challenge current, ingrained belief systems, especially in one's own self.

Current religions, using fear and promise, seem to have made a prisoner of the minds of humankind.

I would soon realise, while taking care of remarkable and famous people, many of them spiritual leaders, that it is not man's knowing of gods and what they order us to do that feeds our spirit. It's the mystery, and our trust, acceptance and love of this unfathomable, unknowable, endless mystery.

Here is one last quote to end the first part of my story:

"I do not believe anything. Most people, even the educated, think that everybody must 'believe' something or other. That if one is not a theist, one must be a dogmatic atheist and if one does not have blind faith in X, one must alternatively have blind faith in not X, or the reverse of X. My own opinion is that belief is the death of intelligence."
Robert Anton Wilson.

Chapter 18 - Malaysia and the Muslim Religion

It is illegal in most Muslim countries to stop being a Muslim and punishable by death in some. This is what the Koran says about apostates: "If they turn their backs, take them and slay them, wherever you find them (Quran 4:89)."

This is an extract from the Muslim holy book called the Koran, which was given to me to read while I was in Malaysia living with a Muslim professor and his family in his home to study this religion.

I had saved enough money to leave Australia. A few friends and my beloved three-year-old daughter came to the airport to wave goodbye. Leaving my daughter for any time was one of the most difficult things I have ever had to do. I told everyone that my farewell had to be a happy occasion in front of my daughter, no matter how we felt. This appeared to work well and she was laughing when I left for Malaysia. I told her that through deep love two people are never really apart and I would visit her often while I was away.

One of the good things about leaving, apart from my seemingly driven search, was that my daughter's mother told me she was happy to get rid of me for a while. She was also loving and enjoying our darling girl.

Malaysia was a predominantly Muslim country and as I had had no contact with this form of religion before, I was interested to learn more about it.

I landed in Kuala Lumpur, the capital city. I had the phone number of a university professor, who was the friend of a friend of mine. When I called him, he invited me to his home and offered me a room. He was most happy to help me learn about Islam.

He had a home like any middle-class three-bedroom house

in the West. He was married, with two beautiful daughters aged eighteen and twenty. They dressed like Western women in the privacy of their home, and were very open and friendly with me. After we quickly got to know each other, we were all often laughing together. Perhaps there is hope in this religion, I thought, if they appear so happy and content.

He was a Muslim but not in the strict sense, he explained to me. He was more like a Sufi, or else I could not be in his home and see his wife and daughter without their veils and long black dresses called abayas. Of course, he would be in serious trouble if the imams (religious leaders like Christian priests) knew that these women did not wear the abaya at all times while I was living here.

Whenever his wife would venture out into the street, she had to wear the Muslim veil and long black frock covering every inch of her body. She dressed this way only to avoid a confrontation with the strict minority clergy who could demand her death if she were seen in the street uncovered.

Her husband, the professor, did not agree with this outdated law but his hands were tied.

I wonder what it is about women that upsets these religious men and their followers? Is it because women are beautiful and this is too tempting for them? Is it because this religion, like Christianity, strictly controls (or tries to) sexuality in their followers? An impossible task!

The professor proudly showed me a large book, beautifully bound, and told me that this was the Koran, the holy book of the Muslims. He read out a few of his favourite passages and then handed the book to me.

During the easy time I spent with this lovely family I studied their holy book and was soon amazed at the similarities between this book and the Christian Bible. Some passages were almost identical and the myth and message were clearly similar. In both books there were definite instructions on when it is necessary to kill, and both have strict rules about sex and morals. Paradise or heaven are only available after death to those who obey all these laws.

It seemed to me that these books contained a lot of the basic wisdom of man, gathered over the centuries, but were changed to suit religious beliefs supposedly given us by different gods. These books are used to control the uneducated masses so giving power over them to the priests and imams.

Had I thrown out the baby with the bath water by rejecting the hidden wisdom of holy books? There was, after all, much of mankind's greatest wisdom, not that of gods, hidden in each book.

Yet, I cannot help but vaguely wonder whether these priests have robbed us of our morals through impossible attempts at controlling humans and their sexuality. Plus, how can humans really take responsibility for all their actions if gods are willing to forgive us for all our sins, including killing – even the disgusting sin of blowing up innocent people including women and children.

In these holy books much of mankind's innate wisdom was there to be found. Writings such as: "As it was in the beginning is now and ever shall be." This statement has the power to bring humanity immediately into the present moment. It was written about four thousand years ago by a wise man. It was not written by an unseen and imagined god in the year of Christ.

One day the family and I packed a picnic basket and drove to a beach in the family car. On arrival at the beach, I was shocked that his beautiful wife was not even allowed to swim, as a matter of respect to Allah. I did not see or feel Allah here, watching whether this woman swims or not.

The women didn't own swimming togs and they would probably drown from the weight of all their clothes if they were swept into deep water. These heavy, hot clothes covered them from the top of their heads to the tips of their feet, and the climate was equatorial and very hot. Women suffer for the sake of sexually weak men.

When the pleasant professor and I went for a swim in our togs I looked up and saw the irrepressible daughters playing in the waves, fully clothed. Their enjoyment should have made me happy but instead it saddened me deeply to see females treated like this in the name of male-dominated religions. To my mind this is simply men controlling women by treating them as lesser beings.

The girls soon became absolutely soaking wet and I could see through their wet tops. I thought it would be much more respectable, and saner, if they were allowed to wear togs. You cannot see through togs.

Don't these imams realise that there is only "one self"? We all come from the same place and to this we return, though we have never really left.

Men must show a more humane attitude towards women and realise they are the same as us and so treat them as we like to be treated.

Day by day it was becoming increasingly obvious that this Muslim-ruled country was set up to serve men. The men have been taught to be misogynistic.

It appears that nearly all the gods that we worship in this world are male, with women treated unequally in all religious countries including India. How ridiculous this is – but I kept my thoughts to myself and suppressed my anger at this blatant and cruel inequality. If I spoke up, I could be jailed forever, or even killed. This was more than a bit bloody primitive.

The professor educated me about the Muslim culture. He told me that some rich men here have as many as four wives and only the older wife is allowed to venture out and do the shopping. She also must always be covered from head to toe while shopping.

These wives were virtual prisoners in their own home and sometimes the youngest wife was only twelve or thirteen, while the husband was often an old man. Why would females accept this obvious inequality, I wondered? Did they even know they had a choice – or did they have a choice? Maybe the choice was life or death if they spoke up.

I swayed between sadness and anger about this cultural inequality but soon dismissed my thoughts when we arrived home. The family was happy again freed from the constraints of their hot clothes and the judging sexual eyes of religious zealots.

During my final day with the professor, he dropped me off at a large marketplace on his way to work at the university.

The market stretched along a wide and busy street with canvas-clad stalls and old wooden shops lining both sides of the road. It was crowded and I soon noticed that all the people who were standing around talking, watching and buying, were men. There was not one woman. I imagined that a place like this could only exist a few hundred years ago in some backward Arab town.

I wandered through this market and bought a few little presents for the kind, generous family I had spent a most interesting time with. Then I sat down at a little table on the street, outside a café, and sipped coffee while I continued to study the Koran.

I was shocked at what I read in the Koran (Quran 4:34): "Men are in charge of women by (right of) what Allah has given one over the other and what they spend (for maintenance) from their wealth. So righteous women are devoutly obedient, guarding in (the husband's) absence what Allah would have them guard. [In Australia we use guard dogs for this purpose.] But those (wives) from whom you fear arrogance – (first) advise them; (then if they persist) forsake them in bed; and (finally) strike them. But if they obey you (once more), seek no means against them. Indeed, Allah is ever Exalted and Grand."

This passage made me realise why women were treated so unfairly in this culture. So, Allah in his infinite wisdom and holy writings was even giving men permission to beat women.

Synchronicity again revealed itself as suddenly I heard a loud noise of excited, angry voices. Looking up from my Koran I saw a crowd of men gathering closely together in the street. They were becoming quite agitated with wild looks on their faces. Picking up the Koran, I pushed my way through the crowd to see what was causing this unrest. In front of this crazed crowd of men I saw two young Western women in their early twenties dressed in long shorts – really quite respectable shorts because they reached to their knees. They were looking frightened as this ever-growing crowd of men advanced towards them, yelling and shaking their fists.

My waiter from the coffee shop was standing near me and I asked him what was happening. He told me that these women were accused of being prostitutes because they were wearing shorts and no burqas, and the flesh of their arms and legs was showing. I asked him what would happen to them and, shaking his young head, he stated quite clearly that they would be stoned and probably beaten. Perhaps killed if they were unlucky.

I was totally shocked and outraged. I looked closely at the faces of some of these men and what I saw was sexual excitement most probably caused by hundreds of years of sexual repression by their religion. Did they hate these innocent women or merely what the women's figures were bringing to the surface in them? Are they angry because they were feeling sexually aroused and this is forbidden by Allah unless one is married?

Glued to the spot by what was unfolding, I thought of the poor woman in Vietnam, and my shock and disgust immediately turned into a raging anger. What could I do? I could not face this whole crazed crowd, as I did not have a machine gun, only a knife. Yet I could not walk away to save myself as I once did in Vietnam and then be haunted for the rest of my life by the possible death of these innocent young Western women. I had vowed never to walk away again from women in trouble and a vow is sacred to me and I could not break it.

Looking around quickly, I saw a taxi on the other side of the street near the girls. Just then one of the girls screamed as an apple hit her. I rushed forward wildly waving the Koran above my head and screaming at the top of my voice at these poor girls. These men were still unconsciously working themselves into a mob frenzy where there are no rules, while I simply took the shortcut and completely acted out the part of their destined madness.

The sight of a totally crazed, 6-foot-tall Westerner waving a large book stopped the men in their tracks, though I knew this would only be for a minute.

Running up to the girls, I grabbed the nearest one by her arm. She was in a state of dazed shock by this time and I had suddenly made it worse. I shook her, saying softly to them both, "I am here to help you. Do exactly as I say. Do exactly as I say. Do you understand?" Wide-eyed, they nodded.

Still madly waving the Koran over my head, and screaming like a crazy man, I said to them, "Jump into that taxi through the door on the other side. Now."

They needed no further encouragement and quickly did as I told them. When the mob realised what was happening, a roar went up. I ran to the driver's door and softly but urgently told him to drive. He cringed and looked at me wide-eyed, as he was also in shock, so I hit him with the Koran and said, "Go, go, go. Imshi, imshi." He needed no further encouragement as objects were already hitting his taxi as the now rabid mob surged forward, realising that they were about to lose their easy, innocent female prey. As the taxi took off, I pretended to grab for the back window to stop it, screaming at the top of my lungs. I then let the taxi bump me and pretended to fall, sprawling to the ground where I rolled away from the rushing mob. The men went racing by me throwing rocks as well as fruit. A stone broke the taxi's back window as it sped away to safety.

I quickly got to my feet and the friendly young waiter grabbed my arm and urgently directed me to the back of his restaurant where I sat down feeling a little nauseous at the risk I had taken. One mistake and I would have also been a victim. I never feel shock or sickness during a skirmish, only after when the mind races in. My heart was still beating at a crazy pace and my mind was in overdrive.

Trying to understand why I was involved with this scenario, I contemplated my understanding that life is like a magic mirror and everything that I am experiencing mirrors something inside of me. What? What does this escapade tell me about myself? Even though I thought I treated women as equals, did I still relate to them as sex objects, and does this, deep in my mind, make them less than me? This may all be hidden beneath the romantic dream of finding my "one and only". My fantasy was still based strongly on sex and desire and how a good woman should act in accordance with my imagination and fading religious teachings, which may have been instilled in me when I was a child and then a boy at a Christian boarding college for four long years.

As my heart pumped madly in this high state I found myself in, I had sudden insights about myself. I realised that this life was going to be a long road to self-realisation with pain and suffering awaiting patiently as my sometimes teachers and saviours.

The waiter tapped me on the shoulder, bringing me back to the present moment, smiled at me and said: "You are very lucky. That rabble may easily have attacked you for hitting that taxi driver and his taxi with the Koran. That is forbidden. That is their holy book. You would be dead for sure."

I smiled back and answered, "Luckily they were more interested in sex and violence than religion."

We spoke for a while and I learned that he was a university student and due to his study, intelligence and understanding, he was no longer a blind religious fanatic. I wondered if the world was now slowly changing and would it happen quickly enough to avoid more, and all too common, religious wars? Or would it become worse for a while as the fanatics cling to their fading beliefs while their religions slowly die as people finally realise not only the good works, but the centuries of prejudice, corruption and the lack of virtues these religions albeit unconsciously invoke?

Rape and mass killing – common in war and now also perpetrated by religious fanatics – have nothing to do with morals whatsoever.

Because the crowd of angry men had turned and was slowly coming back, my new friend quickly rushed me out through the back alley and I made my escape just in case any of these men realised what I had done with a "Holy Koran".

The Holy Koran has a large number of verses giving precise instructions on how to kill infidels. A holy book is not in any way a holy book if it gives instructions on killing and when to make war. Both the Bible and the Koran do this.

It had become obvious that I must find a new way of thinking and connecting to spirit that is not part of any of our ancient, outdated religious teachings.

I was sick at heart, having seen extreme Muslims in action. I decided to leave Malaysia the next day and never return there, or to any other Muslim country, until they outgrew their primitive religious beliefs. The first way they must do this is to treat women as equals.

My pen may not conquer belief but evolution surely will.

When the professor came home, I told him what I had witnessed and he was very sad, almost to the point of tears. He explained that not all Muslims were like this and that the Sufi tradition was a great benefit to mankind. He said it had been bastardised by the mullahs and the religious leaders. Perhaps this is true though I knew nothing about Sufis or their traditions.

The next day the professor gave me a book about Sufis and I caught a train north, safely taking me out of this country.

The book he gave me showed me that the Muslim Sufis were a wise and caring people, unlike the primitive Muslim men I had witnessed in action in Malaysia. I must remember not to generalise about races and countries, in support of my mind's easy judgement, though this is difficult and I sometimes forget.

I wondered if belief in gods, separate from man, creates duality.

After a brief holiday in Buddhist Thailand, shopping, eating wonderful food, and getting excellent massages, I happily flew home to try, yet again, to settle down permanently with my family and fit in to a society based on Christianity.

Chapter 19 - Becoming a Buddhist Monk

I quickly found another job driving a large cement truck. When I had saved enough money, I bought Mary an engagement ring which she lovingly accepted.

For a couple of months, we were all very happy. Mary and I made love nearly every night. I did not realise then that this was a physical over-indulgence until one night after making love, I had not one, but three orgasms in my sleep. Wet dreams as they are called.

I woke up feeling like a nervous wreck with my hands shaking violently. My dreams had all been of strange, unknown women. I realised that I knew nothing about life and sex. Did this happen because I was having too much sex? Obviously!

For the next five months I struggled along in this "normal society" as a "normal man" with a "normal job" and a "normal family". The one thing that kept me happy was being with my beloved daughter every day.

But, as usual, something was driving me, and outbursts of panic were growing with the pressure of what? I did not know. This life did not appear to be mine to claim and direct. It does what it wants, not what I want. Perhaps after losing everything in Vietnam this condition has been set in motion in myself?

One morning I was driving to work and thinking about my problems, so not driving with complete awareness. This disturbed a tradesman driving behind me in a ute with a cement mixer on the back. He accelerated past me and rudely cut me off. We both got out of our vehicles and this fellow came running at me with fists clenched, threatening to teach me a lesson for driving like that. I noticed that he was a heavily muscled man so I turned and ran, not really aware of what was happening or why.

I had only been driving very slowly, during my deep thought, and this had probably disturbed him.

He was cursing me as he chased me around and around our two vehicles. After a few laps, running madly around the vehicles, he began breathing heavily and slowing down. I noticed this and being super-fit myself, I called to him while still running, "You know it is not me who should be running away, it is you." And with that I abruptly stopped running and turned, facing him. He came to a sliding halt and looked into my eyes, which were now reflecting anger at this over-muscled bully who had been cursing me violently and threatening me with foul language and physical retribution.

I then set off after him and he started running in the opposite direction around our two vehicles.

After one round of running, with me gaining on him, this struck me as so ridiculously funny that I began laughing uncontrollably and stopped pursuing him. He did not quite get the joke and jumped in his ute, still cursing, and drove away.

Meanwhile, I did not wish to leave my family again, but the pressure was increasing, and any tablets that the doctors prescribed for me did not help so I had stopped taking them. And, as usual, my body was becoming more and more painful. I could hardly turn my head because of the pain in my neck and back. This intensity was crippling me but still I tried to resist leaving.

I had read a little about Buddhism and had been told of a Buddhist wat (temple) in northern Thailand that took in Westerners, so I discussed this with Mary and she agreed that I should go and enter it. She was probably more than happy to get rid of me again by now, poor woman. It is difficult to live with someone in obvious pain whose moods fluctuate.

Before my neck grew stiff, I had been super-fit as I was playing rugby league for my home city. I also had very long hair including some dreadlocks as my hair was naturally curly.

Each year in northern Australia a competition was held where sportsmen from all sports came together and raced each other for a mile. My neck did not keep me from running even though it was painful to do so. I won my heats against all comers. I then won the final against the best runners in the country, before a packed grandstand of people, all cheering madly.

I was presented with the considerable cash prize for the winner. This was enough for me to be able to afford to go to Thailand plus leave money for my family. I also paid the mortgage on my war veteran loan for my house.

Leaving my little girl, and Mary, was again one of the hardest things I have ever done but I knew, and so did Mary, that something was very different about me. Doctors, psychiatrists and pills could not help.

So off I flew to Thailand and the Buddhist wat. Strangely, I felt something leave my body before the plane landed and my neck became instantly free from pain.

As it turned out I was lucky to be so fit or I would never have had the strength to become a Buddhist monk. In those days it was almost an impossibility for Westerners to become monks, as many foreigners from different countries discovered while I was there.

Bangkok is a city full of temptations but I bypassed them all and headed straight up to the northern city of Chiang Mai where the temple I had heard about would take me in and train me to become a monk.

Chiang Mai was nowhere near as busy as Bangkok and had quite an easy-going feel to it. The locals are whiter and more round-eyed than their southern counterparts and the women there are renowned worldwide for their beauty. At the post office I spoke to a local beauty as we waited in line and to my surprise, she was friendly and we had a long conversation. Asian women are generally shy and not so talkative.

After booking into a hotel by the river I hired a motorbike and rode out to Wat Ram Poeng, which is situated just below the King's palace on the outskirts of the city. It had a high cement wall which enclosed a large temple, plus about forty small plain cement huts. In these huts live the monks and nuns; monks to the south and nuns to the north – both totally separated. And the nuns serve the monks but cannot become monks themselves. They also serve the male trainees who will be taught Vipassana Insight Meditation before qualifying to become a monk but only if they can pass all the tests. Most cannot, including both Westerners and Thais.

I noticed that the monks wore orange while the women were only allowed to wear white.

I was directed to wait outside one of these small concrete huts until the head monk could see me. As I waited, a large pack of dogs strolled past. A couple of them had huge gaping wounds that were crawling with maggots and I wondered why they were not healed, or at least put out of their misery with a quick bullet. I was told later that the Buddhist way is one of non-interference. Not killing any living thing is part of this belief. This seemed cruel and uncaring to me as I do not like to see animals in such terrible suffering when we, as humans, have the power to help them.

We could also assist humans when they are dying in agony and have no hope of ever recovering, but religions also do not allow this mercy.

My interpreter for the interview was an American monk, the only Western monk here, and he told me the form I must follow when I approached the master – the bowing, and touching my head to the ground, plus staying always below the level of the master's head; when I may speak, how to address him, and so on. All these orders made me nervous, as since being in the army I do not like orders, and I found myself sweating as I was ushered into his presence. He was an old man of about sixty-five sitting cross-legged on a raised dais. I proceeded to make mistakes in form as I crawled forward, trying to remember all the bows and all the rules.

The master simply laughed at my mistakes and I took an immediate liking to this easy-going man who was not encumbered by strict form or orders.

During the interview to see whether I was acceptable he seemed to be doing his best to scare me away. He informed me that I would spend all my time in silence except for the daily interviews of around ten minutes with him. I would gradually build up my meditation to nineteen hours a day, alternating between one hour sitting and one hour walking.

I would only be served two meals a day and these must be taken at 7.30 am and 11.30 am. No food must be consumed between noon and daylight of the next day and the only food available was what the monks could beg from the surrounding farmers. Some days there may be no food, he smilingly informed me. In my little cement box with one small window, I must sleep on a wooden bed, with no mattress just a folded rug. I would be allowed no radio or writing paper, and for a bath I would have a cold-water tap and a bucket to pour the cold water over me. He told me that I must buy a bucket. As the month was November it was cold here and I would definitely have to spend Christmas and New Year alone and away from my family in my little cement hut. I may not be allowed, or be ready, to graduate for six months depending on my progress in meditation, he smilingly informed me.

"Do you still want to become a monk?" he asked.

As I had had my fill of the pain of my outside world and was totally fed up with it, and myself, I nodded my head, yes. The master told me that nearly all Westerners last only two or three days, as the pain and the demands of the mind overcome them. In my ignorance I assured him that I would finish the course. He looked at my curly long hair that reached to my shoulders and told me that monks and trainees must shave their head and eyebrows. They must never look at women or touch them, and must take the six sacred vows to be eligible to come here and begin this course.

I rose to the obvious challenge and said, "no worries", so he informed me that I was accepted and to finalise all business with the world as I would be allowed no communication with anyone from the outside while I was here. He also told me to buy a book that would tell me all about Buddhism and to read it before I returned because I was not to bring any books when I reported back with suitably shaved head and eyebrows.

Back in town I purchased a thick old book that looked like a Bible. I was informed that it's the bible of Buddhism. Then I went straight to a barber and asked him to shave me. The shock of looking at a lily-white head with no eyebrows, after such long hair, took a while to get used to. I avoided looking in mirrors.

Back at my hotel I spent most of the night reading these ancient texts and discovered some obvious similarities between the myths of Buddha and Jesus. For example, they were both quite clear about not worshipping any images, neither of gods nor of themselves. Plus, some of their laws were also exactly the same, such as not killing others, although the Bible condones "holy wars", which of course are responsible for multitudes of killings. But the Buddhists take it a step further and vow not to kill any living thing.

A Buddhist monk also takes over two hundred vows, some as detailed as not spitting on any plant. I was pleased to read that tolerance plays a large part in their teachings and realised that this is one of the reasons why Thai people are so easy-going and friendly. Though Christians are taught almost the same thing in the commandment "Judge not lest ye be judged", they do not seem to realise its meaning and are very intolerant of anyone who does not share their beliefs of "the one and only true God" and follow his strict moral orders.

The Buddhist teachings on tolerance are quite detailed and the Thais seem to understand them, and so they do not force their beliefs or personal morals on anyone else. I like that.

While studying I was reminded of a joke about a guy who died and went to heaven. St Peter was giving him a guided tour. The first group of people they came to were all sitting cross-legged very quietly with their eyes closed and the newly dead man asked St Peter who they were. St Peter replied that these were the Buddhists and that they liked to meditate.

A little bit further along there was a group dressed in white, who were dancing, singing, and making a hell of a racket. "Who are this lot?" the new arrival asked. "Oh! These are the Hare Krishnas and this is how they like to spread their message and worship their god."

Around the next corner were a group in various poses, some standing on their heads, others leaning forward with their heads on their knees. "Who are they?" the new arrival asked. "These are the yogis," replied St Peter, "and they love to twist about in the physical body while seeking a future enlightenment."

Around the next corner they ran into a huge brick wall and of course the dead man asked in a shocked voice, "What is this ugly brick wall doing here in heaven?" Placing his finger to his lips, St Peter whispered, "Sshhh, the Catholics are on the other side of that wall and they like to believe that they are here in heaven alone."

The next day I began practising some of the vows I would soon be taking, by not looking at women and not brushing against them in the streets. That night I had a wet dream. Suppression obviously creates its opposite.

Our Christian priests have taught us to suppress sexual desires by trying to copy a fairy tale which states that Jesus was abstinent for his entire life.

Maybe he was but that was his business and his realisation so don't try to copy someone else. Be authentic. The priests and holy men suppress their sexuality for years and it eventually erupts like a volcano as they are often caught having sex with young boys, young girls, or members of their congregations. Yet the Pope and other masters are ignorantly unaware of this maniacal result and the eventual outcome of their anti-sexual teachings. They can't rule nature though they stupidly continue to try.

But I, perhaps because of my Christian-invoked guilt around sex, was disturbed about my wet dream and would have to ask the master about it. I don't know how to control my thoughts about women while asleep or if this is even possible.

My last night of freedom I spent walking around the streets

noticing the nightlife of Chiang Mai. I was quite shocked to see tough-looking Chinese men who seemed to be guarding very young girls who were on display in glass cages. Some of these girls could have been no more than fourteen and this made me angry because these girls did not look happy. I should not have looked at them, but I was not a monk yet.

I thought of my young daughter and was concerned about this incarceration so I could not resist trying to rescue the youngest little girl of about twelve or thirteen years of age who was on display.

I paid the tough-looking Chinese man so I could be with this young girl in private and we were shown to a tiny little room out the back of the cage. The girl was very shy and it soon became obvious that she could not speak even one word of English. She was nervous as I tried to communicate so she pulled up her skirt to reveal that she had no panties on. She lay down and spread her legs. I quickly covered her and lifted her up. I gave her a sum of money. I thought that she could use this money after I helped her to escape.

However, the poor girl did not want to leave when I tried to take her with me. She called out to an older woman and then she began to cry. The older woman told me that the young girl thought I did not love her because I did not touch her. I assured her that this was not true and told her that I was about to become a monk the next day but I would never have sex with any young girl anyway, monk or not.

The older woman was happy on hearing that I was about to become a monk.

My strong moral protector still wanted to save the young girl so I told the older woman to ask the young girl if she was happy or if she would like me to help her escape from here.
The older woman was shocked and told me that this was the young girl's home and her family and she did not wish to leave.

The old woman, frightened now, suddenly called out loudly. Three tough-looking Chinese hoods burst into the room. They were quite angry and I noticed that the little girl looked scared.

My initial instinctual reaction was to calm the men down with a few well-placed punches but the vow of non-violence popped into my head. Also, it seemed that the young girl was happy here so I sadly turned and left. My tolerance level was not like a Buddhist's, I realised; it was more like a Christian's.

I don't like some things about this world and often wish I were superman so I could change them but I am not; and, if the young girl had really wished to leave, those Chinese hoods could perhaps have killed me, as they were armed.

As I slowly walked away, sickened and sad, I was more determined than ever to explore Buddhism to the fullest and see if I could find some peace and understanding in this world of suffering.

But my moral judge and protector was not done yet. Back at my hotel I told the manager about my encounter with the young girl and asked him to ring the police as she was obviously under age. He laughed and informed me that the police or the Thai generals own most of these places. He told me that the girls consider it an honour to work there as it is a way to support their poor families. Monks and these working girls are both respected equally by the Thai people, he laughingly told me, shaking his head at my ignorance and judgements.

This again made me more determined than ever to become a monk as disgust at the "normal" world hardened my resolve.

Brothels can be found all over the world. In the West the women are at least old enough to choose.

I could see as I looked closely at myself in this situation that I was prepared to act as judge and jury even if it had meant violence would erupt. The tolerance I so admired in the Thais was not present in me – neither with the suffering dogs nor these young girls.

On arrival at the temple the next day I was told that my room in the wat was called a cell and a cell it certainly was. There were no windows in my small cement room that came complete with a hard wooden bed about 18 inches wide.

I was told to begin meditating immediately. Sit for twenty

minutes then walk ever so slowly, being aware of each and every little movement, for another twenty minutes for the first few days. I was instructed to do this for nine hours a day.

Within a week I was practising this sitting and walking meditation regime for nineteen hours a day and was allowed to sleep for only five hours a night. I was soon sitting for an hour, then slowly walking for an hour, with no smoko breaks.

The first five days were almost impossible. My hips were bruised from the bed, and sleep was not really possible on such a hard surface that my body was not used to. My body was also in pain from the sitting, which of course I was not used to as I had never practised meditation in my life. I had never even heard of it. This meditation was an Eastern thing to me.

My mind tried to play tricks on me to get me out of there. On the fourth night when it finally became time to sleep, I had just dozed off when I was awakened by someone sitting on the end of my bed. Now I had never seen forms or ghosts of any kind in my life, as I did not believe in the supernatural, but this "form" appeared to speak to me.

He said: "What are you doing here? You belong to me and must serve me."

Due to my Western mind, this form represented Satan or "Old Nick". This invasion made me really angry and I sat up on my piece of hard wood and in no uncertain terms told this form to leave immediately and never come back. He seemed to disappear in a flash and never did return. So much for the mind and its taught beliefs. They are not real – neither God nor Satan – even though these are what Christians imagine they see. Hindus have visions of Krishna or Shiva, while Muslims imagine they see and talk to Allah. The mind can be very tricky, using the archetypes of different religious belief systems to control or terrorise.

Each ever so slowly passing day I looked forward to the ten minutes or so with the master. He always asked me about my progress and if I had any questions. When I told him about my wet dreams and that I was worried that I could not stop thinking about women in my sleep, he laughed long and loud and told me not to worry about what happens when I am asleep. The wet dreams never came back. He had eased my guilty Christian-trained conscience.

Each day he would tell me something new. He said that during the months I was to be here I would compress the experiences of at least ten years into this short time period. He told me that I would experience every emotion with an intensity I had not felt before. He planned to warn me beforehand which emotions would be arising and trying to distract me from my meditations. I realised he could see my path, and me, quite clearly as he was very experienced through many years of watching numerous hopeful monks. He also told me that even the Thai men, who have practised meditation all their lives, fail this course.

For example, one day he told me that I was about to experience love and sure enough this incredible feeling of bliss and compassion came over me that afternoon. I thought, this is it. This is what I have been searching for and now it will never leave me. Of course, like everything else, it did, and the next day I experienced almost unbearable pain, which also tried to stop me meditating.

The master was very supportive of me and told me stories about pain and how the mind easily fools us into believing that we are dying.

One day I was totally convinced that I had appendicitis, or at least a damaged vertebra, and every bit of logic in my mind was screaming at me to leave here, especially when a Thai monk, also doing this course, was rushed to hospital with an actual burst appendix. I was almost convinced that this was a message telling me that I was about to die but some part of me still remained separate and watched my mind at work, knowing that it was not the truth, and then gradually even this pain passed.

At another time I was assailed by fear with the thought that my daughter was sick and maybe dying and I should fly home immediately.

"Compassion" came and went, then a great sadness almost overcame me at not being able to be with my family over Christmas and New Year.

The next day anger arose and became so intense that I thought of ways to kill a barking dog outside my room.

Gradually all these feelings and thoughts lost their power over me and I would sit for days just watching the rise and fall of my breath. I became aware of each part of my body between the still part of the rising and falling of each breath.

My weight and muscle tone slowly fell away as I never had enough food to eat and I was not allowed to exercise. I lost 60 pounds during my time there. I did not really notice this at the time because there were no mirrors.

I saw a woman practitioner go stark raving mad and have to be carried away. Of course, I then had to deal with the possibility of my own madness.

One day the master sent me down to some graves and told me to meditate on death. I kept thinking that a large black cobra was going to slide out of the long grass surrounding these graves and bite me. I had a strong fear of cobras since Vietnam.

After a couple of months, boredom came to visit and this was one of the most difficult tests. I thought of different reasons to tell the master why I had to leave but I could not break the promise I had made to see this through to the end, and so I held firm. Promises were sacred deals for me but how I wished I had not made this particular promise to myself as I am sure I would otherwise have left there a long time ago.

Thoughts of jealousy and possession one night appeared as a huge army of men dressed in black. One of them with a giant penis was having sex with my beloved Mary and then took her away, along with my daughter. My mind tried to convince me for a moment that I would never see them again.

Then came a period of relative calm as my meditation took me beyond time. Only a few seconds seemed to elapse when in truth an hour of sitting meditation had passed.

I thought then that I finally had the mind beaten and from

here on everything would be easy, but this too passed, as did everything else. Hard, easy, love, hate, anger, jealousy, knowing, not knowing – I watched them all come and go with the help of the master.

I also noticed a strange thing happening with the dogs of the monastery while I was there. They were a large pack that lived mainly on rice, and were allowed to do as they pleased because the monks would in no way interfere with them. Some of them were wounded and badly diseased and death was not uncommon. Over the months, I watched dogs all die in exactly the same way. They would go to the front steps of the large temple where the great gold Buddha lay, and here each seemed to die the most painful and excruciating death, which took about three days. A smooth, neat hole had been worn about a foot deep into the hard ground. The dogs, lying on their sides, pushed themselves around and around in circles on the ground as they frothed at the mouth and howled, until eventually they died.

Buddhists have a strong belief that if you do wrong in this life you will come back as a dog in the next life. If you do right you will never come back, and you will go to a place something like heaven.

I watched these dogs for many hours when I finally became a monk and wondered if they were susceptible to the thoughts of humans. This suffering or death ritual does not happen to dogs where I come from.

Over the months quite a few Westerners, both men and women, arrived to do the course, but all except one left after three days. This saddened and disappointed me for a while. I was curious about the one American who stayed longer so I broke the rules and slipped down to his cell one night. He was in a bad state and told me that he had to leave the next day. I did my best to convince him that it would get easier and he should stay but the next day he left. I found out later that he was an artist who was trying to kick a heroin habit. This was not a good place to try and do that and I was amazed that he lasted longer than anyone else. Four days he lasted, to be precise.

By this time, months had passed and the New Year had come and gone. It touched my heart deeply not to spend these days with my daughter and partner.

The weather had turned very cold, as winter had set in. The cold baths were a shock to my system but by now had become a part of my daily ritual.

I asked the master when I could become a monk and he answered, when I was ready. His answer managed to panic me slightly because I was missing Australia and my family and I thought I might be forced to be here for another year. Yet all thoughts passed away eventually and I got back to the job at hand, which was meditation.

Finally, one day the old master handed me three sheets of paper and told me to study them, as this was what was required of me before I could become a monk. I was shocked to discover that I would have to meditate from 4.30 am next Monday morning to noon on Thursday without any sleep. "Determination" it was called and I could see why. I didn't think this was possible when I let my mind think about it. Yet if the master said I was ready then I must be able to do this. He had been right about everything so far and I had grown to respect and trust him, so I quickly put all thoughts about this final requirement – which was to happen in the future – out of my mind.

I was beginning to realise that practising meditation was simply practising how to live each moment and I always had a choice about which thoughts I could linger on and follow. It was all up to me to choose and be master of my thoughts. The witness was growing stronger in me.

Ah, to be grateful and silent. I wondered later if it is possible to use words to describe this state of being that all religions seem to be aiming for in one way or another. I guess I'll know when I get there. I have no doubt that this is my heritage as it is for all humankind. I called it "consciousness" at this time in my life.

And I wondered if the poor animals would be happier? They would certainly be more relieved as I imagine we would not be so cruel to them or treat them as any less than ourselves when we become softer, meeker, and more conscious.

This is a copy of the script for determination that the master gave me:

The First Day of Determination

1. Finding mercy for one's self recite the following passages in Pali:

"AHANG SUKITO HOMI, NIT DUKKO HOMI, AVERO HOMI, ABHAYA BHACCHO HOMI, ANIKO HOMI, SUKHI ATTANANG PARI HARAMI." (which translates):
May I be happy, free from suffering, free from enmity, diseases, and grief, free from troubles, difficulties and dangers, and be protected from all misfortune.

2. Extending your friendship to all beings recite the following passage in Pali:

"SABBHE SATTA SUKITA HONTU, AVERA, ABHAYA BHACCHA, ANIKA HONTU, SUKHI ATTA NANG PARI HARANTU."
May all sentient beings be happy, free from suffering, free from enmity, diseases and grief, free from troubles, difficulties and dangers, and be protected from all misfortune.

3. May gross perceptions of three characteristics of phenomena cease and may more subtle realisations of these three characteristics be attained during twenty-four hours.

4. Having determined as above, perform the walking meditation first, then sit down and acknowledge the rising and falling, sitting, touching. Perform the two exercises in alternation throughout twenty-four hours without sleep.

The Second Day of Determination

1. Finding mercy for one's self, and extending your friendship to all beings as described the previous day.

2. Perform the walking meditation first, then proceed as follows.

a. Make a wish that in a period of one hour, the phenomena of arising and ceasing shall appear (at least five times).

b. If within this hour the phenomena of arising and ceasing appear distinctly (at least five times and possibly as many as sixty-five times) reduce the period of the exercise down to thirty minutes and make the wish that within these thirty minutes the phenomena of arising and ceasing shall appear to you several times.

c. Make a wish in the same manner and reduce the period of the exercise down to 20 – 15 – 10 – 5 minutes. Within five minutes the phenomena should appear at least twice (but they may appear as many as six times).

3. Repeat in alternation to complete twenty-four hours also without sleep.

The Third Day of Determination

1. Finding mercy for one's self, and extending your friendship to all beings as described the previous day.

2. Perform the walking meditation first; then in the sitting posture make a wish that you attain a steady concentration for five minutes. Next acknowledge rising, falling, sitting, touching. The resolution is fulfilled if the mind abides in concentration and becomes unconscious of outside phenomena for five full minutes. Keep a check carefully on the time and if this exercise cannot be continued for five minutes, repeat it until you are successful. Then try to increase the period of full concentration.

3. Make a resolve that you obtain a steady concentration without consciousness of outside phenomena for ten minutes. If this cannot be achieved yet, try again until you are quite experienced. Then practise further for 15 – 20 – 30 minutes to one hour, 2-3-4-5-6-7-8 up to twenty-four hours.

4. The number of minutes and hours is to be reckoned from the point of steady concentration with non-consciousness onwards. In such a condition, we do not experience any feeling. The period wished for being fulfilled, consciousness will return of its own accord as in waking, but this is not waking.

Perform the two exercises of walking and sitting meditation in alternation throughout twenty-four hours without sleep and bath.

On the first night of this final task, I was assailed by the thought that this regime was not possible. The pain, starvation and deprivation of these difficult months, combined with lack of sleep, were together taking their toll. Yet like all thoughts, no matter how powerful or painful, they passed away and were overcome by the witness.

The mind can be such a trickster in its efforts to protect the body, and this practice was definitely extending the limit of my now skinny body's power.

By the second evening the hour of sitting would sometimes seem to pass in a second. No sooner had I closed my eyes than the hour appeared to be over. The master was very pleased about this as I looked up once after an hour to see him standing at my door smiling, so I relaxed more and more as the hours continued to turn into seconds.

Passing the halfway mark was quite a relief. I did take some refuge in the thought that I would definitely sleep for at least two days come Thursday afternoon.

Even extreme tiredness and drowsiness came and went, just like everything else. I noticed that excitement or its opposite, boredom – any emotion in fact – had little hold over me for very long.

Once when sitting, I toppled over and landed on my back, but escape into sleep was not possible so I immediately sat back up and resumed the meditation position.

The weaker my body became the more powerful my mind became. Once I knew I could save the world and I knew how to do it. So what? I let even this pass.

At another stage I opened my eyes after another hour of sitting that seemed to pass in no time at all, to see the master's face beaming down at me, nodding his head. I wondered what he saw.

It seemed to me that I had stepped out of time and had no feeling of a body or of being anyone or anything.

Thursday finally arrived and the mammoth task was done. Everyone seemed happy that I, a Westerner, had passed the final test.

The American monk offered to take me to the King's palace temple on the mountain nearby and then to see one of the most famous monks in Thailand who was visiting Chiang Mai.

Contrary to my fears and expectations, I was not in the least bit tired. Surprisingly, I was bursting with energy. Life is a mystery.

The King's wat, on the mountain, is steeped in history and was a pleasure to visit, especially as this was my first time away from Wat Ram Poeng in many months. Freedom is a potent elixir.

My revered friend then took me to town and bought me a meal. This seemed to be the best tasting food I have ever had. Thai food is simply wonderful and doubly so when half starved, as I was.

After the satisfying meal we walked to the public square and sat and meditated with the famous old monk. After a couple of hours of meditation, which again appeared to pass in no time at all, I opened my eyes and watched the old man until dusk when the sun finally kissed the horizon. The monk was thin and looked to be about seventy years old. Dust, flies, insects, mosquitoes, rain, nothing could make him move even an eyelash. What incredible self-control, I thought. I admired him and felt proud to be a monk.

Then I received my first shock about Buddhism. A limousine drove up onto the pavement to pick him up. Two strong-looking monks stepped out, bowed to him, and then bodily picked him up and gently placed him in the back seat of the limo with his legs still crossed. I turned to my friend and asked him why he did not walk to the car himself, and was nonchalantly informed that he is crippled from sitting cross-legged for so long, all his life. He told me that the old monk is greatly revered and admired by people from all over the world, as it is said that he has overcome the body, pain, and this earthly life.

As I watched, it seemed to me that he had simply crippled himself and this seemed a sad act against nature and the body. There is no reason, sacred or otherwise, that could convince me otherwise. I felt so sorry for the poor man! Was he trapped in his beliefs, at the cost of his temple, the body? I could not imagine the enormous pain he must have suffered to become so crippled.

However, this one flaw was not enough to shake my faith in my newfound religion even though, at this moment in time, I did not think of Buddhism as another religion. Before, I had nothing; now I had something. I belonged, and I had earned my hard-won acceptance as a monk. Yes, I had earned my robe and I was still at peace with Buddhism.

Back at the wat, I felt incredibly free after my ordeal of the past months. I was now allowed to walk around the wat, to talk to people, to sit in the temple, to eat with the other monks, to meditate much less, and even to write. I read more books on Buddhism. I also ate more, and left the enclosure of the temple grounds when I felt like it.

During the prayers and chanting with the other monks, when we gathered together in the temple, I suffered through a plague of mosquitoes, always remembering my vow not to kill any living thing. During meditation a monk should not move even if bothered by mere mosquito stings, so my "believer's hands" never even brushed them away.

Being an ex-boxer, I made friends with a monk who was once a very famous and successful Thai boxer, a revered champion no less. He was about forty years of age and had become a monk at the end of his boxing career. He was a strong, happy man with very few scars, which told me just how good a fighter he must have been. He had been watching me in the temple while I was trying to ignore the mosquitoes and one day, after the chanting and meditation, he approached me with a big smile. We then walked together out of the door of the temple. He held a closed fist very close to my face. I looked from his fist to his twinkling eyes, wondering what he was going to do with that fist, and then he turned his fist over and opened his hand. On his hand lay many dead mozzies. For some reason I began to count them and he closed his hand and whispered, "Twenty-one." I found this overwhelmingly ironic and funny and had to hurry away to laugh behind a tree. It was a shock, however, as I thought everyone was incredibly devoted and followed the vows to the letter. It was also a relief as I realised that I did not have to sit and let the mosquitoes bite me. Also, the rebel in me was aroused from its slumber by this laughing monk.

In my studies of Buddhism, as I continued to read their books, one thing kept bothering me. Buddha was very clear and stated quite often that mankind must never worship him as a god nor worship images of him. Jesus was also quite clear about this worship though one would never know it in the Christian religion these days. The poor man hangs, with nails through his hands and feet, from crosses in churches around the world and the masses worship this image.

It seems that the teachings of great human beings like Buddha and Jesus are changed by lesser men, over the centuries, to serve themselves and give them more power over easily led people.

Anyway, quite often the local people, especially the poor uneducated farmers in this area, came for three days at a time and slept on the floor of the large temple which housed the huge golden Buddha. They brought gifts of considerable value to these poor peasants and laid them at the feet of the golden image. They also brought offerings of food for the monks. Then they prayed and chanted to the statue, just as Christians do with the image of Jesus.

Day after day I watched these beautiful people and I realised that they were doing what the Buddha, in his wisdom, had specifically instructed them not to do. I decided that I must ask the master why this was allowed.

The next day I was given the opportunity as one of the revered head monks of Thailand was visiting our wat.

The master had taken my passport about a week before and when I was summoned before him, and the powerful man by his side, he returned my passport freshly stamped with a five-year visa. The master explained that I could stay in Thailand for the rest of my life and now they wanted me to take my begging bowl and travel the entire country with this head monk as he spoke to powerful people and other monks.

A surge of pride went through my chest as I realised that this was a great honour.

But first I just had to clear up this misunderstanding about the peasants and their obvious worship, which Buddha forbade. I explained all this in detail to the two older men. They nodded and spoke to each other in Thai, which I could not understand. Finally, they turned back to me and the powerful monk said that it was all right for the peasants to worship a statue of Buddha as they were at a different level of understanding than I was.

I wanted to stay in Thailand and be a travelling monk for a while. I admired and respected the Thai people and their country, and I was very grateful to my wise and compassionate master for his teachings. However, I was concerned about the status of separation, placing myself above other people, even if they were simple farmers. I could not even consider living in such a way. Right there and then I said, "No! This is not the way for me. I must leave here now for I do not wish to separate myself from other people by using my mind and imagining myself to be above them."

The men were surprised, and briefly showed it on their faces, but they quickly composed themselves. I admired the way they immediately took everything in their stride and even though the disappointment showed through for a split second, it passed by in a flash. What wonderful living examples of their meditation practice these men were.

They smiled and gave me their blessings and I left immediately. I suppose, in retrospect, that I was more than anxious to see my daughter and partner again and this may have influenced the fact that rather than discussing it with my beloved master, I made this quick decision.

I was intending to return home to marry my daughter Cheyanne's mother. I wanted nothing else but to be with my family. Surely this was not too much to ask and would be possible for me to do. I had become sick of this driven search for answers where I did not seem to have had a choice to simply stay in one place, in one job, with one woman and be with my darling daughter.

A new determination replaced the one I had just endured. Surely after all this practice on dealing with the mind I could finally go home and fit into a normal society.

Chapter 20 - The Dream Destroyed

Arriving home into the arms of Mary and my daughter brought me a happiness that threatened to overwhelm me, even after all my monk training of self-control.

For two days I remained enraptured in this love.

Then suddenly I was brought back to earth with a thud and I was once again awoken from my perfect dream of a happy marriage and a normal life.

Mary told me something that I would have preferred never to hear, but she became overwhelmed by her perceived guilt and needed to share it with me and ask for forgiveness.

This may well have been the result of the Christian practice of confessing to a priest to relieve oneself of guilt.

I was away a long time and Mary was a red-blooded woman who was obviously free to do as she pleased. And nature has its own calling.

I realised with a shock that I still had many lessons to learn in this lifetime. It amazed me how life seemed to ferret out my weaknesses and misconceptions and then hand me the perfect lesson to rid myself of these fears.

Mary and I have a lot of mutual friends. One of my friends was a New Zealander. When he first arrived in Australia, I helped him find a job and settle in. I also protected him like a brother. Once when four men took a dislike to his accent, long hair, and the way he dressed, I stood beside him and convinced them, with closed fists and strong words, not to attack him.

In the past when we went swimming with our friends, none of us wore togs. We swam and sunbaked in the nude. Everyone noticed that this particular man, let's call him Colin, had a huge penis. We joked about it and asked him if his body had enough blood for him to get an erection, as it hung down his leg for at least 8 inches.

In my dream at the Buddhist wat in Thailand I remembered seeing Mary having intercourse with a man with a huge penis. I wondered about the special psychic connection between lovers.

While growing up on a cattle station I would hear the men talking about big penises and who had been at the front of the line when God was handing them out. The common perception by the men was that the bigger the penis the better a woman liked the man. Of course, this is ridiculous but I knew no better at this stage of my life. So here I was, about to suffer again from my teachings and misconceptions.

Mary told me that she had slept with Colin while I was away. I reacted by going into our bedroom and starting to pack my suitcase. I was not mad at her. How could I expect her to wait for me at home when I spent so much time away? I was fiercely angry with Colin and planned to go around to his home and give him a terrible beating. These thoughts of violence did not sit well with me for long. Because of my Buddhist master's teaching, and my recent training, I knew I had a choice. I became still and focused on my breath and let all thoughts of violence, as I watched them continually rising and falling, slowly pass harmlessly away. My time spent as a monk had not been in vain. However, I continued packing. I was sure Mary would now wish to be with Colin.

Mary came into the room and asked me what I was doing. In my cowboy naivety I told her that obviously she would not wish to be with me anymore as Colin had such a big dick.

She laughed at me and sat me down. She told me that she did not enjoy sex with him and only did it once. "And anyway," she said, "he is married to Anna."

I suddenly remembered something my father once told me when the men were talking about big penises and how much the women loved these men.

He had pulled me aside and said, "Put your thumb in your ear, son, and wriggle it around," so I did. He then said, "Now put your little finger in your ear and wriggle it around," so I did. He then asked me which one felt the best. Stunned, my mouth dropped open, so I tried it again. The little finger's movement felt far better.

"It is the same with penises," my father said.

Mary laughed and hugged me when I told her this story and then said, "I know who I want to be with. Only you!"

Being a man is not easy at times because of our teachings and perceptions, and even if you have a bigger than average penis, someone else will have a bigger one.

I told my dad what had happened the next time we were mustering cattle on the family station. He laughed and said, "It's not the size of the nail, son, it's the hammer that drives it."

My father's sexual education of his son. Not bad really.

Anyway, the lesson was learned and I have not had to worry about penis envy ever again. Pain, shock and suffering are competent teachers.

And jealousy is a strange curse, inflicted upon us by our religious moral teachings. I have often wondered at how sex, which makes loving partners so happy, can cause one of the partners to commit even murder if the other dares to enjoy this act with another? Madness!

The ancient Tahitians did not suffer from possession and jealousy. Why? Because they were not taught by Christians or Muslims. They shared sex with anyone and, strangely enough, they suffered from no sexual diseases until the ships laden with Christian sailors from the West first landed on their shores. Soon the innocent Tahitians were dying from syphilis and other diseases brought to them by God-fearing Christians.

One thing that I did take note of is that guilt can be a strong love tonic. Mary was much more loving towards me and much more attentive, cooking me beautiful meals, while being the perfect housewife and lover.

Mary and I got on with our lives but soon I was again being plagued by the restlessness and search for answers. I shared this with her and we decided to work really hard for six months and then go to India in search of gurus and truth. We would also take our darling daughter with us, as she was old enough at five years of age and she loved to travel.

I worked for the government during the day, and at night I again got a job as a bouncer at a hotel I used to run. I worked with one other man who was, thankfully, very big and muscular. We figured out a way to avoid most fights by working together. I would distract the aggressive drunks and he would pick them up from behind and we would take them outside and convince them not to fight, without offending their egotistical manhood.

I was pleased with my personal transformation and healing after my training as a Buddhist monk. I had finally stopped going to the hotels where I was not welcome, waiting for men to attack me.

I also continued my meditation practice.

The money rolled in. Mary, however, was not earning very much in her job as a hairdresser, so a Maori friend of hers convinced her to fly up to the Gulf of Carpentaria and work on a prawn trawler. The prawns were plentiful and crew members were earning huge amounts of money for just a few months of work.

My mother agreed to move in with me and take care of our daughter while I was working. This was a satisfactory arrangement because my daughter loved her nan.

Mary often rang from the Gulf and told me how much money she was earning. We decided to leave for India in September because my darling daughter was not enjoying school, and her report card showed this clearly. It stated that she was not mixing with other children and was not interested in learning.

Because I had been a schoolteacher for a year before Vietnam, we decided that I could easily teach her by correspondence in India as she was only in grade one.

During this time, I had been gathering information about the spiritual side of India for our trip, and was particularly interested in gurus.

There was also a religion called Tantra that taught about the satisfaction found in proper lovemaking.

Also, there was a famous yoga teacher named Iyengar in the same city as a famous guru, so we decided to go to this city of Poona, 90 miles from Bombay. Here they had what was called an ashram where we could go and learn about spiritual matters. All we needed to do now was another two months of long hours of work to pay for the trip. Then we could afford to travel "five-star" with our combined earnings.

I was as happy as I had been for a long time. I still had mood swings and occasional nightmares of the war, but they were less intense and I could meditate my way through them and use my witness to overcome them and any thoughts about war.

This feeling of having a family, like normal people, was important to me. Maybe I could fit into society at last, if destiny approved.

Mary rang one day and said her boat was in port for a few days so she was flying down to see us. It was wonderful to see her again and we took Cheyanne to a famous restaurant in a nearby tourist resort town.

Mary seemed very distant during lunch and appeared nervous to see me. Probably because she had been away for so long, I thought.

On the drive back to the city, she broke the news that she had fallen in love with her ship's captain and she was going to pack and leave both her daughter and me.

I stopped the car near a cliff, and because I did not want Cheyanne to see me, climbed down about 30 feet, where I vomited in reaction to the shock and pain. I couldn't understand why love hurts so much or how to protect myself from such pain. I did not know how, even after all my training, but I was to find out. One must discover the value of surrender and the falsehood of thinking that we alone control our lives in this human form. Life is a mystery to be accepted and lived.

Mary eventually told me that her new lover was aged in his

forties. As we were both in our twenties, he sounded like a very old man to me.

When we arrived home, she immediately packed and left. I think what shocked me the most was the fact that she was leaving her little daughter Cheyanne, not just me. I told Cheyanne that Mummy was going back to work and would be home when she finished as I did not believe our child should think badly of either of her parents or worry about her mother leaving us. How could I tell my daughter that her mum was leaving both of us for some drunken old sailor?

For the next three days I could not work and I lay on our bed vomiting until there was nothing but bile left in my stomach to pass.

I asked my mother to take Cheyanne to my grandmother's home as I did not want either of them seeing me in this much pain.

In my suffering I watched my mind. I imagined all sorts of strange things as I flipped through my emotions of sadness, love, hate, anger and hopelessness. I imagined, during the worst times, shooting Mary and her old drunken boyfriend. I realised that this was just the mind doing its dance to try to alleviate my sorrow and battered ego, and I quickly let these thoughts pass. My recent Buddhist training once again served me well.

I wish to make a point, which I eventually realised in this sad state, and that is that women are in touch with their feelings and with what love is, whereas men of my generation are taught to be macho and strong, while never showing weakness. I had been told over and over that big boys do not cry and a man must never cry like a little girl. This denies us feelings and makes us hard.

It seemed to me at the time that I had no understanding of how to love a woman, so that is why Mary left me. My feelings must be repressed and dormant, so how could I feel love after a lifetime of suppression in the name of manhood?

Also, as men in this flawed society, we are congratulated by other men if we can have sex with many women, whereas women are considered to be sluts if they do the same thing.

It appears that somehow, I – and probably most other males – see sex as proving love, and use sex in place of feelings. Is this why some men embrace homosexuality so they can bypass this male masochism and finally express their feelings? Is that why many of our greatest artists are gay? Have they conquered their unfeeling male side and embraced, in relief, their feminine and intuitive side? I must find the answer to these questions. Not through becoming a gay man as I do not have a gay bone in my body. Perhaps I would find the answers through spiritual masters who lived in India.

Eventually I could face the world again and I went back to work to earn more money. I was more determined than ever to go to India and take Cheyanne with me.

I gradually forgave Mary, as I knew I had not always been the easiest man to live with in my sometimes withdrawn state of being. In the past when my moods overcame me, I would withdraw from both Mary and the world, and that is a difficult psychic thing for any woman to endure. Even though I was never physically violent with her or any other woman, my withdrawals were wounding and seemed to bring out the wrath in every woman I was with for many years to come.

Yet my emotional pain from losing Mary only slowly receded. It would be a long time in the future before I could get close to any woman again. Whenever I did get close to a woman, my memory of this pain of abandonment would arise, and I would soon run from any poor woman I met who was interested in me.

In retrospect, the complete lesson about sex and love had not been learned and the opportunities would be handed to me time and again until I finally understood. The lesson of romantic love and the longing to marry and make a relationship last forever, like in the fairy tales, always eluded me. Was I deluded or was the fairy tale an illusion?

Existence is indeed a kind teacher, even when it seems to be relentlessly cruel.

I did see a doctor about this deep heart pain and he, once again, wanted me to take some new tablets called anti-depressants. I took two and they made me start grinding my teeth, so that reinforced my determination not to ever take tablets made by pharmaceutical companies. The usual medications freely fed to vets do not work for me or, I suspect, for other vets or depressed people. These tablets seem to be a ticket to a slow downward spiral after the initial relief. I am sure that one day this obvious truth will be revealed and more control will be exercised over prescribing these tablets, for both adults and children, than it is now by doctors and the AMA (Australian Medical Association). Pills do not fix everything.

The doctor did, however, tell me about a rehabilitation centre in Melbourne that specialised in helping war veterans. He said it was a beautiful converted mansion on a large forested estate and it was run by the leading psychiatrist of the entire Armed Forces. He told me that this man was highly esteemed and had written books on the subject of sick veterans and was quite respected in his field. The doctor convinced me that the only way to break out of this depression and the pain I was feeling was to enter this fancy "nuthouse".

Fortunately, one of Cheyanne's grandmothers lived in Melbourne so I reluctantly decided that I would seek help for myself, yet again, in the medical field. Desperation is a strange driving partner.

My best friend at the time was a solicitor and he and his father were wealthy and had influential friends in government. This home in Melbourne only allowed residential patients. I needed to be a day patient because of my daughter. I also carried a deep distrust, by this time, of authority and the medical profession, so I had my lawyer mate phone the head of Veterans' Affairs and procure me a letter saying that I was to be entered as a day patient, a volunteer, and could leave whenever I chose. I was assured that this was a legal document and it was signed by my lawyer, myself, and the vet administrator. I was making sure that a psyche hospital could not detain me. Psychiatrists appear to be convinced that all Vietnam vets are crazy in one way or another.

I gave notice at work, rented my home, and flew to Melbourne with Cheyanne. The plan was to spend a month or so at this establishment to see if they could help me conquer this pain and the nightmares of Vietnam that had returned in my weakened, broken-hearted state. Then I would take my darling Cheyanne to India.

If you have never seen a movie starring Jack Nicholson called One Flew Over the Cuckoo's Nest, I would urge you to see it, as it is a classic. Then you will understand what I mean when I say this place was definitely a "cuckoo's nest" and I was about to land in it.

Chapter 21 - The Cuckoo's Nest

A grandparent's love for their grandchild is a blessing to behold. It was a relief for me to be in a stable home with people who had been married for decades, even though they did have hidden secrets which I found out about at a much later date. It seems that most married people compromise their freedom for security, money, the children, for respect from their peers, or for some other reason.

The grandparents' lounge room was well lived in and felt snug and warm. Every available space was filled with something – books, a fish tank, photos of family, plus mementos from holiday places that they had visited in their long lives.

There were magnificent old sunken lounges that seemed to hug me when I sat in them, with ingrained wood on their arms, all adding to the ambience and feeling of warmth and safety.

And to see my daughter, the light of my life, laughingly happy with her doting grandparents, brought tears to my eyes and eased my nervous tension a little.

I spent a week at this welcoming home allowing my girl to settle in while also exploring parts of Melbourne with her. Places such as the magnificent botanical gardens, plus various museums and art galleries opened a whole new world for her.

Finally, on a Monday morning I drove into the beautiful gardens and grounds of the expensive asylum. Huge fir trees lined the driveway while flowers and lush green lawns abounded in every direction. At this point I did not think of this place as an asylum. That soon changed.

The huge red brick house was obviously once an estate for a very wealthy person. Its intricate and unique architectural design reminded me of old English mansions. I was pleasantly surprised.

This asylum for vets wasn't the usual drab army barracks or army hospitals that I was familiar with.

The Admissions office, just inside the massive front doors, was typical of any hospital. After introducing myself they reluctantly registered me as a day patient but only after some discussion after I presented all my paperwork. I did not show them the letter from Veterans' Affairs and my solicitor.

A large muscular male nurse then guided me to a comfortable chair in a waiting room next to a magnificent, closed mahogany door. I was informed that the director, who was also the renowned psychiatrist, would see me soon.

I felt hopeful that maybe here, in this sophisticated city of Melbourne, such professional people would be able to help me. I was desperate to be able to sleep without the nightmares, which had arisen once again, constantly waking me during the nights. I needed to be able to relax around large crowds of people and to conquer my intense anger when it arose.

Of course, the Buddhist training had helped me conquer these emotions as they arose but I could not control the nightmares or the pain of lost love.

My enthralment with this place would not last, however. I got my first of many shocks in this exquisite madhouse where the game being played was as intricate as chess, yet as deadly as an ancient duel. Here, I soon had to fight against becoming a mindless zombie. This situation was much worse than being shot in a war battle, which would actually be much kinder than what was being done to the chosen few in this place.

After waiting an hour, the usual long army wait, a thickset, tall, no nonsense looking matron, dressed in a pure white immaculately ironed uniform, opened the psychiatrist's door. With no introduction or friendly hello, she ordered me to "come in". So far, these people seemed anything but kind and considerate. I wondered why they had me wait nervously for an hour when no one had come or gone from the doctor's room during the entire time I was sitting there?

The shock that confronted me as I entered this huge room

went deeper than the mere physical. It outraged me that someone could be this insensitive, as men who first arrive here are obviously already nervous from their war experiences or they surely would not be here. We come seeking comfort and help.

As I approached the psychiatrist's huge teak desk, two white dogs which had been hiding beneath it, suddenly attacked me. They were poodles and they ran straight at me as if they were going to bite. My first instinct was to kick them in self-defence, but my second instinct was to give them the benefit of the doubt that they would not bite me. My next instinct was to get them to stop by overpowering them with a strong command and physical dominance. My mind worked at lightning speed and I realised that any of these reactions would not be in my best interests in achieving anything in this mansion. The elderly man with the thick glasses was studying me intensely from behind his desk and he was the obvious owner of these neurotic dogs, so I simply chose to freeze.

They came very close and after at least a minute of constant, loud barking, he who holds power called them off with one command. I wondered if he thought that this would impress me, or if he was just studying my reaction for his future reference? He disappointed me. I found this whole sorry episode a childish and insensitive action against men who are suffering from the stress and shock of war and are obviously nervous when entering an asylum, even one as well concealed as this one appeared to be.

These two, the doctor and the nurse, may even have thought that they were conducting an experiment. They stared at me intensely, perhaps looking for fear, aggression, or whatever negative things that the psychiatric profession takes so seriously but seems to take such joy in discovering. But all it shows me is their disconnection from their hearts and their limited intelligence. Well, this is a great start to my therapy, I thought.

My initial instinct was to turn and walk out of this place but I ignored my gut feeling and still hoped that these fools might be able to help me.

Was I naive, or just stupid, living in hope? Perhaps that wise man Jesus was talking about places like this when he was purported to have said: "Give up all hope ye who would enter here." That saying simply means to me that to be present in the eternal now, the mind must be free from any hope.

After a month of enduring this place, my initial awareness and natural instinct to leave would reveal its truth in a way too horrible and shocking for me to even imagine at this early stage as I stood there still clinging to a slight hope.

What a pity that I did not put my trust in myself, for I would have quickly brought these dogs to heel and spoken my truth to these two people and chastised them for their gross insensitivity and for playing flawed mind games with wounded soldiers.

The great and noted psychiatrist did not shake hands or introduce himself to me. He simply handed me a form to fill out and informed me in a bland voice of the rules of his little kingdom. No drinking, no aggression (I wonder what he thinks a dog attack is?), attend all classes, and take (ALL) the medications I am given.

He then casually dismissed me without even looking up from his desk, and Nurse Ratched, as the men not so fondly called her, escorted me into the bowels of this lovely home and introduced me to the permanent, residential, inmates.

In the central area where everyone met for "classes", announcements and tablet taking, was a huge lounge obviously once used by a very wealthy family, who previously owned it, to relax and socialise. As I marvelled at this luxurious setting, I pictured comfortable leather lounges, huge TVs, and well-stocked bookcases. Now it was a stark meeting and marshalling area with none of these amenities but, strangely enough, still as luxurious a place as I had ever been in.

The polished wooden floor, probably hard oak, was dark and bare. I wondered what had become of the rich Persian rugs that had added to its oaken splendour and warmed the feet and hearts of its former, somewhat luckier, occupants.

Now the environment was similar to an army parade ground as the thirty, much older than me, inmates and I stood together for roll call.

After we all answered "present", two trolleys were wheeled into the room. One had tea, coffee and cakes while the other carried medications. I was handed a yellow, full strength, 10ml Valium tablet and was told I must take three of these a day. I put this in my mouth, as Nurse Ratched watched me with her eagle eyes, and pretended to swallow. I wondered how they expected me to drive home at night loaded up on 30ml of Valium, and why were they giving me this numbing drug when I had not even seen a doctor and been asked what medications I was already taking? They must have actually assumed that all veterans need Valium.

Nurse R then proceeded to tell me my schedule. First, I had to see the resident doctor (a little late) and then it's woodwork in the carpentry shop followed by a general gathering to discuss treatments and how we are coping.

The doctor gave me a quick physical exam, asked me what medications I am on (none, but obviously they should have asked me this before handing me strong Valium), and then quickly dismissed me.

A lot of the people who worked in this institute had been in the army, and some still were, so they carried their rank of captain, or whatever, and ordered me around as if I was still in the army. This annoyed me, but I decided to hide my annoyance for now.

After the woodwork class where I was expected to make a table (I did this in school a long time ago), we were called to lunch. Here I met a younger man in his twenties, about 5 feet 8 inches tall, with light blond hair and pleasant features. He was obviously fit and in some type of training. This was the first person who worked here who was gracious and used his manners. "Hi, my name is Peter and I am the physical education instructor."

He was the first person who worked here to actually shake

my hand and properly introduce himself in this cold place. He would turn out to be one of my saviours.

During the following week the older residents slowly warmed to me and finally offered their hands. They then gradually revealed their fascinating secrets of living here permanently and it became obvious why they were slow to trust anyone.

Peter, the physical education instructor, asked me if I played any sports.

"Yes," I answered, "I play rugby league football."

"Professional?" he inquired.

"Yes."

"Are you any good?" he asked.

Just average since Vietnam," I answered.

"Why is that?" he responded with obvious concern in his eyes.

"I think it was the three-year layoff, the operations on my face, and especially the Agent Orange poisoning which nearly killed me."

He asked about the Agent Orange and how it nearly killed me so I told him about the stay in hospital and the swelling of my entire body, together with the heat, the pain, and then the long, long road of recovery to just a semblance of my former health.

Peter then asked me how good was I at this game of rugby before Vietnam.

I answered, "I played for my city at age eighteen and was also picked for the Australian Army rugby union team. I was good enough."

Shaking his head, he said something I was happy to hear. "Bloody armies."

"You have sure got that right," I laughed.

"Do you play any other sports?"

"I run and I practise self-defence."

"Are you a good runner?" he asked.

"Pretty good. I recently won the finals of the combined sportsmen's golden mile in North Queensland."

"That is great because I love running too, so we can train every day here together if you like?"

"I would love to," I smiled, "and it might keep me sane in this strange place."

"You have sure got that right," he laughed.

If it were not for this kind man, the outcome of my story here in this place, and indeed my entire life, may have been much different.

So every day Peter and I ran for half an hour, and every day during the slow warm-up, Peter told me things about this place that surprised me, to say the least. The rest I found out from a couple of the good old Second World War veterans who had eventually taken me under their wing after they befriended and trusted me.

Most days we did a couple of hours in the woodwork building and this was where I got to know these inmates. Some of them were pretty good at woodwork and they helped me out as I did not have the angles in my head to be a good carpenter. Eventually, I managed to make a fair bedside dresser which I gave to my daughter's grandparents. This carpentry helped to make the time here pass while I waited for the help which had not yet been forthcoming. I was beginning to doubt that it ever would.

It was actually on my second day during lunchtime when I became really concerned about the viability of this place. Two long tables had been set up in the big lounge and it was here that I first saw the man in the wheelchair. A large male orderly wheeled him into the lunchroom. After he was spoon fed, with a bib tied around his neck, he began yelling at the top of his voice.

"Incoming, incoming. Everybody take cover. Man the guns. Fire, fire. Enemy sighted. Kill them, kill them."

He continued screaming these words as we all tried to relax and eat.

I went and sat next to Peter who was eating with us, and

said to him, "Poor bugger. Has he been like this since the war?"

Peter looked at me for a while with obvious sadness in his eyes. Nurse Ratched was hovering nearby and only after she left, he told me, "No, he has only been like this since his treatment."

"What treatment?" I asked.

Peter looked around to make sure Ratched was not listening and finally said, "Come outside."

I was happy to follow Peter into the garden because I suddenly had no interest in my food. I was glad to escape the screaming war madness emanating from this poor disabled man in the wheelchair. What on earth was he doing here, I wondered? Apart from disturbing the other patients, was he perhaps a warning?

Surrounded by the landscaped beauty of these trees and gardens, Peter told me about this man-made horror.

This old man had been in the navy and was quite normal, if a little withdrawn, when he first came here. In the respected opinion of the "boss man", the armed forces leading psychiatrist no less, this old navy man was considered to not be responding well enough to the medications he was being forced to take (he probably spat them out like me) and so needed further "help".

Peter told me that this once physically healthy but now broken man was brutally transferred to the grounds of a nearby hospital where there was a large room standing alone, with a veranda. Like a prison, the veranda had steel bars around it, as did all the windows. It was, in fact, a prison, Peter said.

In the room this man was injected with daily doses of Valium. After six long weeks of this chemical therapy, the boss man psychiatrist did not think that the navy man had responded well enough yet again. I wondered how this famous psychiatrist could possibly know this, if the poor man was in such a drug-induced haze?

Unfortunately, they transferred this drug-induced man to another barred prison a little further away. For another six long weeks he was given electric shock treatment. At the end of this six weeks of "treatment" the boss man judged him to still be in need of his expert "help". So, he finally ordered him to be given a "lobotomy". To say that I was shocked to hear what a lobotomy actually was is an understatement. A lobotomy is where the bone around your head is sawed right through. This bone is then raised to allow access to the exposed brain. Then parts of the brain are removed.

The outcome of these expert healing procedures was here for all to see. This poor war veteran in the wheelchair, who could walk before the lobotomy, was now more of a vegetable imprisoned forever in a wheelchair. Obviously, from the nightmare he is trapped in, none of the above inhumane methods worked. The procedures only benefited the mad psychiatrist with the angry dogs, experimenting with a live patient so he could finish another book. These heartless people will obviously need to experiment again. Not with me, I thought.

I asked Peter if Jim (that was his name) used to relive the war at lunchtimes before his treatments ever occurred.

Peter told me that he had not.

"Why on earth do they keep him here now?" I inquired.

"He is paraded here as a warning to you and all the other patients."

How sick is this? Do all psychiatrists go mad because they are constantly looking for the negative in people and this becomes the mirror they live in? Why can't they look for the positive in people and teach them how to think positively, not direct them to concentrate on past so-called wounds?

This place was strange and shocking to me in many ways and I was fast being cured of ever seeking help again in places like this, or with psychiatrists anywhere.

Some of the old men who used this place as a home because they had no other had put their carpentry skills to good use and built brilliant hide-a-ways in the back of their wardrobes. Here they hid alcohol and other banned contraband. They also had skeleton keys to the back doors of this locked down night-time prison, which they used to let themselves out so they could go to the local pub. I met them there one night when they invited me to join them for a drink.

They tried to explain to me that living in the institute was like a game and one had to understand the rules of the game and know how to con the boss. I was impressed by their cunning and their ability to live here even with all this obvious impending danger waiting to strike them, instigated by this mad doctor.

Some funny events took place in this veterans' retreat, probably one of the most luxurious "nuthouses" in the world.

One day we were all taken outside onto the lovely lawn to learn and practise archery. The bows were as tall as me and the arrows were steel tipped and could easily bring down a deer or a human. I thought, yet again, how dumb is this?

The large targets were set up about 50 metres away at the base of the huge fir trees. We were then ordered to step forward, in groups of three, and shoot an arrow, while all the others stood behind us and watched.

After my shot, I stood behind one old Second World War vet and watched him draw back on his large, tight-stringed bow. As he struggled to draw back the string on the bow, the effort forced his bow slowly upwards until the arrow was pointing straight up to the sky. Just before he fell flat on his back, he let the arrow fly, straight up into the heavens.

Amid the yells and the screams, we all turned as one, running helter-skelter for cover. The arrow eventually landed harmlessly but I fell down with old Al and another vet because we were laughing so hard.

Archery was cancelled and never took place again while I was there.

Another time one of these old wags set off a smoke bomb in a room adjoining the kitchen. As the huge amounts of smoke filtered into the room where we were participating in group therapy, the cheeky old bugger yelled, "Fire, fire, fire! Run for it or we are all cooked meat."

All the patients and staff ran flat out onto the green grounds, some yelling and screaming. Once outside I was again overcome with laughter at this hilarious, mad stampede. Lying on the grass in hysterics, I looked up to see the boss, and his barking dogs, staring cold-eyed down at me. Nurse Ratched was by his side and her look was even worse if possible. It seemed I was caught laughing, which was not playing their stupid game, and perhaps they thought I had something to do with this prank.

The next day Peter and my old war friends quietly told me that I had been chosen to follow Jim, the lobotomised man in the wheelchair. They advised me to run for it as quickly as possible.

Running was not my way out of this. Now I was angry and did not intend, or need, to control it so I calmly waited for my summons. When I was finally ordered to appear, my anger at this craziness and crass stupidity, to say nothing of the heartlessness, had built to a hurtful feeling in the pit of my stomach. This foolish psychiatrist was about to experiment with another human, me. Why? Because I was the youngest and fittest and he could, possibly, write another book? He had obviously failed with his last human experiment. People like him do not admit failure.

The summons had come after another disturbing lunch with poor Jim present and screaming accordingly about "incomings", killing, and bombs exploding.

As I entered the boss's large office, and as I knew what he had planned for me, I had nothing to gain by being polite. I knew by now that I certainly would never receive any sane help here, so I had nothing to lose.

When the dogs duly ran at me, barking loudly, I took a commanding step towards them and yelled loudly, "Shut up and sit down," while pointing my finger under the boss's desk. The dogs yelped and retreated immediately, hiding under his legs. Funny how dogs often become like their human masters.

I looked around the large room and quickly saw everything. Two large orderlies, probably from the nearby hospital, were standing in the left corner, near the door. They were dressed in white with their arms folded. They were on my left about four paces away. Nurse Ratched stood about another four paces to my right, near the huge bay window. She was always dressed in well-pressed whites but for the first time I saw her mouth drop open in amazement at someone taking control of the mad psychiatrist's dogs.

One of the orderlies was trying hard to suppress a smile and it took all my self-control not to burst out laughing again, but there was work to be done here. The look of astonishment on the psychiatrist's face was worth more to me than you could imagine.

Slowly recovering his composure, he finally directed me to sit on the hard wooden chair at the front of his desk. Once I was seated a look of unconcealed joy came over his face as he began to speak with Nurse Ratched, now standing by his desk to his left. The two orderlies moved to the door with their arms still folded. Subtle bastards, all of them! They acted as if they were about to take a prisoner.

The psychiatrist looked at me with what I imagine he considered a kind smile on his face (the first time he had smiled at me) and said, "You do not seem to be responding to the medications we have been giving you."

I felt like telling him that was because I was not taking them but I held my tongue and let him continue digging his own hole in his smirking way.

I had chosen not to take Valium because it is an insidious drug that has caused more eventual breakdowns and hospitalisations of veterans than any other drugs we have been given. We feel relaxed, dopey, bleary eyed and calm after taking the tablet but gradually, over the weeks and months, it leaves the nervous system in a state of distress and anger easily arises, then all types of misplaced emotions follow as the effects of the drug wear off each day.

I have seen a few of my mates taken away and placed in straitjackets after being on Valium for over a year. In the padded cells they were then fed "Wobbly Oblies" (the serviceman's nickname for Rohypnol), a ridiculously strong sleeping tablet that sends legs rubbery and wobbly, and which was soon to be banned as it eventually encourages suicide which is caused by its accompanying deep depression.

Valium, however, has never been banned. Its name just keeps getting changed by the big pharmaceutical companies and most doctors still prescribe the newly named tablet quite liberally. Diazepam, I think it is called now, but that name will also change.

I stared intently at this foolish, gloating creature sitting before me. In his kindest voice, never used before now, he informed me that I would be taken to another hospital where I will be very relaxed. He did not even bother to ask me if I wished to go or what he intended to do to me there.

That was quite enough of playtime for me so I spoke freely for the first time. "Oh! You mean the lovely relaxing room with the steel bars where I will first be injected with Valium for six weeks, followed by shock treatment for another six weeks, and then kindly given a lobotomy, like you did to poor old Jim. That did him a real lot of good, didn't it?"

I was pleased to note that the kind expression on his face was immediately replaced with a frown and an expression of anger and consternation. He looked to the two big men by the door and nodded, obviously warning them to get ready to grab me.

He slowly recovered his composure and said, "Yes, that is right. You will have no troubles at all to worry you after the treatments as you are different to Jim, and younger."

This man was either an idiot or a monster. A bit of both probably. How could he possibly do that to someone else after what he had done to Jim? He was fortunate that I am not a violent man because I would dearly have liked to get up and slap him very hard.

Instead, I ever so slowly reached into my pocket and handed him the letter from my solicitor and the head of Veterans' Affairs, stating that I was a voluntary day patient and could leave this establishment whenever I chose.

On reading the letter his frown returned and he handed the letter to Nurse Ratched.

I stood up and gave him my parting shot. "If I were you, I would leave those barking and ridiculous attack dogs at home.

This is no place to have them attacking nervous soldiers when they enter. And you should be ashamed of yourself for what you have done to poor old Jim. You definitely should not be planning on doing it to someone else."

Then I did something I should not have but this fool obviously needed at least a little shock. On his desk, right in front of me, was a metal, 8-inch letter opener, similar to a knife, protruding from a heavy stone. As I turned to leave, I pretended to accidentally knock it off his desk. Before it hit the floor, I snatched the knife out of its stone holder and turned back to the boss with it in my right hand and the rock in my left, both held high as if to attack. The look of horror on his face, as he cringed in his chair, was quickly followed by the colour in his face instantly fading. This was a joy to behold and I said, while placing the letter opener and stone back on his desk, "By the way, there are not enough of you here to put me in your sick prison."

I turned and walked to the door. The orderlies hastily stepped aside and one of them was now grinning openly so I said to him, "I don't think this work suits you, mate."

In the adjoining car park, I unlocked my car and then looked back inside through the large bay window. The lame psychiatrist and Nurse Ratched were standing at the window staring at me with a look that seemed to say, "Well there is one that got away."

With a laugh directed straight at them, I climbed in and drove away with great relief. I could not resist spinning the back wheels of my car.

As the adrenaline rush died down, I contemplated what had just taken place. One of the things that I learned the hard way in Vietnam is that every person has good and bad in them and even nice guys can kill. My anger and actions could have landed me in serious trouble. I realised that I am not a soldier and do not wish to be one as I do not want to kill or hurt anybody, not even that sick psycho. I must un-learn what I was taught in the army and not act like a soldier, always running towards trouble and trained to kill in a split second. This anger and urge to fight is no longer of any use to me. I do not want to be a physical fighter.

I realised that I may always be a warrior but I would like to think that I can be a spiritual warrior. I had discovered that Christianity, Buddhism, Islam, and their current offshoot which is Western psychiatry, could not show me the way. I must now continue to seek elsewhere for the answers to my confusion and questions. I must search for the answers to the riddles embedded in this strange life in this world I find myself in.

To the East I shall go. India is known as the Motherland and I am also told it is the land of the spirit.

Chapter 22 - The Spiritual East

My daughter and I left Melbourne and returned home to stay while I earned a little more money. She went back to school but missed her mother terribly, often asking when she would return.

My heart was filled with love for my little daughter. I often thought that if it were not for her on this earth, I would not bother to carry on and would have gladly died. Sometimes it all seemed too difficult but I was not ready to give up. Yet because of my daughter I have a goal. It is to leave this world a better place than it is now.

She was excited about travelling to far-off lands. Her teacher again told me that she did not mix with other children and was not interested in her lessons.

However, she played happily on her own in her own little world. Of course, because she is like this, the other kids bully her. When I found out about the bullying, I sped to the school prepared to find these bullies, and also to berate the teachers for allowing this to occur.

Cheyanne met me at the gate with a big smile and a hug. I told her what I was going to do and she said, "Dad, I can take care of my own fights, and anyway we are leaving here."

I nearly burst with pride at my tough little girl.

I was hoping that travel and new experiences would change Cheyanne so that she could one day enjoy school and mixing with friends.

We soon flew to India and then set off by train and bus to the nearby neighbouring northern area of Kashmir. It was still open to tourists at this time in the 70s, but not many people went there, as India and Pakistan were about to fight over ownership of this beautiful small territory. Religious beliefs, as usual, were at the basis of this war. Hindus against Muslims.

After spending a few relaxing and fascinating days on a houseboat on the large lake that adjoins the capital, we caught another bus to the far north-west corner of the country where we were booked into a ski resort for a few weeks. Cheyanne loved the thought of snow, which she had never seen, and was excited about learning to ski.

The mountain resort had not been opened for very long and was in a small valley. The ski slope was not very long and I was sure that professional skiers would have found it terribly boring, but for us it was heaven.

In Australia before we left, I had begun drinking and smoking, even though I do not really smoke. I had also been taking the odd Mandrax tablet to calm my sad heart and troubled nerves which had been made worse by the shocking Melbourne asylum. However, travelling with my daughter I gave all that up "cold turkey", including the barbiturate Mandrax.

On arrival at the ski resort the friendly desk clerk handed me a large block of hashish. I put this hash in the drawer next to my bed just to enforce my intention. The use of hashish was not yet outlawed here so was obviously included in our hotel costs.

The ski slopes were covered in snow and surrounded by trees. They were only about 30-degree slopes and not of an international standard, so we were the only skiers there.

Cheyanne was a joy to behold as she skied, screaming and laughing, down a long run.

There is a type of flying fox that one sits on which takes you to the top of the ski slope. Cheyanne was too small to reach it so the attendant lifted her on. At the top of the slope, she was too small to step off so she would fling herself onto the snow and land flat on her back with her skis protruding into the sky. She was a tough one, my daughter.

I, on the other hand, was just a nervous wreck, coming down cold turkey from tobacco, drugs, alcohol and opiates.

I soon broke out in an itchy, horrible looking rash that covered my stomach, chest and arms. My mind kept telling me to smoke some hash and I would be fine, but I had set my course and I was not going to deviate from it. Mandrax are, I discovered, a strong drug, related to opiates, and difficult to give up. Of course, so are the legal drugs of alcohol and, especially, tobacco.

There was a resident doctor at the resort so I made an appointment, because the rash was getting worse.

I felt terribly nervous, also sleepless, and my hands were shaking.

The doctor was a small Kashmiri man, with light brown skin, about thirty-five years old.

It was one of the most amazing, satisfying and shortest visits I had ever had with a doctor. One of the first times, also, when a doctor did not quickly write a prescription for yet another pharmaceutical drug.

He looked closely at the rash and me, then sat back down behind his ancient, dark wooden desk, looked me straight in the eye and said, "You do not need me as you already know what is causing this nervous rash."

He then took up his pen, entirely dismissing me, and began writing.

In that very same instance when he told me "you know", I did know. It was simply a nervous reaction caused by my body as it came off drugs, tobacco and alcohol.

Amazed at the insight of this doctor, I walked out of his office and neither of us said another word.

For two weeks I skied every day with my beloved daughter while the battle inside me raged. The mind gave me every excuse imaginable to drink or smoke. Thankfully, the insight meditation training in Thailand kept me strong. Witnessing and awareness are a great power and I had the ability to watch my own demanding mind. I also, once again, had the understanding that there was a watcher in me separate from the mind.

Each day, nothing was pleasurable in the least, except the

joy of watching my little daughter laughing after her falls, as she gradually became an accomplished skier.

Of course, I hid my nerve-racking condition from her and acted as if I was having as much fun as she obviously was.

When the skiing was over for the day, I then taught her the school lessons she needed to learn. I was pleased to see that she was a quick learner.

I was always up early in the mornings as sleep was difficult. I shall never forget the sunrise that revealed itself to me on the fifteenth day. As I watched it creep slowly above the horizon, life seemed to flow back into my veins. I stood open-mouthed, in awe of its colours and smooth rising sparkling magnificence. The dark cloud of withdrawal lifted and I seemed to really see the snow-covered valley for the first time. I also saw the eagles soaring in the clear blue sky, plus the smiling friendly people as they walked past, and the snow-topped trees on the beautiful bright white slopes.

Breakfast was suddenly a mouth-watering delight, as food once again became a pleasure to eat. The rash, of course, disappeared with no medical help apart from the supporting words of that brilliant young doctor.

I dropped the hashish, which I had never touched, into the rubbish bin.

It was soon time to leave this wonderful place. I renamed it "Resurrection" for what it had done for me and my daughter.

We spent another night in the still peaceful capital on an affordable yet luxurious houseboat, and the next day we flew to Bombay on our way to the beachside resort of Goa in the south.

On the plane ride I read the magazine that airlines supply in the seat pocket. In it was an article about the controversial so-called Sex Guru named Bhagwan Shree Rajneesh. He had thousands of Western followers who lived with him in an ashram in a city named Poona, which is about 140 kilometres east of Bombay. This man resided in a palatial palace living a life of opulence and luxury made possible by the donations of his wealthy followers who supposedly indulged in sexual orgies.

Sex obviously sells. I found it interesting that people were so taken in by this man that they showered him with money, jewels and expensive Rolls Royce cars. How did he do this, I wondered?

It sounded a bit like a sect and I was no follower of any religion or sect so I replaced the magazine and forgot about this colourful fellow. There would have to be other gurus who I could seek out in India. I was not interested in sexual orgies or giving this man my money.

In Bombay we spent one night in one of the world's great hotels, the Bombay Taj Mahal. It was out of my price range but still very cheap compared with Western prices and I wished to give Cheyanne a taste of this luxury.

The foyer of this great hotel was magnificent and was alive with self-important, rich-looking people of many different nationalities.

I sat on a lounge in this opulent foyer with Cheyanne, discreetly observing the people because I liked to know what was going on around me. There were some obvious hustlers working and also some very good-looking prostitutes. I watched a beautiful blonde woman approach a man. Her mark was a rich-looking Arab and she soon led him away, probably to his room.

Surprisingly, I was to meet this same woman a month later in an ashram.

Our room in the Taj was, surprisingly, not much different to the rooms of most motels. It was just a little larger. The glitz, and beckoning splendour, was only on display in the foyer.

We ventured out that night into the bustling streets of Bombay to eat at a famous restaurant I had also read about in the magazine on the plane. It was only a short walk from the Taj Mahal to the restaurant and even though I was holding onto the hand of my daughter, I was offered hashish, speed, women and boys. I dismissed all these hustlers with an angry "no". A definite, strong "no" is powerful in India and was the only thing that would stop these hustlers in their tracks.

The streets were alive and these Indians all seemed to be selling something. Shoeshines, food, yo-yos, dolls, anything you could possibly want.

The common people make up at least 90 per cent of the population in overcrowded India and they are often terribly poor. The government is corrupt, as baksheesh is the way of doing business in India, and these "under the table" baksheesh payments can buy you anything from a car licence to freedom from arrest for the worst crimes imaginable.

I bought Cheyanne a colouring-in book and a yo-yo. After a great Indian meal, we made our way back to the Taj to have an early night. The next day we would be catching a rough and ready passenger/come freight boat that sailed to Goa. We did not have a sleeping cabin as I had been informed that the few available in first class were booked out six months in advance.

Early the next day, carrying my backpack and Cheyanne, I hurried to the pier. We took up position in front of two 10-foot-high wire gates and waited with a huge crowd of mainly Indian people, all anxious to catch the boat. We had been told that we must get on this boat as quickly as possible as it would be overcrowded (typical in India) and it would be difficult to find a place on the deck where we could sit and also sleep, as it was an overnight sail.

I placed Cheyanne on the ground next to me and held her hand, ready to move quickly.

The large gates were finally flung open and an immediate and dangerous stampede began. We found ourselves being pushed and shoved, with Cheyanne soon being trampled underfoot. She screamed, "Daddy," in fear, so I pulled her in front of me and proceeded to quickly clear the space around us. I did this with my hands, elbows and feet. Anybody who rudely ran into us I softly hit, in order to protect my daughter. Suddenly, almost like magic, there was a clear space around us for a few feet on either side. My little daughter was no longer terrorised and trampled by the madly rushing crowd. She looked up at me and said, "Thanks, Daddy."

We reached the ship in one piece with my darling daughter calm once again. Unfortunately, we could only find a small place to sit on the deck and I wondered how we would be able to sleep as there was no room to stretch out. This ship was definitely way overcrowded.

As we sailed away, an Indian family next to us began force-feeding a screaming, obese child who was only about eighteen months old. They just kept forcing food into the poor child's mouth against his obvious wishes.

Cheyanne tugged me by the arm and said, "Daddy, stop them hurting that little boy," so I intervened.

I slowly, in halting English, explained to them that the child had definitely had enough food. I discovered later that this over-feeding happens often in India because the parents sometimes grew up starving, so if they can afford it, they think that lots of food for their children is healthy.

Indians are usually not an aggressive people, and this couple smiled in an eye-open, shocked sort of way, but thankfully stopped force-feeding the poor child.

Cheyanne then began playing with him, much to his delight.

Due to the relentless racket going on around us I decided to use some good old Indian "baksheesh" (accepted bribery) to get us a bed upstairs.

I was quickly informed by a purser who was guarding the steps that this was not possible as all cabins had been booked out for months but I managed to bribe him so he let us walk upstairs. Just so my daughter could have a look, I explained, as she had never been on a big ship.

The people we met as we walked around this roomy top deck were mainly rich-looking Indians.

Then a strikingly beautiful, exotic looking, blonde Austrian woman, about thirty years old, made friends with Cheyanne while I was speaking to an Indian couple who I thought could help us find a place just to stretch out on the deck somewhere.

I finally turned around and found my daughter and the Austrian in an animated conversation. It was immediately obvious that my daughter had this lady enthralled.

I slowly strolled over to them. Cheyanne introduced me and we all sat and talked. When I explained our predicament, this generous woman insisted that we sleep in her luxurious cabin. She informed the purser that she was expecting us, so we retrieved our luggage from the bottom deck and spent the rest of the trip in first-class luxury with this auburn-haired beauty named Inga. That night Inga and Cheyanne slept in her bed together and I slept on the floor.

In Goa we got to know Inga a lot better and so a year later our Austrian friend came and stayed with us for a month, as our guest, in our home in Australia.

Goa, in those days, was unspoilt by the later influx of drunken young tourists from around the world, all wanting to get high.

At that time in Goa, I soon discovered that a large array of drugs was being sold freely on the beaches and in the restaurants. The well-off hippie looking types that lived there openly smoked hashish in the many cafés. I did not.

The beach towns were plentiful and fresh sea produce was rowed ashore, then immediately cooked for us on the beach at ridiculously low prices.

Accommodation was also abundant and a comfortable beach hut cost us about one dollar a night.

We made friends with a bunch of crazy Italians who enjoyed cooking us excellent spaghetti and Italian meals, and also liked offering me all types of drugs, which I again refused. One of their friends, a good-looking young Italian man, became psychotic, probably from the drugs, and would talk to himself all day while performing intricate dances. They explained to me that this was acceptable behaviour and he would pass through this psychotic stage, if they ignored him, and would eventually bring valuable information back to the group. I had my doubts, but sure enough, after a week of this seeming madness, he greeted us one morning, quite sane again. In the West, he would have quickly been put in an asylum and pumped full of Valium and sleeping tablets. If he continued to act this strangely, he perhaps would be given shock treatment.

Where I come from everyone is expected to act normal like everybody else. If you do not you are placed in jail or a hospital.

Another embarrassing incident that occurred happened on a beach where everyone swam and sunbaked nude. Busloads of Indian tourists would stare in disbelief, from a high embankment, stunned by such a casual freedom.

I was sitting on my towel watching my daughter play in the sand and sea when two women walked down the beach and spread their towels quite close to me. One of them was among the most beautiful women I had ever seen. However, her face seemed to hold a sorrow and sadness that touched my heart and I felt like hugging her and telling her that whatever was bothering her would pass. Her friend was the polar opposite. She was a huge, angry looking woman and not pretty at all. Her hair was cut short and she looked like a man. She was nothing at all like the beautiful blonde woman she accompanied and I wondered if she was guarding this delicate beauty.

Cheyanne saw them setting up their towels next to me and came to investigate these new people.

The women undressed and sat on their large towels.

Having been mesmerised by the beautiful blonde's face I finally managed to glance down and noticed her perfectly formed, upright breasts that were of generous size.

Cheyanne, who was heading towards me, veered off to my right and kneeled down before these two women and stared, with her mouth hanging open. She turned to me with a shocked look on her innocent face and called out loudly, "Daddy, come here."

Happy to come closer to this angel she was staring at, I jumped up and walked over.

Cheyanne cried out as I stood in front of these two opposite looking people, "Daddy, look, look, this girl has a penis."

Glancing to where she was pointing, I saw that the gorgeous blonde indeed had a perfectly formed uncircumcised penis, plus two generous balls hanging down to the towel.

For a moment I was too stunned to speak and everyone was looking at me. I recovered and said, "Yes, I know, darlin, and it is quite okay."

"But Daddy, she is a girl."

If there was a hole in the sand, I would have crawled into it. "Yes, she is," I answered. Smiling at the woman, because I realised that this was a woman trapped in half a man's body, I said, "Sorry, we are from a remote place in Australia," as if this would make everything all right.

The beautiful blonde smiled at me and said in a guttural German voice, "That is okay; your daughter is very beautiful."

"Thank you, and so are you," I answered, feeling much more than I possibly should have for this sad person, seemingly trapped between two worlds.

I scooped Cheyanne up, ran down the beach and jumped in the ocean with her where I explained that there were some people in the world called hermaphrodites who were both male and female at the same time. This explanation finally appeased her curiosity and I managed to escape without further embarrassment.

The next day I was beginning to wonder where we would go next to find an Indian holy man, as I thought it was high time (forgive the pun) we left Goa.

While we were walking hand in hand to our favourite breakfast restaurant, I saw a man and a woman gliding across the sand dunes dressed very strangely. They were wearing long orange robes which almost dragged on the sand. Around their necks they wore strings of wooden beads, malas, with a photo of a bearded man hanging suspended from the malas in a round wooden disk.

Curiosity got the better of me and I asked them why they were dressed like this and who was the man in the photo.

These two ignorant snobs totally ignored me and walked past with their heads in the air. I thought this was strange and rude, but held my peace.

At the restaurant we met my Italian friends who had come to share a last meal with us, and I asked them who these people were and why would they not speak with me?

They explained that they were arrogant German sannyasins and because I was not a sannyasin, they would not speak to me or anyone else who was not one.

"What is a sannyasin and who is that man hanging around their neck?" I asked.

"That is the Bhagwan. He is a guru in Poona, a city to the north, and sannyasins are his followers, like disciples. He can make love with any of his many thousands of followers."

I had read a little about this fellow on the plane and in Australia, and each time decided to forget about him, but this news intrigued me. I thought, what could this old-looking man with a long grey beard possibly say and do to make people give him lots of money and expensive gifts such as luxury cars? And why would any of these many Western women who followed him gladly have sex with him at the drop of a hat? On further questioning the Italians assured me that any of these women would. He was an old, short, grey bearded man. What power did he have over these women?

I decided then and there that I had to visit his ashram in Poona and find out his secret.

We said our goodbyes to our many new friends and then caught the plane to the northern city of Poona to meet this fascinating enigma of a man.

Travelling by motorbike in India. My escape!

Chapter 23 - The Ashram

I reminded myself to focus on the teachings, not the teacher or guru.

Poona is a large industrial type city about 90 miles inland from the huge city of Bombay, which was later renamed Mumbai.

After a bumpy flight we checked into a nice, clean, well-run hotel just around the corner from the Bhagwan's ashram. After a meal and a short nap, we set out on foot to the ashram.

Turning right, and strolling down a small street directly in front of the ashram, we passed a few Indian magicians sitting on the side of the road performing their magic for any passerby. What these magicians did was beyond belief and so I gave them some rupees after each magical performance, which fascinated and amazed us both.

The ashram was enclosed by a very high cement wall that offered protection and also kept out unwanted visitors. The entrance in the middle of this wall was called the "Gateless Gate". It was a cement archway with two large, heavy swinging wooden gates of considerable beauty and weight.

As Cheyanne and I, tightly holding hands, were about to enter these gates a procession of Western people dancing and singing, laughing and yelling, came out through the Gateless Gate, forcing us back onto the street.

As we stood watching these mad, dancing, seemingly happy people, I wondered what the occasion was and what they were celebrating.

Suddenly, in the middle of this loud procession, I saw a dead man lying with his face and body bared for all to see.
He was being carried shoulder high on a bed of flowers spread over a crude wooden litter.

I found out later that this was the German Prince of Hanover and he was one of Bhagwan's many bodyguards who died while practising karate in the ashram's dojo. They were taking him down to the river to burn him. The way they were laughing and celebrating, I thought they must be happy he died.

After the procession had passed, I said to Cheyanne, "Come on, darlin, let's get out of here. These people are crazy. He has not got even one friend who is sad he died."

I turned to leave. Cheyanne gripped my hand tighter and pulled me back and said, "Look, Daddy, children; let's go in there."

I turned and looked through the gate and there were about a dozen children, some around her age, playing happily inside. Now Cheyanne had been travelling with me for quite a while and although she played well by herself and had enjoyed the snow and the beaches, she had not had other kids to play with.

I really did not wish to enter this seeming madhouse so I stood hesitatingly at the entrance and thought about it.

Quickly making up my mind, I said aloud, addressing this Bhagwan, "Okay, old man, we are coming in. You show me your tricks and I shall show you some of mine."

As soon as we entered, Cheyanne let go of my hand and approached a half black, half white girl from America. After a quick chat, and as only children can, they began playing happily together. For our entire stay in Poona, they remained firm friends and would often have sleepovers at each other's place.

The mother of this little girl came over and introduced herself to me. She was a light brown coloured black/white American and very beautiful. We eventually became firm friends and often shared the care of our two beautiful children.

Stepping inside this ashram and off the dirty Indian street was like stepping into a paradise. The place must have covered four or five acres. It used to be a run-down mansion but now the gardens were beautifully kept. The trees and plants were like a lush rainforest. All this was maintained by Western gardeners who were not paid a penny for their hard work. All were volunteers and followers of the Bhagwan, and they had come to Mother India from all corners of the earth. Nice con, I thought at the time. Free labour. Holy work.

On my left stood a massive circus tent that was capable of holding thousands of his followers. This was called "Buddha Hall" and this was where the man gave discourse in the evening.

Discourse consisted of him telling stories, sometimes for hours, often with a spiritual message, interspersed with crude jokes.

The ashram also had cafés, restaurants, accommodation, a swimming pool, and lots of therapy rooms and chambers. Exploring these chambers, I saw that the floors and walls were lined with rubber mattresses, and I would soon enough discover why these were necessary.

Nearly all the many people who wandered around this paradise wore long maroon or black robes. The ones visiting from all corners of the world wore the maroon robes and these people were usually partaking in the many group therapy sessions on offer here for a handsome price. The ones leading the group therapies wore the black robes and there were not so many of them.

Whatever I would think about this Bhagwan over the years, this trickster, flawed, crazy genius of a man, the phenomenon of this place in India was a miracle to behold. What a creation! Gathered here were many famous people, some extremely rich, together with amazing artists and musicians, doctors, psychiatrists, politicians, dancers, famous actors, etc. They were all drawn here by this enigma of a man.

A famous orchestra once played in Buddha Hall and a thousand of us danced to the wonderful Austrian waltz. People from practically every country in the world were here. What a gathering. Palestinians with Israelis, Croatians with Serbs, sworn enemies all meeting in peace. Amazing what this Bhagwan had put together, to put it mildly.

Once Cheyanne had settled into a routine, I began to join in. Of course, I refused to wear a robe and a wooden mala hanging around my neck with his picture on it so a lot of the German sannyasins continued to ignore me. They seemed to be a very mind-orientated race, these Germans. Very cerebral.

Also, if you were not a sannyasin, dressing like a girl, you could not sit anywhere near Bhagwan in Buddha Hall or enter his home in the mansion now named "Lao Tzu".

Also, I could not partake in such intense groups like the famous "Encounter" or "Primal" groups.

I was amused that the "serious" followers, many of them disillusioned Catholics and Muslims with believers' minds, simply ignored me as a lesser being because I did not wear a mala and an orange robe. It seemed to me that they had transformed their longing for heaven or paradise into the same longing for something in the future, found only in the East, called "enlightenment". A sad waste of "now time", I thought. The same longing mind, just a different longing, and always to be found somewhere, unknown, in the future.

Here the ex-Catholics hated the Pope, while the cured Muslims hated Islam.

The powers that rule in this particular paradise told me that I could, and should, enrol in a group named "Centering". This was a group where eighty newcomers to the ashram (forty men and forty women) would meet each day and play games while getting to know each other and learning the many rules of this seeming paradise.

I signed up and joined the group because my new mother friend, who promised to take care of Cheyenne and her daughter, insisted that I did. She was also one of the "Black Robes" who led groups, but not this particular one.

The group started out with games which were as simple as hopscotch and musical chairs. Some games were obviously psychologically based – simple stuff that did not impress, or have any effect on me, though others broke down crying.

By the third day I relaxed in the group. I had met a beautiful blonde woman from Vienna in Austria, where Freud, the celebrated psychologist, once lived. She was also a psychologist who had come here to study different and new forms of psychology which were practised and experimented with in these many unique groups. Breath therapy, body therapy, sex therapy, mind therapy – it was all happening here, and it did not take me long to come to the conclusion that Vietnam was the frontline of a physical war with guns and bombs, while this place was the frontline, in the beguiling East, of a mental war. This war was being fought by using minds, not mines.

I would later discover further frontlines of consciousness based in the West, in the good old USA, at a magical, wonderful place called Esalen Institute. I shall tell that story later.

On the fourth day of this Centering group, I had finally found the courage to ask Marie, my now "potential" Austrian girlfriend, to dinner that night. A very big Dutchman had befriended me and he was standing next to me while we were on a coffee break halfway through the morning's group entertainment.

This Centering group was taking place in the highest building in the ashram which was an old wooden building three storeys high. We were standing talking, while looking down at the entrance to Lao Tzu, the house where Bhagwan lived in complete luxury in a room with air-conditioning, and marble floors. I would visit the grounds of his house at a later date.

And I would only visit his room after he died.

Suddenly, a man holding an old suitcase was set upon by three large, robed Western men. These men were some of the Bhagwan's karate guards. A heated argument ensued and one of the robed men, a redheaded man who I later learned was the head of the guards, hit the undefended man with a karate chop. The other two grabbed him and began dragging him out of the ashram with the man screaming loudly. It was a pitiful sound. The people mingling in the ashram stopped to look but no one interfered to help him.

They dragged the little man out through the front gate and took him about 30 metres down the road. From our position on the third floor, we could watch this confrontation unfold.

Two of the karate guards began punching the pitifully wailing fellow who was not fighting back, while the other one stood with his back to them in a karate pose and made sure nobody came near them to interfere.

The non-resisting little man fell and they started kicking him on the ground. Still, no one helped the man though some Indians had gathered and were watching. Bhagwan's sannyasins just kept walking past.

My new Dutch friend stood 6 feet 6 inches tall and must have weighed at least 15 stone.

I simply could not watch another human being kicked mercilessly on the ground by obvious karate bullies. My blood boiled with indignation, so I turned to the Dutchman and asked him, "Can you fight?"

"No," he replied, "I have never been in a fight."

"Well, I can," I answered. "Will you come and stand behind me and look as if you can fight so I can stop this slaughter?"

"Yes," he answered immediately, without thinking, and I admired him for his courage.

We ran down the steps and out the gate in time to see this poor man struggle to his feet, crying. One of the bodyguards karate kicked him in the ribs and he fell back to the ground.

The man directing people away saw us coming quickly towards them and yelled, "Stop, go away."

I instantly noted that he was standing with his feet more than shoulder width apart and each leg was placed side by side with his knees bent in some sort of weird karate pose.

Now any street fighter or boxer knows that you must place one leg behind the other for balance and speed of movement in a fight and you definitely do not hold your hands so far from your body, as he was doing. I didn't know what these fellows learned in that dojo but in this instance, I knew it was not going to help him.

I came to a stop just out of his reach and said to them all, "Leave him alone. He has had enough of this."

The other two brave men were kicking him again. Is this what they teach in karate? Brave bastards! Obviously, they were not taught the ways of the honourable warrior, therefore I did not think they could really fight. I was about to test my theory.

The man facing me reached forward, off balance, and with his finger pointed at me, tried to push me in the chest. It took all my self-control not to flatten him with one punch, but I thought I had better save the punches for the other two. I simply grabbed his outstretched finger and bent it painfully backwards, forcing him to his knees. I then yelled loudly at the other two, "Stop this fucking slaughter now."

These words commanded their immediate attention and they swung away from their prey and took in the situation. The big redhaired man, who had started this attack, had the glazed look of blood lust in his eyes. He hesitated and must have thought better of attacking me because of the big Dutchman, looking suitably mean, now standing next to me. Good on him.

The redhead growled, "This is none of your business. Fuck off."

In situations of adrenaline-fuelled confrontations such as this, it amazes me at how quickly my mind can work. The wild, pumping, angry side of me said to break the kneeling man's finger, putting him out of action, and then teach these other two a well-deserved lesson about real fighting. Cowards like this cannot beat me. Even if they got lucky and knocked me down, I would simply pull out my razor-sharp knife and cut their kicking legs. Cowards are afraid of blood.

The intention with the use of a knife is to make sure it has a point like a needle, is as sharp as a razor, and has a hole drilled in one end with a leather loop attached. The loop is wrapped around the wrist so you cannot lose the knife and then have it used against you. So, either way these sadistic animals would lose and learn a valuable lesson. I was sure of this.

Yet my mind was still working at a fast rate. A beautiful European woman was waiting for me back in the ashram and Cheyanne loved playing there with her friends. If I fought these men I would not be allowed back inside. If I were forced to cut them, I might be thrown into an Indian prison and where would that leave my darling daughter?

So, instead, I gave the kneeling man in front of me a little extra pressure on his bent finger until he cried out, and then I released him to lie there nursing it.

The other two must still have been worried about the size of my Dutch mate because they had not moved. I weighed 90 kilos and was very fit and neither the Dutchman nor I were restricted in movement by an ankle-length robe.

Just then a large Indian tuk-tuk driver, who had watched all this unfold, came and stood by my side and said to the redhead, "Enough, baba."

This Indian man would soon become a firm friend of mine. I would meet his wife and child and Cheyanne and I would eat with him in his hovel of a house. I would also give him what money I could afford. A very funny episode occurred with a famous man and I when I was with this tuk-tuk driver one dark night.

Veeresh who owned and ran the famous Humaniversity in Holland was one of Bhagwan's favourites. He came to India to see his master and then hired a huge hall where all his followers came to listen to him speak. It was by invitation only but I was dragged along by my friend and sat in the hall on a dark night with about 300 other sannyasins. Veeresh was an interesting man who I would stay with at his university at a later date. At the end of his talk he deftly rolled the biggest joint I have ever seen. He smoked a little and then passed it around the room.

An announcer asked us to all remain seated when Veeresh left. I did not wish to smoke so I stood up and crept out. My tuk-tuk driver, Prakash, and many other drivers were about 50 metres from this hall waiting for their fares to return. I began playing with Prakash and we were soon wrestling on the ground. Suddenly Veeresh appeared alone, and seeing what was happening, he panicked. He thought I was being attacked and would probably die because there were so many Indians. He ran into the centre of them all, pushing and screaming. He raised me off the ground and pushed me into the back seat of his waiting taxi, jumping in after me and screaming at his driver to speed away. I was laughing so hard he thought I was in a panic. When he saw that I was laughing, and looked at my Akubra hat, he slapped his head and said, "Good lord, I've just tried to rescue Crocodile Dundee." He began laughing with me and from that moment we became firm friends.

But back to these guards beating up this man. Prakash's added presence had caused these fools to hesitate further, and I decided to use this opportunity to talk them out of a fight with me. This was their lucky day because I now badly wanted to pay them back for what I had witnessed them do but I am not violent unless attacked and they had not yet attacked me.

So I said, "Listen, fellows, you will kill this poor man if you keep kicking him and is that worth the three of you spending twenty years in an Indian prison for murder with all these witnesses around?"

The leader looked down at the man who still lay crying on the ground feeling his wounds, with the contents of his suitcase strewn about the road.

"Don't worry about him; we will get rid of him for you," I said, making the poor man the common enemy.

By this time a large crowd of Indians had magically gathered, as always happens with a disturbance in India, and it was obvious that they were there to support the taxi driver because he was an Indian and, as I found out later, a leader of his people.

With this extra threat, the cowardly redhead, who had been fingering a whistle around his neck, obviously used to summon more guards, dropped his hands and stomped off, followed by his two brave minions.

We immediately helped the broken and bleeding man to his feet. His face was bloody and his ribs probably broken.

The taxi driver, who was named Prakash, had kindly gathered the man's belongings, put them back in his battered suitcase, and placed them in his taxi. We helped the man into Prakash's vehicle and took him around the corner to a café called the German Bakery.

This was a wonderful coffee and bakehouse run by a German and was the favourite meeting place of sannyasins until it was blown up by mad Muslims who, like Christians, objected to any belief that was not their own.

Beliefs are responsible for so much killing everywhere the world over.

In those early days of old India, the traffic was not so crazy and thick, so sitting outside the café on the footpath was a pleasurable experience. This would change dramatically over the next twenty years, making sitting outside in the dust and smog no longer an option.

I bought the man, my Dutch friend, and Prakash, a coffee and we sat down and listened to the beaten man's sad story.

He said that he was an actor – not a particularly successful one – but had managed to save US$80,000 during the past twenty years of his work. This was his entire life savings and quite a lot of money in those days.

A couple of Bhagwan's female followers had told him that if he gave his life savings to the Bhagwan then he could live in a room in Lao Tzu House for the rest of his life, close to the master. Everything would be supplied free, including food and board. Indeed, he had been living there for two weeks to see if he liked it, before being instructed to fly back to America and transfer all his money into one of the ashram's many overseas accounts.

This poor man then came back with his life's possessions stuffed into one ancient suitcase, looking forward to his idealistic future at the master's feet. He told us that he did not even have a return ticket to the USA and now he had been refused entry back into his new, promised, and paid for home. No wonder he was shocked and upset, and more so now, as he found it difficult to swallow and agony to cough or sneeze. Obviously, they had broken his ribs with their cowardly placed kicks.

My ribs had once been broken while playing professional

rugby league so I knew the pain he was in and quickly warned him to avoid sneezing at all cost. If you pinch your nose with your forefinger and thumb, at the top where your eyebrows meet, squeezing as hard as you can, this will stop a sneeze. A sneeze causes shocking pain when one has broken ribs.

I explained to the actor that nothing could be done for broken ribs but he should still go to the hospital and have himself checked and x-rayed in case a rib had pierced his lung.

I paid my new Indian friend Prakash to take this poor worshipping fool to the hospital nearby, and I gave him some money to pay the doctor. Then I sat back down with my new Dutch mate and worked out our next strategy which we would need to get back into the ashram without being identified and banned for life.

Because we were still dressed in our Western clothes, I bought two orange robes, two pairs of reflector sunglasses, and a red cap each. I then arranged for him to walk around to the back entrance of the ashram, and I would wait five more minutes and enter through the front, which would be the more dangerous entrance where the guards were now gathered.

As I walked down the same street where the confrontation took place, I passed many guards, waiting and watching. None recognised me in my new outfit and I sailed effortlessly through the Gateless Gate.

Then I found myself thinking in exactly the same way as all worshippers tend to do. I made excuses for the Bhagwan, thinking that of course he knew nothing about this actor, his robbery (because that is what it was), or the cowardly and violent assault by his guards. He would have been in his room meditating and heard nothing, and the women who had arranged to take all the actor's money could not have possibly told Bhagwan. Ah! The tricky mind, diluting the truth, in its mad rush to be part of a group.

Satisfied with this assumption, I rejoined the frivolity and enjoyment of my first group experience. I was also excited about my first date since my split with my daughter's mother.

That night we went to one of the excellent restaurants in Poona. Marie was an interesting dinner date and told me a lot about Austria. When I dropped her off, we kissed goodnight. She asked me if I would like to stay the night, but I said that I had to pick up my daughter. This was not the truth as she was having an overnight stay at her friend's place.

I did not know if this was a common reaction, but ever since my break-up, when I tried to get close to a woman my body filled with pain. I was still contemplating why this hurt lingered within. It must be something to with a misconception about love and my inner beliefs. Was this burning away my romantic idealism through sorrow and suffering? Was I learning what not to do in my life so I could be free to live life in a different way?

Many years later, beyond the hurt, I realised that this intense pain had served me because it had burnt away thousands of years of teachings that reside in all cultures. It had arisen into the Westerner culture from what was called "The Unholy Alliance", which seems to still exist as a living history. In centuries past, the Christians had been at war with Mordred, King Arthur, and the knights of the round table because their way of living for hundreds of years was different to the new laws of the Christians.

The Christians, and the mystic loving knights elect, after centuries of fighting, had finally met and made peace and then drew up a treaty. This treaty stated, among other things, that the Christian ideal of marriage would be united with the Arthurian way of romantic love. This type of love had been adopted from the love legend of Iseult the Fair and Tristan, which we in the West still live by to this day. Eastern people, apparently, do not suffer from this romantic induced love combined with Christian marriage and its strict moral teachings.

This treaty also stated that all churches in England would thenceforth be built with a castle turret on one end. This can still be seen in the churches in southern England.

So, our consciousness has been infused with the belief in marriage and its possession of another through romantic love and this is how we live.

Obviously, this ancient cultural belief system must be what was being burnt out of me through almost unbearable pain of the heart and disappointment of the mind. Perhaps the pain would disappear when the job was done and I could make love again, yet in a different way. Sooner rather than later I hoped as my body was again reacting quite sexually at times. Evolution, combined with nature, must know what it was doing.

I had often thought, in the past, that a good alternative to marriage would be free love and lots of sex. This ashram preached exactly this way of relating and most of these sannyasins fucked like rabbits.

At one point, Bhagwan had even decreed that if anyone asked you to have sex you must agree and say yes. I met one woman who confided in me that she had had sex with eleven men that very day. I asked her if she was sore.

As I watched this sexual phenomenon unfold in the ashram, strangely enough it turned my stomach and I refused many offers from women, and as usual I ignored the Bhagwan's orders. I don't like orders.

I quickly learned that my dream of a society indulging in free and random sex was just another silly dream.

One young German woman demanded angrily that I have sex with her after I had refused her first offer. She said in an accusing tone, "But you must – the master commands it."

I laughed and said, "That is only an order for the German mind," which was a bit nasty of me to say to her.

The poor girl fell down crying and I had to pick her up and hold her as I apologised. I did wonder, however, if the truth had cut through all this ill-thought out mental and emotional nonsense.

It was patently obvious to my logical mind that sexual diseases such as crabs and gonorrhoea would soon break out. Eventually, after a few weeks this happened and the hospitals became over-crowded with sannyasins.

Bhagwan quickly withdrew the order of obligatory, rampant sex, which he had wrongly yet hopefully reasoned would cure Westerners of their misguided sexual and romantic desires. Not a bad aim, I suppose, even if obviously flawed.

The Bhagwan, like so many of the famous new age spiritual teachers, was a flawed genius of the highest order. The man was simply amazing but he was a man, not a god.

I would later realise that there is no spiritual separation in the entire universe, so we are all a part of what man currently dreams of as separate gods. I prefer to call that unknown, evolution, though even that little "knowing" of it is already too much. It is the mystery that keeps life wholesome.

Over the coming years when I would return to Poona to pick up my Enfield motorbike and ride off into old India, I would witness some funny things in the ashram before I rode away. Like the time the Bhagwan had to go to the dentist and was given laughing gas to stop any pain. He thought this was a most wonderful invention and so he arranged to take a full bottle of the gas home with him. Then he proceeded to use it every day. This gas, of course, was eventually detrimental to his health and the gas had to be taken from him.

At a later time, in Oregon in the USA, he thought it would be fun to drive one of his many Rolls Royces by himself, instead of being driven by his French chauffeur. Each day he would fly down the road and ended up with his Rolls in a ditch, laughing his head off. I wondered if he had found the laughing gas again?

I heard that the keys to his Rolls Royces, all 90 of them, were taken away from him as he had to be stopped from driving for his own safety.

Meanwhile, back at the ashram in India, it was the last day of the Centering group, and we all had to attend satsang that evening. As we stood in line waiting to enter Buddha Hall, "sniffers" (chosen sannyasins) sniffed us all over to make sure we wore no perfumes and had washed our heads with a non-smelling shampoo that they only sold in their shop. They said that this sniffing was because the master was allergic to perfumes, even though we were sitting metres from him because he was sitting on his throne-like chair on a raised stage.

That night, in the year 1979, this man, the flawed genius, shocked me to the core and thankfully exposed even my longing to have a follower's mind so I could be like everyone else here

Buddha Hall had a large marble floor inside and a marble road running around it. Bhagwan was driven the 80 metres from his house to the hall, in one of his new Rolls Royces. Then about four of the world's most beautiful women imaginable, who waited patiently outside the tent, greeted him lovingly. Then a dark-haired beauty, who was once a famous international model, opened his door and escorted him into the tent. He was greeted with a roar from his many disciples, nearly all Westerners in those days, and he acknowledged them with his open arms waving in time to the music while they all danced. He directed everyone like a big-band conductor, with a huge smile on his face. The dancing would become frenzied as his hands quickened the pace. Amazing!

This place still seemed like a lot of madness to me. My beliefs were destroyed in Vietnam and so I was wary of beliefs of any kind. Here the disciple surrenders to the master. Surrender to a belief, or a man, to my present way of thinking, showed weakness. Why do people sell their souls for a nice feeling and to fit in with other believers? Just like any old religion. Surrendering to someone, or to an imaginary god, was not an option for me. Once again as in the war with my platoon in Vietnam, I felt left out and alone and I had dreams in my sleep about this isolation from all others.

My instinct, again, was to run and leave this strange place forever. Yet I was feeling a strong attraction to this blonde Austrian woman standing next to me in Buddha Hall holding my hand and my little daughter was having a lot of fun with her newfound friends. Escape was not about to happen.

This guy, Bhagwan Rajneesh, might be a freaky fellow, but he appeared to be conducting one of the greatest shows on earth. There were things about this place that were unbelievably good.

Some of his wise and intelligent words, for instance, plus the great food, the experiments in consciousness, and some interesting seekers who I would meet.

All life, everywhere that I have been, seemed to be made up of both the good and the bad, love and hate, happiness and sadness, the meeting of opposites, the duality. What causes this extreme duality, I wondered?

During the evening's discourse he actually spoke about the altercation with the actor. He said that some people had interfered with ashram business and his guards, and if anyone did this again, they would be banned forever from ever entering his ashram. He reinforced this statement by saying that he always knew everything that happened here because it was his ashram.

My mind went into immediate reaction as I maintained a silent conversation with myself. "Well, old fellow, you certainly do not know that it was me who stopped your guards, and you never will. That was one of my tricks, now show me some more of yours. Just do not make them violent please, as I do not wish to hurt anyone."

Yet again I noticed how the longing mind makes allowances. I still believed that this nice looking, kind, spiritual old man with a long white beard could not possibly have known that his disciples kicked someone on the ground. My instinct was to jump up and tell him what had happened, but my concern for my daughter, and my distrust of his minions, held me back. I kept silent, as usual. We were not even allowed to cough in discourse, let alone speak. But I finally had had enough of this place and planned to leave soon with my daughter.

Eventually, I bought my first new Indian Enfield motorbike. They were dirt cheap compared to the price of bikes back home in the West. I had a mechanic change the Indian carburettor for a good German one and also had him make a few other changes to make the bike go faster and be more reliable.

Cheyanne and I would soon use this bike to move around India and visit other famous gurus as we explored this ancient country.

I had noticed that some Indian men, probably because of

their religion, have little consideration for the rights of women, and they appeared to be overtly sexual around them. It was not really their fault as all the male-dominated religions teach these males that women are less than men and sex with them is sinful unless they are married. All the religions such as Muslim, Christian, Hindu, even Buddhist, are guilty of this crime against women, yet women still follow and support them. Why, I wondered? Do they need to believe, and fit in to their societies, so badly?

As a Buddhist monk I had to take some of the strangest vows, such as never looking at or even brushing against a woman when I was walking down the street. And Buddhist women are not allowed to become monks, just as Christian women, for centuries, have not been allowed to become priests. And of course, Muslim women cannot drive, cannot leave the house without being covered from head to foot, including the face, and only with their husband's permission. And they have no rights, as a man can have them killed in some countries if he simply says, even without proof, that they have committed adultery. And he can divorce any one of his many wives by saying "I divorce you" three times. Crazy, primitive stuff that women are forced to suffer on this poor backward earth of man.

And when will these male-run religions understand the reason why their priests and holy men have sex with young boys and girls, or, indeed, anyone they can? The ridiculously obvious reason is simply this. You cannot copy anyone, and just because Jesus, Buddha and other great humans of the past are purported by the priests to have risen above the act of sex and nature, it does not mean that we can copy them. A person must come to this state of "no sex" after his own experience and understanding if this is what is meant to be for that particular individual. Sex is not a bad thing when done with love and consideration for the other. And why do all the religions wish to control sex? Do they believe that this gives them more power over people? Most likely.

When will mankind realise that we cannot control nature?

So, when these popes, priests and imams, these (wise?) holy men, order their priests and followers to copy Jesus or someone, and never have sex again, they are setting these priests and followers up to eventually become sex-starved maniacs who take advantage of even children, and will fuck even a twitching muscle it seems. Nature cannot be controlled in this way. I wonder how many children must be assaulted before the overlords wake up to this obvious fact and give up their lust for power by trying to control humans sexually?

And it is not only sex. Try telling these priests that they must never think, or dream, of, say, elephants. They will of course think and dream of elephants. It is a strong reverse command. Duality in action.

Before leaving the ashram, Marie, Cheyanne and I were on our way to another fabulous Indian restaurant. We had climbed out of my friend Prakash's tuk-tuk and as I stood talking to him, I noticed a lot of Indian men standing at the entrance to the restaurant drinking alcohol and smoking beedies, which are an awful, strong Indian cigarette. Two of these Indian men suddenly roughly pushed Marie aside and grabbed a handful of Cheyanne's lovely blonde hair. They were obviously enthralled by her beautiful hair and were laughing in amazement as they pulled it and lifted her onto her toes. My adrenaline kicked in immediately as the urge to protect your young is one of the strongest feelings on this planet, and by now Cheyanne was screaming in terror. The man holding her could not care less that she was screaming and he was smiling drunkenly as he lifted her even higher.

I rushed forward and spun the man around roughly. He let go of her hair and I grabbed him tightly by the throat with one hand and lifted him in the air and slammed him hard against the brick wall of the restaurant.

I said, "You must never touch her or any other little girl again."

He nodded with a stupid grin of amazement as his face slowly changed colour.

It is very unwise to touch an Indian aggressively for any reason, and I had reacted without thinking to protect my little girl.

A crowd of Indians appeared immediately, as if by magic, and many were drunk. Now Indians are a pretty peaceful race but if a crowd forms like this they can riot, attack and kill, like a bunch of crazed animals. I have seen them rioting and it was a frightening sight.

They advanced towards me and I told Marie to take Cheyanne inside as I prepared to fight for my life.

Suddenly, big Prakash and his taxi driver mates stepped in front of me and he yelled at the crowd in Hindi. They snapped out of their killer, crowd mentality and melted away. Prakash had come to my aid once again.

After that, Prakash and I became even closer. I would take Cheyanne over to his home more often and we would eat with his family. Prakash was big for an Indian. Not quite as tall as me but more thickly built. He liked to play with me, so after dinner, or even when we met in town, we would wrestle. In the evening, while at his home, we could often be found rolling around on the floor with the children joining in and jumping on us. These were good times with good people which sometimes brought tears of happiness to my eyes when the children laughed hilariously.

I must repeat that there were some great things about India and this amazing ashram. Many seekers from all corners of the world came here to find something new and to break out of the normal world order and their conditioned way of thinking and acting.

My daughter benefited greatly by being there. She was socially withdrawn in Australia. Her teachers had told me that she was a loner, not particularly interested in her schoolwork, and would not join in dancing with the rest of the class.

One night in Buddha Hall, after discourse, a famous band played great rock music to a packed audience. Now one of the things that these sannyasins can do is dance uninhibitedly like no one is watching.

I eventually joined in, as I like to dance, but my gorgeous Cheyanne just stood at the edge of the dance floor watching these wild people giving themselves up to the dance. I watched her on this particular night, from the corner of my eye so she would not know that I was watching. She was gaping with an open mouth at this wild explosion of so many dancing people, including children.

Suddenly she just let go and joined in, dancing madly with the same abandonment as these sannyasins.

I don't think I had felt anything so wonderful in my life. My self-contained beloved daughter had broken through whatever chains had bound her and I knew her life would be different from that moment on. She looked so beautiful in the abandonment to her own unique dance.

When we finally returned to Australia, her life reflected this transformation. She always came around the top of her class in exams, and she also joined in everything, while becoming the social leader of her group of kids.

Once again, I realised that in any large organisation, just like life, the good comes with the bad. The rush for power, sex or money creates a certain madness in individuals, and people at the top often act badly to protect their strong, newfound addictions and desires. People in government often blatantly display this characteristic.

Spiritual leaders such as this man Bhagwan, and the others I would soon meet, have the added nectar of "worship" offered to them by their followers. This they must also resist. I was glad it was not me being tempted with this money, power and devotion. I think I would also fail as the mind is a tricky master in its headlong quest for what it sometimes thinks it wants.

Bhagwan, who would soon announce that he had changed his name to Osho, was a master of the spoken word and I was happy I had found someone who had some interesting answers that I could agree with.

For instance, he was once asked in satsang what his ten

commandments were. In reply he said, spontaneously, that it was a difficult matter because he was against any kind of commandment but "just for fun" he would set out the following:

- *Never obey anyone's commandment unless it is coming from within you also.*
- *There is no God other than life itself. [I like that one.]*
- *Truth is within you. Do not search for it anywhere else.*
- *Love is prayer.*
- *To become a nothingness is the door to truth. Nothingness itself is the means, the goal, and the attainment.*
- *Life is now and here.*
- *Live wakefully.*
- *Do not swim – float.*
- *Die each moment so that you can be new each moment.*
- *Do not search. That which is, is. Stop and see.*

And this man, who was also a self-confessed madman and proud of it, said that he had seven circles of people surrounding him. I could not remember all the circles but the seventh was the outer circle made up of traders, here to make money, selling everything imaginable from food and clothing to drugs and gold.

The sixth circle was made up of the curious, here for a quick look.

The second circle was made up of the believers, who he said were often ex-Catholics or Muslims.

It was patently obvious that these people were the disciples, and in their minds, they had transferred their indoctrinated beliefs of heavens and paradises to another future belief, which was "enlightenment".

I cannot remember the other circles but I had already figured out the people of the second circle and was impressed that he saw so clearly and was not afraid to speak what he saw.

Another quality that impressed me about this man, and made me smile, was that he would say one thing and then three days later he would say the exact opposite, as if each was the utter truth. He got delight in confusing the grasping minds while also freaking out the true believers.

Oh, I remembered. The last circle was called the "Inner Circle", and these were the ones, Osho said, who had the potential to be the most dangerous because they wielded enormous power, controlling the money and running this ashram together with Osho's huge worldwide businesses which generated untold millions of dollars. The holy business made more money than the oil business.

So naturally, the slick, the greedy, the criminal and the pleasure seekers are all going to join the innocent and the true seekers who are in the majority here. Yet all these people had difficult life lessons to learn as this was a tricky place to be.

I delayed leaving this ashram, after conferring with my daughter who wished to stay.

After two months of this sometimes surprising and exciting life, I decided to partake in the famous and notorious "Encounter" group. In order to attend this and other popular groups I had to become a sannyasin. For this and no other reason, Cheyanne and I both took sannyas. I did not take sannyas to become a disciple.

In those heady days, sannyas was given in his private house as the hordes of seekers had not yet arrived, but were soon to come in large numbers.

That night Osho placed his hands on my head and a mala around my neck. I was given a new name, the female name of a famous dancer, because he told me that I was blessed to be both a mother and a father to my daughter. Cheyanne, when he asked her, chose to keep her own name. Smart girl.

This was what he said to me as he touched my head:

"This is your new name: Swami Sahajo. Sahajo means natural, spontaneous.

Man can function in two ways. One is out of the memory. Then whatsoever he does is a kind of reaction. He is not responding to the reality. He is functioning out of his past. He is not spontaneous.
He has some principles about what to do, what not to do. He has a

certain ideology to be followed

Rather than responding to the reality that he is encountering, he is responding to his past ideology. He will always be missing the point. And that creates great frustration in life, because you always go on missing the target, you are always falling short. You always reach to the station when the train has left.

One has to be spontaneous in life, one has to be responsible. By responsibility I don't mean that one has to be very dutiful. That is a wrong meaning of the word responsibility. Responsibility simply means capacity to respond not according to the past but according to the present moment, reflecting the reality that is in front of you. Act out of your presence, awareness, act out of your totality. Then your action is a response, it is no more of a reaction. And every response brings a deep fulfilment.

And if life becomes responsible in this sense, then each moment brings more and more maturity, growth, fulfilment, contentment, and you start feeling grounded, rooted, centred, integrated. And to be integrated is to be reborn. That is the beginning of a real life. Before that we are only in a kind of womb, a psychological womb. One has to come out of it.

Socrates has said: The function of the master is that of a midwife. He is right. The function of the master is exactly that of a midwife. He helps you to come out of the psychological womb".

He had given me a female name. How interesting, as females have had such a big effect and influence in my life.

Now that I had a mala and an orange dress, I could mix freely with the members of the "inner circle" and also get closer to Osho and further study this amazing human and listen to his sometimes powerful words.

I soon discovered that the inner circle crowd had great parties complete with Dom Perignon champagne costing $120 a bottle together with wonderful feasts of exotic foods fit for a king. Some of these inner circle people whom I met, now that I dressed like them and could easily act like them, counted their wealth in hundreds of millions of dollars and their family products such as Bata shoes and Revlon perfumes were sold the wide world over. These people were called the "jet set". They flew around the world in their own jets. I liked them for they were all new age warriors rebelling against their confusing, normal world. Their vast wealth had not made them happy. Like mystics, they intuited a brighter future, yet here and now they were all as fucked up, feeling as lost and needy as the therapists, the broke, the broken, the dropouts, the criminals, and me.

Yet all these different people were here and in the frontline of consciousness in the East.

A place called Esalen, in California, was the Western frontline of consciousness that heralds impending change. Esalen was a place, as I said, that I would soon visit and almost accidentally become a part of.

Here, in this Eastern ashram, all beliefs were challenged and the normal mind was given no place to rest and say, "I know". This was a war of the psychic kind and it was trying desperately to clear an opening for a new human being to appear on earth. Like all evolution it was a painstakingly slow process. Yet this battle will rage until the final moment when the human psyche shifts and is set free from the chains of the old order of institutionalised historical belief systems – beliefs that demand separation through thought control, sect isolation, and, so, extreme duality.

I often tired of this war of the psyche in this ashram so my little girl and I took many long trips throughout India on my new motorbike.

Soon it was time to return to Australia.

My ever-loved daughter had been changed by this ashram and our travels, and that was miracle enough for me. All the negative things I had witnessed about the Bhagwan were forgiven. However, I was still not a believer or a follower and never would be.

Worshipping, whether it be of a live person or an imagined god, is not my way. I was glad to leave. My money was almost gone so I needed to find work immediately on my return to Australia.

I would return one day with my daughter when she became

older.

Before departing, I had one last date with Marie. She had been very patient with me, unaware of this strange pain I had continually felt when trying to get close to women.

We had a beautiful dinner in our favourite restaurant and then returned home and got into bed. Even though the fear and the pain were present I closed my mind to these negative feelings. Because I was leaving India the next day, this seemed to make it easier.

I remembered what my family taught me on the vast cattle stations. "When you get thrown from a horse, get straight back on."

Marie was a strikingly beautiful blonde woman with perfect breasts and body. Once I began making love, I did not want to stop as it had been a long time without the magic of a woman's love, touch and feeling. As a man I realised that I was fascinated with the depth of a woman's ability to feel.

Sex is one of my most potent remedies for stopping thought. Intense feeling is stronger than the mind, and the night seemed to pass in an instant, as time disappeared along with thought, while we were making love.

Women are love.

Marie was amazed that I had waited this long. She had come to believe that I was either impotent or gay. I suppose I was impotent for a time.

I told her that I would visit her in Austria and some years later I did. Marie lived near Sigmund Freud's house in Vienna and she took me there and also to the fabulous Austrian palace, which reminded me of a large dolls' house.

Of course, as usual, I did not stay long with her in Vienna. I went off in search of the hidden answers to my questions. Why the world was like it is, and how to change myself while in it.

After six months in India, Cheyanne and I were finally coming home.

Chapter 24 - A Miracle

I think India benefited my daughter more than me. After we arrived back in Australia, what I saw as a minor miracle occurred. Cheyanne had received her first report card since returning to school. She was now in the top three in her class and the report card also stated that she had gone from being a loner to having lots of friends. She also enjoyed dancing.

After school she would bring kids home to our house and play happily with them. I was overjoyed to see my little girl like this, though she sometimes told me that she missed her mother who was still on the boat with the old sea captain.

Months later her mother suddenly reappeared, done and finished with her drunken sailor boyfriend. Yet it was too late for us to become a couple again as she was not interested.

If I had been a smart and vindictive sort of man with less trust, I would have legally claimed my daughter, and my home, after her mother left us, but when my daughter saw her mum, they ran crying into each other's arms.

That night I asked Cheyanne if she would like to live with her mum for a while. Cheyanne immediately answered yes.

Next day I told her mother that she could move back into my home, which I had purchased with my war loan and savings. I gave her permission to live there for a year and then I would move back to be with my daughter for a year. We could share my home and my daughter, as this is what I thought would be best for her.

I asked her mother to promise to give my daughter back in a year, and also to promise that she would not try to take my home from me.

A promise is something that I had been taught is a sacred

oath so I trusted people who make promises. Cheyanne's mum obviously did not share this belief and at a later date when I said to her, "but you promised", she blithely told me that she had had her fingers crossed behind her back so the promise did not count. I eventually lost my house and my daughter. I was shocked and devastated at this further betrayal. Yet everything happens for a reason and though it took a few years, I eventually learned the true meaning and value of forgiveness. How to forgive other people's actions even though they did not agree with my moral code. What I believe about promises is not necessarily what anyone else has to believe. At least the mother loved her daughter and now wished to be with her. This outcome was actually a blessing for Cheyanne because I was not the world's most stable person when trying to live in a "traditional" society.

I did not fight this woman for the legal rights to my home. Instead, I was able to continue to see my darling daughter and spend time with her, but never again in my own home. I have never been able to buy another house to live in because of the painful feelings the concept of home brings up.

Everyone has to live with their own actions for the rest of their life. It is not up to me to judge my daughter's mother, or anyone else for that matter, though it is sometimes extremely hard not to do so. She had her reasons and I am certainly not always right.

I decided to leave town for a while and go mustering on the family cattle station with my beloved uncles, Boof and Jimmy, who were half Chinese. I loved them both dearly and often wished one of them had been my dad as they were happy and easy-going, unlike my father, who was serious and sometimes brutal.

I was raised in a fighting family where two things that received a lot of attention were fighting and sex. We were not a Christian or religious family, but sex could still not be discussed openly with women. Subconsciously, Christianity still held sway in our society, and sex was always hidden. Normal society with its mass consciousness has a strong effect on everyone. And because sex is a taboo subject, people were obsessed with it. This is also why this denial of sex, and the ridiculous pressure on priests to abstain, eventually drives them mad.

But fighting! My family discussed that a lot and we sometimes fought others with fists and guns.

I had ridden my motorbike the 300 kilometres out west to the original family cattle and horse station, which was near the mining town of Kidston in west Queensland.

I was gratefully kept busy helping my family muster cattle, brand, and take the big calves off their mothers.

It did not matter what work we were doing on this particular property, whether it was fencing, breaking in horses, or branding our cattle – when it started raining heavily, everyone would immediately stop work, grab a bottle and a pair of tweezers, and head to their favourite creek or gully. The rain would quickly wash the topsoil away, revealing gold pieces. These pieces were put in the bottles and stored, usually under the bed.

I was never interested in finding gold and so I usually rode my horse into the deep and quiet bush country to find "bush tucker" while the family looked for gold. The bush tucker was what interested me and I collected many books on the subject as well as receiving the teachings of my Aboriginal "grandfather".

I simply loved riding my horse deep into the bush by myself, as the bush emanated a stillness and peaceful energy that enthralled and calmed me.

After weeks of healing in this unspoiled natural environment I felt much better. I said goodbye to everyone, jumped on my motorbike and headed home to Cairns where I had to find somewhere to live, plus get myself some work and make some money.

Both these tasks I found easy to do and quickly found work during the day in an office job. At night I worked as a bouncer at the largest hotel in town where the best bands attracted thousands of people from all walks of life.

I got to spend time with my daughter but only on the weekends for the next six months. She always jumped into my arms when I arrived to pick her up. I wished that her mum and I could live together for her sake.

Once again, I was saving every penny I could to return to India and the ashrams. I felt a strong pull to return because I intuitively knew that there was a mystery I would uncover there. I needed to understand this life. Life still did not make sense to me and I was not happy for a lot of the time so I knew that I had to find the answers.

I thought that it must be possible to find the sweetness of my breath and my aliveness, free from the pursuits of pleasure, or romantic dreams of an intimate other, or the longing for money – the seemingly endless desires of my present worldly way.

I came across the writings of an ancient sage, which inspired my search:

The Great Way is not difficult for those who have no preferences.

When love and hate are both absent everything becomes clear and undisguised.

Make the smallest distinction however, and heaven and earth are set infinitely apart.

If you wish to see the truth then hold no opinion for or against.

The struggle of what one likes and dislikes is the disease of the mind.

When the deep meaning of things is not understood the mind's essential peace is disturbed to no avail.

Rest and unrest derive from illusion; with enlightenment there is no liking and disliking.

However, I had no idea what enlightenment was. Was it real? What did it mean? For me, "the deep meaning of things is not understood", as the sage wrote.

It was again time to leave for India. I would search out wisdom and teachings, other lessons, but I could not search for enlightenment. Enlightenment was a foreign Eastern concept to me.

I always liked to wait for existence in everyday life to present what was needed for me and then I could react accordingly. Life then becomes all "grist for the mill" as someone once said.

"There is a mystery beyond the capacities and powers of the mind."
J. Krishnamurti

On reading these words it was time for me to surrender into another mystery, summoned by life itself.

Chapter 25 - Back to India

I arrived, alone this time, at the ashram in Poona, after a long and tiring journey.

After I picked up my motorbike, which had been kept in storage, I attended a satsang with Bhagwan Shree Rajneesh, now renamed Osho, in Buddha Hall. He put me to sleep.

The next day I decided to participate in a dance group.

This group had twenty participants – ten men and ten women from all corners of the earth, as usual. On the first day I was so nervous and withdrawn that I could not join in the dancing. It hurt me to feel so left out as everyone else was having fun. I had forgotten that I was always nervous in a room full of people I have never met. It takes me a while to study them, feel safe, and then I may relax.

But this time the intensity of my isolation reached a new level, until after two long days I suddenly rose above my intense fears and soon became the dance. All the feelings of isolation and "I can't", or there is something wrong with me, simply fell away and the feeling of pure joy that passed through me caused a strange phenomenon to occur. After the group disbanded for the day, I was walking back to my hotel and was suddenly struck by the amazing beauty of the trees, the birds, the sky and the sun. I froze in mind-numbing wonder. For hours I stood under a tree and everything appeared to be perfect in my world, as joy kept flooding through my body. When it finally became dark, many hours later, my mind crept back in and reminded me that I must move as I could not stay there all night with the mosquitoes biting me.

Some people would refer to this experience as a satori ... a deep insight in which someone sees and understands the true nature of the universe and of reality itself. Or, as I have grown to understand, a moment in which the experiencer and that which is experienced are one.

Phenomena similar to this would happen again in my life and in a much more intense and wonderful way. This would come later in America.

After a week of nightly satsangs, I was bored beyond belief. Osho, as well as being a madman, was also a genius and had some interesting things to say, but he often seemed to take forever to say them. He was so boring he often put me to sleep. He hissed when he spoke and this had a hypnotic effect.

My boredom was simply a sign that I must move on.

I always used the ashram as my India base before I rode off in search of other Indian marvels and men.

At a much later date, years later, I remembered going to satsang one evening at 6.30 pm, passing the mad sniffers, and sitting on the cold marble floor as I had forgotten to bring my cushion. I calmly sat and waited for this giant of a man to appear at 7 pm.

He was driven up slowly in one of his luxury cars. A stunningly beautiful German woman, who was the same ex-international model as always, opened his door. She escorted him to the front door of the hall where another stunningly beautiful woman, a blonde, took him to a place where he stood in front of his huge chair, which looked like a large throne. I wondered why he did not allow some not so beautiful women to closely serve him.

As a crowd we thankfully rose, the loud music began, and he directed us with waving arms and we all danced.

After five or so minutes this came to an abrupt halt as he sat on the polished throne, crossed his legs, folded his hands, took some silent time, then began his discourse.

This night his discourse was a terribly long one about America and its evil ways. He went on and on and on. Sitting on a cold marble floor for four hours with my legs crossed in a crowded temple became painful beyond belief, especially when what he was saying sounded like nonsense to me. He was raving on about how the American police and the CIA planted a bomb under his chair in a prison where they were holding him. Somehow the bomb failed to explode. He then said that they gave him poison. The poisoning I could possibly have believed if it wasn't for the ridiculous story about a bomb, which then failed to detonate, being placed under his chair. The powers that be in the USA would simply not be so stupid, or daring, as to blow him up with a bomb in their own prison. Why was he carrying on about this? Was he stoned or on some drug, I wondered?

Meanwhile, the older wiser sannyasins were sitting on cushions and flat chairs with a back support so their discomfort level was obviously much lower than mine.

When you are sitting in satsang you cannot cough or sneeze (the germs may reach himself) and you most definitely cannot talk. If you do any of these things you will be thrown out by the many guards.

Also, you cannot rest by lying down as I tried to do. A large German woman slapped me and pulled me back up. Germans should never be police.

And finally, it is totally forbidden to stand up and leave while the master is speaking.

To keep all these rules in place, his karate guards lined the outer area of the large hall.

By 11.30 pm I had had more than enough. Rules like this are for fools, I thought. I had been sitting in pain, becoming angrier and angrier, as five hours monotonously and slowly had passed by. Five hours that seemed like fifty.

I was in the middle of the hall, lost in a bunched sea of people. I stood up, turned, and stepped my way precariously through the sea of packed bodies, heading slowly towards the back exit. Osho just kept talking. Some of these rule-following sheep hissed at me like snakes, too afraid to even speak and say "sit down".

At the exit a guard stood up and poked me in the chest,

whispering, "Sit down."

What was it with these karate men? Didn't they teach them anything about the danger of fingers poked in bodies in their training? As usual I simply grabbed his finger, which was jammed rudely in my chest, and bent it painfully backwards, forcing him to slowly sit down. He had a horrified look on his face as he could not call out for his mates in holy satsang and this gave me an edge and a little amusement. I then released his finger and walked out. No guards followed me because they, also, were not allowed to leave and disrupt satsang.

I decided then and there that I was not attending any more satsangs.

Osho did indeed sometimes reveal wonderful spiritual lessons and interesting gems in his long speeches but if I felt the urge to listen, I would sit outside the hall where large TV screens were set up for outside watchers who did not pass the sniff test.

This listening would not happen often for me I was certain.

Anyway, way back before this incident happened, and Osho had not yet flown to America, I decided that I had already had enough of the ashram with its discourses, so I rode my motorcycle across India to a desert state in the north called Rajasthan. I like deserts because there are not many people living there and the silence is loud and uninterrupted.

In India when travelling, however, there is only one road rule. The largest vehicle has right-of-way and it may be driving on either side of the highway. The traffic is awfully thick near Bombay and other large cities so mindless driving chaos often occurs.

When I saw a truck or a bus coming straight at me around a corner, or hurtling from a side road, I was forced to ride my Enfield motorbike straight off the side of the road and sometimes into ditches or over farm lands. Once I had to lie the bike down going about 70 kilometres an hour as a huge truck shot out of a side road. Luckily, Enfields are very strong and sturdy and the sidebars I had fitted protected my legs and body when I was forced to slide the bike along on its side.

After days of riding in the Rajasthan desert I arrived at a city called Jodhpur and I booked into a maharajah's palace which was still inhabited by the royal personage and his family. They rented me a room fit for a king and at a reasonable price.

The maharajah's son had just returned from a prestigious college in England and at dinner we soon discovered that we all shared a common love of horses. When I told them that I came from a cattle station background in Australia and that my first job was breaking in horses, they excitedly informed me that they owned a white Arab stallion that was very difficult to ride and could I ride it for them?

No worries, I said, as this challenge excited me. Their horses were all expensive and magnificent animals.

They took me to their large stables and past rows of beautiful horses to this fiery, foot-stamping Arab.

I spent the next two days getting to know the horse and letting him get used to me, as I knew these royals were great horsemen and if they said that this horse was difficult it would probably be pretty bad. So every time I came near him I gave him a little treat. I never looked directly into his eyes and never threatened him. He was a magnificent animal whose spirit had never been broken. He quickly got to know me and looked forward to my arrival and the games I played with him before I gave him sugar or sweets.

On the third morning I told the family that it was time to ride the horse into the desert at the back of their palace, so they all gathered on a large white balcony to watch me.

The horse had not been ridden for a long time and when I mounted him in a small yard he pig-jumped (bucked softly) in nervous anticipation. I held him tightly and kept pulling him in circles, to stop him from really bucking hard. I spoke calmly to him to let him know that I knew what I was doing and thus gain his respect and confidence.

When the gate to the mounting yard was finally opened, this magnificent white stallion took off at a flat gallop. I had a hold of his head in case he tripped, but did not try to slow him down as this would have made him crazy. We galloped at full speed past the large balcony and the entire family erupted in a cheer. I shall never forget that ride. That Arab horse was a wonderful "stayer" and galloped for miles before he slowed down and I could take full control of him. After he had used his pent-up energy, we slowly explored the desert and enjoyed ourselves together.

That night I was invited to a great feast with the family and they begged me to stay on and be their well-paid "horse man". It was tempting, even though well-paid in India is not a lot of money, but I was anxious to continue my journey to a primitive place near the Pakistan border called Jaisalmer where, I had been told, I could take a week-long camel ride through the Thar Desert.

I took my leave from the kind family the next day.

Jaisalmer was basically a palace and fort combined into one huge building. It was made a long time ago from ancient rocks piled high on top of each other. It looked like it might collapse at any moment. Inside this fort/palace there were shops, accommodation, and everything you would find in any large town. Outside this fort there were also a few shops and cheap accommodation.

After a couple of lazy hours spent inspecting rooms for rent, I finally booked into one of the old palace rooms in the fort. It cost next to nothing.

I then spent the next two days inspecting camels. I would look at their teeth and hooves, the condition of their backs and the condition of the saddles. I also spoke to all the camel owners.

Some of these animals were treated very harshly and the poor condition of many quickly revealed that.

I finally found two fat well-kept animals with good hooves, belonging to Muslim men. One man was about twenty years of age and the other about thirty-five. I explained to them that I did not wish to ride, perched behind one of them, with no control over the camel, as all Westerners were forced to do. I explained that I wished to ride alone on my own camel and they could both accompany me for a week, on the other camel. I explained that I would pay for both camels and both owners. They only agreed to this after watching me ride the camel of my choice and seeing the control I quickly exercised over him.

For the next seven days we would ride through the desert, visiting temples, a deserted city, and other wonders of this area. They would cook our simple meals each day and also be my guides as I had no map or compass.

At night we camped under the stars on the red desert sands where we ate our simple meals. It was similar to Australian deserts but there were no ants, scorpions, snakes or insects to bother us at night.

I would also read to them from an interesting book titled Zarathustra, A God That Can Dance – a book written by none other than Osho.

At the end of our trip, they no longer wished to be strict Muslims and were busy planning a trip to Poona to see this enigma of a man.

One evening, sitting under the vast array of stars, my contemplations took me back to about twenty years of age. In Vietnam, as a Special Service soldier, I was on a mission alone to read signs and see who had been leaving any tracks during the night. I was walking slowly, with thick jungle on both sides of me, while carrying a cocked sub-machine gun with the safety catch off.

Stepping around a corner on the small track I was following I was confronted by a huge black cobra with a white neck. This cobra was reared up with a flat hooded head in the full strike position. Its head was about 3 feet high, held in the air.

Without thinking I immediately opened fire. Now the weapon I was carrying fired hundreds of rounds per minute and as I held my finger on the trigger, I watched the bush that the cobra was poised in front of entirely disintegrate.

Obviously, I surmised, the cobra was blown to bits and I advanced slowly forward, after re-loading, to view the remains of this unnaturally huge snake.

To my absolute amazement there was no sign of any dead snake. Impossible, I thought. My heart was pumping energy spectacularly through my body and I was covered in goose bumps.

In this state of highly enlivened awareness, I continued along the path expecting to be discovered at any moment by the Vietcong after that infernal racket of an emptying machine gun. With all my senses awakened, just 50 yards further on I saw where the ground beneath me had been slightly disturbed. On closer inspection it revealed a "jumping-jack" mine. This is a terrible American mine that explodes under your feet when you first step on it and then flies into the air and explodes a second time in front of your exposed body.

I thanked the cobra, which was obviously not really there. Without it I would not have been in such a heightened state of awareness waiting for a sudden attack. I would have probably stepped on this terrible mine.

Back in the Indian desert, on the fifth day of riding – after having walked through an eerily deserted city, plus one night, while camped in the dry riverbed near this ghostly city, listening to an absolutely wondrous flute player whose flute sounded like it was coming from three directions at once – we finally rode up to a simple chai tent perched on the top of a large sand dune – the first café of any sort we had seen on this ride.

I do not drink chai because it is made with buffalo milk and is too milky and fatty for me. This was all the café served.

On the other side of the ancient looking tent/café, at the bottom of this high desert dune, stood an old Jain temple.
A short way past the temple was a large village constructed of stone.

My camel men happily drank their chai while I tied my camel ("trust in God but tether your camel") and went down the steep dune to inspect this beautiful and intricately carved temple.

The ancient temple was completely enclosed by a large 10-foot-high stone wall and had a small front gate also covered in intricate Jain carvings. I wandered through this gate, crossed the courtyard and entered the temple.

There was nobody else around.

Inside, the temple was divided into three rooms. As I stepped into the middle of the first room a blinding light shot out of the ceiling and overwhelmed me, frightening me almost senseless. Springing backwards, I looked up and saw a tiny hole in the roof of this room. I found out later that at a certain short moment during the day the sun is directly overhead and lights up the centre of this small room.

Collecting myself, with my heart still beating wildly, I slowly explored the rest of this empty temple.

The next room, which was the largest, was bare and had more carved and amazing stonework on its walls. The floor was smooth and well used so I assumed that people sat here to worship.

To my right was a smaller room which appeared to be the heart of this sandstone temple, complete with a statue of what I assumed was a Jain god. On a stand in the middle of this room lay a very large book, beautifully bound in thick leather.

I did not enter this room. I could see that it was their holy place and because there was no one around I still had my boots on. One does not enter a holy place with boots on. I suppose that I should have taken my boots off before I entered the temple but it was a little eerie and dark inside and I thought I may have had to move fast.

I wandered outside and walked slowly around the tiled stone courtyard while still inside the 10-foot-high stone walls.

In the middle of the right side of the temple I noticed a large hole about a foot square that seemed to extend under the temple. My curiosity was piqued so I approached this hole, knelt down and looked in.

To my absolute shock, in front of my face was poised a huge black cobra with a white neck with its head flattened and ready to strike. It looked exactly the same as the one in Vietnam so long ago.

Instantly, adrenaline and explosive energy coursed through my body and I threw myself violently to the side, rolled, and sprung to my feet. I then sort of danced, jumped and ran, yelling madly at the top of my lungs, all the way around to the other side of the temple.

To this day energy can still enter my body when I think of that huge snake, ready to strike, right in front of my face.

Hearing my loud yells, a Jain priest, the holy man of the temple, came running from the town with some village people hot on his heels.

My two camel men arrived with them as it must have sounded like I was in dire need.

The priest and the villagers did not speak English so my camel men interpreted for me.

I quickly told them of this large black cobra with the white neck in the hole under the temple.

"Show me," the suddenly excited priest said to me in his language.

I slowly edged my way back to the other side of the courtyard, peeping around the corner and ready to flee at any sign of that huge cobra. Standing far away from the hole, still at the corner of the temple, I told them that it was in that large hole and I was definitely coming no nearer.

"Be careful!" I yelled as they all hurried, seemingly unconcerned, to the hole and bent down to look. They had no fear of deadly cobras ready to strike, I thought. Crazy people!

Seeing no snake, they asked me to look again.

Shaking my head, I said, "There is no chance in hell of that."

So the priest then asked me to follow him back into the temple after I had removed my boots. We walked through the first small room, which was now dark, as the sun had passed the hole in the ceiling, then through the large worship room and finally into the small holy room.

The priest gestured excitedly for me to come over and stand next to him while he began leafing through the huge holy book on the stand. The rest of the crowd knelt piously in the other room.

When he reached a certain page, I pointed and cried out, "That's it, that's the cobra."

There in front of me was a perfect painting of the cobra I had seen here. It was also an exact copy of the one in Vietnam. My skin tingled and was again covered in goose bumps and I unconsciously stepped away from this perfect depiction of the cobra.

The priest and the people became very excited when my camel men told them what I had said. They all began talking at once.

The priest ordered them to be quiet with a raised hand and then told me, through my interpreter, that this snake was an image of their god and their legend was that a man would come one day who would see this snake.

Astounded, I asked if anyone else had ever seen this snake.

"No, never, and this has been our legend for many hundreds of years," he replied, "and now you must come into our village and stay here with us."

My camel men told me this and also that I could have anything I wanted in this village for I was now, according to their legend, their god man.

Good God! I immediately felt a strong fear of being kidnapped and held against my will. I quickly explained to the now large jabbering crowd that I had a daughter in Australia who I loved very much and it was my sacred duty to return to her. I told them that one day I would return and come back to them.

Over their disappointed protests, I quickly escaped to my camel and took off at a fast pace into the desert with the camel men hot on my trail.

That week in the desert was peppered with more wondrous events. One day we rode through a small village on the banks of a very dry river. In this village I noticed incredibly poor and starving people. There were no cars here, only animals for transport.

I asked my men why the people were starving. They told me that there had not been a drop of rain for four years.

It angered me to see the billion-dollar war planes flying overhead when just a few million dollars from the government could easily pipe water to these desperate people and alleviate their many deaths and long-term suffering.

"War, what is it good for, absolutely nothing," as someone once sang.

Riding on into the deep desert, we finally stopped for lunch. I told my camel drivers that because there was a god man present here, it would rain tomorrow.

They laughed and called me crazy. "And who is this great enlightened man – you?" they asked.

Every night after dinner they had eagerly waited for me to read them a chapter out of Osho's book Zarathustra, A God That Can Dance. After the reading they would question me about this strange man.

"He is the god man, not me," I told them.

Next day as we wound our way on a slow, almost hypnotic, camel ride through the dunes, dark clouds began to build up on the horizon. I smiled as they pointed and excitedly spoke to each other in their own language.

By late afternoon it was raining heavily.

Now as I have stated earlier, I know nothing about enlightenment nor god men. It is simply not part of my belief system. However, I can be a bit of a trickster. Because in the past I have anxiously awaited rain on our cattle station during long periods of drought-like conditions I can now sense and smell rain long before it appears. Australian Aborigines can also easily do this and they have taught me many of their ways.

The two Muslims became even more fascinated with Osho's writing after that day and, to my relief, stopped the loud prayer wailing that they usually did in the very early morning and night.

In the afternoons when we stopped for the nights, we would now wrestle and play and they would still call me crazy. They were a lot of fun and I enjoyed playing tricks on them.

The older one had what I told him was a magic cane. It was a beautifully carved cane that he carried with him everywhere. For days I had been trying to buy it from him. I had offered him my expensive knife which I carried in a pouch by my side and which he admired greatly. I had offered him more and more money every day for his magic cane. He would just laugh and call me crazy. He liked that cane.

The day after the incident at the Jain temple followed by the rain on the starving village, this camel man engaged me in a solemn ritual where he presented me with the magic cane and flatly refused to take anything at all for it. I was deeply touched.

One other thing I would like to add. If you are having trouble with feeling sexy, take a long camel ride. The younger camel rider told me how many Western girls he had made love to on these rides. He was not very handsome and his face was slightly pockmarked from a disease he had when younger, so I did not believe him and put it down to a young man's sexual boasting. I laughed and told him this so he pulled out a small album. Inside this album were photos of him with some lovely young Western females. He then showed me the love letters they wrote to him. I was astounded. There were Australian, American and European girls.

Three days later I understood why they had sex with him. When one finally, totally, relaxes into the mesmerising rhythm of riding these "ships of the desert", the pelvis and lower back move constantly back and forth in an easy manner in complete alignment with the camel's movement.

On the third night I had two amazing wet dreams. My sexual energy had been intensely raised to an almost unbearable level by this constant moving of the pelvis back and forth, to and fro. It is the same movement that occurs when engaged in the sexual act.

We arrived back in the palace city Jaisalmer and I decided

that this was enough of gods and spirituality for me for a while.

I was reminded that night of the same experience I finally had of god in Vietnam. I read in a magazine left at my hotel something Peter O'Toole said when a reporter asked him: "When did you know you were God?"

Peter answered, "When I was praying to God, and suddenly realised that I was talking to myself."

Or as Barry Long, the one and only world-famous Australian guru, always said at the beginning of his well-attended discourses: "I am God, do not put your I on me." I shall speak more about Barry and my time with him later.

I handsomely tipped my newfound friends and forced the camel cane man to accept my knife.

The next day I began the long ride back to Poona and the Osho ashram. When riding long distances, I seem to be able to think things out. In regard to that cobra at the Jain temple, I do not think I am some special man who has finally come after hundreds of years. No, I just tuned into some psychic ability I sometimes have, but which I have no control over. I probably just picked up the longing of these Jain people's thoughts.

After many days of long riding, I finally arrived back at the ashram where I dressed accordingly, changed my long hairstyle, and kept a low profile.

I soon made friends with a wonderful Japanese man who was Osho's personal gardener in the extensive grounds of Osho's house, Lao Tzu. No one was allowed in there without permission and there were guards everywhere, protecting the great man himself.

I took Avi, the Japanese man, to Australia at a later date and drove him across our glorious country. What a good man he was. At that time, however, I was a zealous vegan and would not let him bring meat into our camp when his beautiful meat-eating Japanese girlfriend joined us. My forceful thinking that veganism is the only way to be on this planet was embarrassing, egotistic, and unfairly forceful, in retrospect. Please do not let me be a zealot with belief of any kind again. I wish I could apologise to him, but they left and I have never seen him again.

Back in the ashram, one day my friend Avi arranged for me to accompany him into the bowels of the Lao Tzu gardens to help him shift a large, heavy statue of a Buddha. After we finished, I took my leave from him and headed out of Lao Tzu along a tiled pathway.

Rounding a corner in this beautiful man-made forest and garden I ran smack bang into the man himself sitting on a high-backed garden seat with a woman named Vivek sitting beside him.

His close presence, suddenly directed right at me, was quite strong and commanding so I put my hands together in front of my chest and bowed my head to him.

He said, "Hello, Sahajo."

I was impressed that he remembered my name and who I was because we had only met once when he gave me the mala and he had given out many thousands of these to new sannyasins.

"How are you enjoying the ashram after your trip into the desert?" Again, he surprised me. I wondered how he knew this as well.

Now Osho had commanded all his sannyasins to wear the orange robe and the mala with his photo on it when returning to their home countries. The week before, a brave Englishman told him that he could not do this in the job he had and Osho sent him packing, telling him never to return.

He said to me, "Any questions, Sahajo?"

I suddenly had an overwhelming urge to speak my truth. If I was to be banished forever, so be it, as Cheyanne was not here with me on this occasion.

"Yes, I do have something to tell you. Back in Australia I sometimes work on the family cattle station, and at other times as a crowd controller in very rough venues. If I wear this robe and mala, I do not think I will live for very long."

I thought I had come this far so I might as well jump right in.

"Also, I have no understanding of this master–disciple thing and I cannot surrender to anyone right now."

There it was. Why did I do this, I thought? Usually I keep my thoughts hidden from people of authority with different ideas to me as it seems the saner thing to do.

Osho acted in a way I never could have imagined. He laughed out loud and long, then said to me, "Sahajo, do not be concerned about those things. This ashram is here for you and you are my sannyasin. Participate in any group you wish. And you need not join the long lines waiting for meals. Just go to the head of the line. And here in my ashram everything is free for you."

With an open mouth I thanked this mysterious man and then asked for one more thing. "Osho, one day I wish to write a book and I would like to include something you say about 'love' in it. May I do that?"

He laughed again and said, "Of course, just go to the office and they will give you my written permission."

I later did this and received written permission to include a sermon on love he gave from his book Sermons in Stones.

Looking at this man I suddenly realised many things about my mind and myself. He is an enigma, a riddle. This ashram and its many wonders exist only because of him and it is an amazing phenomenon at the forefront of consciousness. I often judged him and my judgemental mind can be a burden to only me, it seemed – the judge. He was just doing his thing. He had no "velcro" of the mind, which simply means he did not get stuck for long in any petty thoughts of the constantly thinking and demanding mind. He was just meandering along in freedom and obviously may react differently in any moment. He really was a genius and a madman all rolled into one. He often said that he was a madman.

And he purposely does not let the mind of his followers rest anywhere for long. Just when you may think you have "got it" and understand the intricacies of the universe, he tells you the very opposite. Just when you think you know what he will do next, he also does the opposite. He is a mystery and right then I was flooded with gratitude. Perhaps he saw something in me that I am completely ignorant of. To do something like this for me? I was stunned, and brought my hands together in front of my chest and bowed my head to this phenomenon of a human being. I took my leave by walking backwards. Luckily, I did not fall over anything.

Next day I tested his words and walked to the head of the long, long line of people waiting to get their meals. I do not like lining up for food since the long lines of the Christian boarding school I was forced to attend and later, the food lines in the army.

Some people, however, took exception to me walking to the front of the line and asked me what I thought I was doing, and who did I think I was?

A woman I did not know stepped forward and said, "This is Sahajo and Osho has said that he does not have to get into line for meals."

No money was then charged for my food by someone ticking a meal coupon.

After the meal I was fascinated by all this so I went to the registration desk, in the middle of the ashram, to sign up for a few groups. Once again, a woman appeared and wrote a large sum of money next to my name in a book, to be used as payment for any group I wished to experience.

At the time I thought this was a great gift from Osho but here things were not always as they seemed.

The first group I did was called the Breath group. A well-known German female therapist supported by four assistants ran this group. What they got us, the participants, to do was basically forced breathing until we were sweating, cathartic and could take no more of this induced pain. Then we were encouraged to have an emotional explosion and breakdown of some sort.

All around me this group of easily led people were falling apart, catharting, crying and falling down.

I found all this a little too obvious and ridiculous so I extracted myself from the screaming mob and went and sat on the mattresses piled in the corner, where I simply watched what was going on.

The German therapist (why are so many of the therapists here German?) took exception at her lack of complete control and loudly ordered me to leave.

I refused because I found it amusing to see her having a catharsis like everyone else. She angrily stomped out of the room to get the guards to throw me out.

I was told later by a woman close to Osho that she eventually went to him, and he said, "That is Sahajo. Let him do as he pleases." This again amazed me. Osho was full of surprises.

She returned, more subdued, and tried hard to ignore my presence. Why do people so readily take orders here, just like in an army? Belief can be a mind- numbing tonic it seems. I'd have to be aware of that.

During one of the meal breaks, when everyone else was directed to leave the room, I stayed and listened to her speaking privately to her assistants about a tall American friend, who often rode motorbikes with me. This young, good-looking man was participating in this group. This mind-dominated therapist told her assistants that he was obviously twisted in the birth canal and needed a lot of therapy or he would never overcome this trauma.

I was astounded. This just seemed illogical and impossible to know and an unconscious way for her to feel superior (the life-saving therapist) and hold power over my friend. She was planning to persuade him to pay to participate in more groups and individual therapies of her choosing, to save him.

It seems no accident that the word therapist should be lengthened in this case to spell "the rapist". This would indeed be raping someone's mind by burdening them with this controlling, unfounded information.

What could I do? When John, my friend, came back from lunch I insisted that we leave this group immediately and take a long, overland ride up to a hill station, which had upmarket accommodation. He agreed after I explained the seeming futility and madness of this group behaviour and we both immediately walked out.

Of course, I never did tell him what she had said about him being twisted in birth.

I eventually chose another free group to participate in. It was called "Encounter". This group had received much publicity in the newspapers and became notorious worldwide. In this group there were absolutely no rules. People fought, broke bones, raped, and did whatever they felt like, supposedly to free themselves from any ingrained hang-ups they were carrying.

The group was run by a German doctor named Fritjof, and strangely enough it turned out to be one of the most satisfying groups I would ever participate in. Fritjof, unlike other so-called therapists, had no agenda or form to follow. He allowed everything to occur in its own time and its own way, then dealt with the situations as they arose. This also seemed to me like an intelligent way to deal with life itself. Surrender to what is and then act accordingly.

The group was held in an underground soundproof room with mattresses on the floor and all the walls. The twenty-six participants, half men and half women as usual, waited outside until we were told to enter. We were warned to expect anything once we entered this room. This was unusual and exciting, I thought. A challenge.

I hung back and entered last. When I walked slowly and warily into the underground cell, a tall, strong looking Frenchman took immediate exception to me as he thought I was English. He simply strutted up to me, said nothing, gave no warning, and then threw a hard punch aimed at my head. I leaned back, causing him to miss with his telegraphed, round-arm haymaker of a punch, and said to him softly, "Don't do that."

This infuriated him and he threw another punch which just grazed my chin, and called me "an English bastard".

I turned to Fritjof and said, "That's twice. You had better tell this man if he throws another punch, he is going to spend some time in hospital and miss this group."

I said this without taking my eyes off this raging maniac of a Frenchman. I found out later that he had seen me talking to his cute girlfriend and became jealous. I had no idea who his cute girlfriend was. Jealousy is something I strive to understand in myself and not react to, as it can produce deadly over-reactions like this.

Fritjof smiled, looked closely at me, and then said to the man, "I think you would be wise to listen to him."

Fortunately, he listened and stomped away.

By the time this group, which had lasted five days, was over, there was a peace, love, acceptance and bonding present among us all. Even the bloody Frenchman.

People catharted and screamed, as they seemed to like to do that here in the groups, and Fritjof handled it all with intelligence and helpful direction after it occurred in individuals. He never responded with any judgement, or led them to cathart.

A married couple had a terrible row and the woman was angry, sobbing, and abusing her husband. Fritjof surprised me by telling this poor woman to go and sit in front of me. Ignoring me, he said to her, "He can see you. He does not know what he is doing, or how he is seeing, so take no notice of him. Just sit in front of him and do not talk and he will see you as you really are."

I had no idea what he was talking about and I was nervous about this highly emotional woman being placed so close to me with her angry husband nearby, but I trusted this doctor and his intelligence, so I quietly sat cross-legged as she placed herself, cross-legged, knee to knee, in front of me. I sat there in a neutral, mindless zone, just looking at her as she sat there still sobbing and shaking. Gradually her sobbing stopped. Looking closely at me, she began laughing, and then stood up and sat on my lap with her arms around my neck, still laughing. She kissed me and I hoped that her husband would not get jealous and attack me. He did not. A wise man, perhaps freed from the tentacles of jealousy.

Wonderful things like this happened during the entire group.

On the first day of this group two of the men held a woman participant down and were about to force her into having sex. I could not watch this so I picked up one of the rapists and threw him against a mattress on the wall. His friend quickly slunk away. The young woman hugged and thanked me. She had no interest in being forced to have sex with these men.

Near the end of the group, Fritjof asked me if I had a girlfriend here. I said no, the girls do not seem interested in me. He laughed and said that this was not true and I could have a date with any woman in the room. I laughed and said, "Yeah! As if."

Fritjof laughed again and said to me, "Do you want to prove me wrong?"

"Sure, why not?" I said.

He answered, "Okay, ask any three women here for a date."

I thought this was most amusing, and though it was alluring because there were some beautiful women from foreign lands sitting in the room, I thought that my trickster self would easily prove this man wrong. Just for fun.

One of the women in this group was a confirmed lesbian (not that there is anything wrong with that, as Jerry Seinfeld would say). She had previously told the group that she had been a lesbian for her entire life. I had also met her beautiful young partner during one of the coffee breaks. This woman was quite beautiful, generously proportioned, and also had a commanding strength, which I like in a woman.

Smiling, and sure that I was about to be proven correct, I turned to her and said, "Would you have dinner with me tonight?"

She answered with a smile, "I would love to."

This surprised me but I thought, it's just a meal.

"Two more," Fritjof said.

These two I would make count in case he really was right. I took my time looking around the room and picked out two of the most beautiful women. They both nodded happily when I asked them for a date.

I was amazed at how well this man saw things. I wondered why I was so shy with women and thought so little of myself.

The group came to an end with us all hugging and many of us exchanging addresses and phone numbers. This was the best group I ever attended in the ashram. Its lack of structure was so liberating, and anything was then possible. Could life be lived like this without all of the governing rules and regulations? I bet it could because people would realise that they alone are responsible for their actions and there is no God to forgive them for their sins.

I would participate in more groups in the future, both here and also in the West at a wonderful place called Esalen Institute in California, which I write about later.

That night I took my lesbian friend to dinner. After dinner, much to my surprise, she insisted that we make love and so we did in a slow, intimate and loving way. For three nights she stayed with me and then invited me to attend the "White Robe" satsang with her on the fourth night.

We dressed in our white robes and walked together, hand in hand, to the Gateless Gate at the front of the ashram.

This area was packed with sannyasins, all waiting patiently to get into Buddha Hall to listen to the master.

In the middle of the crowd at the front of the gate, my lesbian friend's lovely young lover erupted out of the crowd, flung herself at my friend's feet and howling loudly, begged her to come back to her and leave me.

This outburst was obviously something my newfound lover would have to deal with. Feeling embarrassed I turned to her and said, "This is your life which you must deal with yourself. It is none of my business." Embarrassed, I hurried away to let them sort out their problem. I missed satsang.

I never saw either of them again so I imagine they made up and left the ashram together.

My other two dates held no such dramas and we ate,

danced and had fun.

I had, because of this Encounter group, become a bit of a group groupie and tried other therapy-based groups based on childhood and trauma. They all bored me and seemed like a waste of time. All were based on psychology and all assumed that something was wrong with every person and only the group leaders could magically fix it. This seemed like wrong thinking and wrong teaching to me. Life is difficult enough without being told we are fucked up because of something that happened to us in our past. Luckily, I had never told anyone here that I had been a soldier in Vietnam. They surely would have had a good old negative feed on me with that information.

My last group, which cured me of doing these free groups, was a torture indeed. About thirty of us were locked in a room for five days and we had to sit in front of another person and say, "Tell me who you are?"

For five minutes they would have to tell me their answer to this question, then it would be my turn to answer the same question.

Now when you have done this non-stop for three days straight it becomes infernally boring. "Thou shalt not bore God" could be a new commandment, I thought.

At the door of the entrance to this group, a guard was always positioned who was there for the sole purpose of stopping anyone from leaving. Obviously, people would wish to leave, perhaps thinking, like myself, that one could die from boredom in this very room.

Now this is a type of Zen question that is supposedly impossible to answer and as I talked and talked, trying to answer, it seemed that it was indeed an impossibility. The monkey mind came up with all sorts of rubbish as an answer.

After three gruelling days I was ready to pay a lot of money just to get out of this room. However, I did not wish to confront the large karate guard at the door.

When it was again my turn to answer the question, "Who am I?", a satisfactory answer finally arose and I said to my partner, "I am nobody and nothing. Thoughts arise and fall away. I am not my thoughts. Nothing is permanent, including me."

That was enough of an answer for me and I quickly stood up, picked up my bag and headed for the door. The group leader, who must have also been infernally bored by now, asked me where I was going. I told him that I had answered the question to my satisfaction and I was done.

The guard folded his arms and confidently said, "You cannot leave." He definitely should not have been standing there with folded arms if he had any chance whatsoever of stopping me from leaving this group. These guards were allowed to hit people and often did. I dropped my bag, put my hands under his folded arms, keeping them tightly folded, lifted him and placed him to one side. I quickly unlocked the door and walked out. Thankfully he did not try to stop me again. And luckily for me, Osho likes a rebel so nothing more was said about my leaving so abruptly and forcefully.

The immediate sense of glorious freedom was a delightful tonic indeed, and I was finally cured of being a group participant even if all the groups were free.

Perhaps it is worth mentioning one other of the many groups I was a participant in. It was called Tantra. It was a group that promised to train us in the art of true sexuality. There was no mention of love or respect, just sex, and this group was quite famous and difficult to get into by then.

In those early days the AIDS virus was not a problem and no one needed a blood test declaring them AIDS free before being allowed to enter the ashram. That all came later.

Thirty of us in this group were made up of, as usual, half men and half women. We entered the underground dungeon and began three days of sexual torture. We were forced to rotate and push our hips, plus move our bodies in various ways for hours on end, sweating until we dropped. We copied rituals of African tribes such as the magnificent Zulus, no less, plus other tribal rituals from around the world that are performed before sex or sexual orgies. On the fourth day the lights were dimmed and the men were told to indulge in sex with each other, while the women were told to do the same. This horrified me, to put it mildly. The group leader was a homosexual man and it seemed obvious that he was trying to live out his weird fantasies through us.

I had come to know everyone in this group. There was a big strong man from Israel who had recently been released from the Israeli army. I whispered in his ear and asked him if he was gay. He said no so I told him to lie next to me and we could protect each other because gang rape was accepted and allowed in these early ashram groups run by sometimes kinky "group leaders".
I truly wondered again why people follow orders and go against any moral fibre they possess simply because of their belief in a guru or a god.

During the next hour or so the Israeli and I would often physically pick up any male bodies creeping towards us and return them to the centre of the circle.

We all, of course, were participating in this group totally naked.

The next day the tall black American group leader, who was about a quarter cast (three quarters white), picked me out and accused me of being a closet homosexual. I patiently tried to explain to him that I had no homosexual tendencies and if I did have, then I would have no problem acting them out. He said that he could prove that I did have. He had me stand in the middle of the group of participants who were all seated in a circle. He then asked the group if there were any men here who fancied me sexually.

One man stepped forward who I knew was an obvious homosexual. The other man surprised me. He was a good looking, very masculine, Spanish guy who had become my friend. We hung out together, had lunches, compared our motorbikes, etc. I would have bet a lot of money that he was not gay.

To put it mildly I realised once again that I was a little naive when it came to these worldly matters.

The leader told me to stand there and let them touch me. They began stroking me. The Spanish man kissed my neck, and the gay guy touched my penis, which had absolutely no reaction whatsoever. This was not a turn-on for me in the slightest. I turned to the group leader and said, "See, I told you." He then left me alone, even though he was obviously disappointed.

The next day the lights were again turned down low, after more intense hours of sexual exercises, and we were set loose on each other. Men and women came together in a rush.

I crawled into a corner, this time by myself, and when anyone came to me, I simply placed him or her back in the madly "fucking" crowd of moving bodies.

I realised that I did not like orgies and would never participate in one again. It all seemed stupid and dangerous because of diseases. And it was slightly disgusting to me, devoid of love and respect. Besides, too many of these penises had obviously been in strange places and had not been washed.

I came back to India again with my daughter when she was seventeen years old. This time she had no interest in staying at the ashram so she went to Goa with her girlfriend from Australia.

I stayed on for a while in Poona because young women do not want their father hanging around 24/7. I participated in a couple more groups. One was a dance group, which was okay because I like dancing.

Fortunately, there were many dance parties here at night that had fantastic music played by international DJs and travelling bands of worldwide fame. It was amazing, the number of famous and talented people who passed through these ashram doors in those heady days when Osho was alive.

One day I was summoned to the office and told that Osho would like me to be his bodyguard.

I wondered if this man, who says that he knows everything that goes on in his ashram, knows what I have done to some of his ashram guards?

Anyway, I thought, why not? Something new to do for a

short time plus he was surrounded by an inner circle of the most beautiful women, all hand-picked by him from the hundreds of thousands who flock here. Perhaps, exactly like I had hand-picked the two women in the Encounter group.

"Good," said the woman who asked me to be the bodyguard, "but first you must clean the toilets for the next six weeks to prove your loyalty and after that you will be his personal guard."

These people were full of unbelievable surprises. When I had participated in any live-in group, I was forced to use the ashram toilets and showers. When there was no group occurring, I avoided these toilets like the plague. Why did I avoid these toilets! Osho had often said that we have been influenced negatively in a sexual way by religious teachings that control all normal societies. We need to be sexually free, he said. Part of his freeing process was lots of sex plus the use of these toilets and showers to rid us of our inhibitions. The showers had no walls or doors and were all placed side by side. Directly opposite these showers, about six feet away, are the toilets, also with no doors or walls. The toilets were the Indian type. They are little holes in the floor which you have to squat over. When you go for a shower or a shit you wait your turn. In front of you are both women and men, pissing and shitting. In front of you while you squat are people showering.

The smell was almost overpowering as a lot of Westerners were often suffering from Indian dysentery – "bad belly". A lot of the shit missed the small holes. The worst thing I did was to look at the people on these toilets when I first entered. Only once did I look, then never again.

So I almost laughed when the woman told me I would be cleaning these toilets all day for six weeks to prove my loyalty. Was this meant to cleanse my shitty ego? Ha!

I quickly told the stupid woman that I suddenly remembered that I had to go to Goa to pick up my daughter and take her back to Australia. I would become the bodyguard of other gurus over the next years, never Osho's, and so had no intimate access to his beautiful women and chosen "Inner Circle". The "Inner Circle" were the ones that he called the 21 Coconuts.

I did, however, have an affair for some years with his most beautiful assistant – the woman who escorted Osho into Buddha Hall after opening the door of his car each night at satsang. She had not only been a successful international model; she was also a German psychologist. She had been in a relationship with Osho's French chauffeur for many years and assured me that she had sex with no other sannyasins and she did not partake in any of the groups. She was a strict meditator and stood on high moral ground.

Unfortunately, I knew nothing about German psychologists. Naive once again. Germans should do themselves a favour and avoid being guards and psychologists. Too heady!

I learned many things from this beauty while I travelled with her in India, Australia, the USA and Europe. When she felt good, she was heaven to be with. However, when she felt bad our relating was an abusive hell of psychological proportions. Ah, that German mind – and this particular mind was also well trained in psychology. She was also a true believer in her "enlightened master" Osho. She often reminded me that he was enlightened and therefore had the authority of a god. Now how can people say that someone is enlightened if they are not enlightened themselves? Common sense dictates that if a person was enlightened, he would never know it as the "I" supposedly disappears with the mind. And a person would never allow others to worship them.

When I took her to Australia, I introduced her to my mum. After a week my mother said, "Son, you must have robbed a Chinaman to be with this one." Ever the joker is my dear mum, but she had quite clear vision about people.

Life often seems to me like an intense softening process of both the mind and the heart. The meek shall inherit the earth, but only after we drop all the baggage we are presently lumbered with.

Osho and his huge following would soon leave India, bound for Oregon in the USA where they had purchased a huge cattle ranch. I would be summoned to America to take care of the horses and cattle on this vast ranch. Things do not always turn out as they are supposed to. I soon became aware in America that I had no interest whatsoever in going to a place where believers and spiritual devotees carried machine guns and pistols.

If they had ever discovered my background, they would probably have put me in charge of their motley army.
No thanks, I had had quite enough of armies in Vietnam.

Nevertheless, in Osho's intense world of opposites I did receive many gifts and words of wisdom. One of Osho's better sermons on love, taken from his book Sermons in Stones, I liked. He had given me his signed permission to use it in a book I always planned to write. Here it is:

> *Beloved Osho, This is an excerpt from a conversation with Gurdjieff; these are his words: "With ordinary love goes hate: I love this, I hate that. Today I love you; next week, or next hour, or next minute, I hate you. He who can really love, can be; he who can be, can do; he who can do, is. To know about real love, one must forget all about love and must look for direction. As we are, we cannot love. We love something because something in ourselves combines with another's emanations. We allow ourselves to be influenced. We project our feelings upon others. Anger begets anger. We receive what we give. Everything attracts or repels. There is the love of sex, which is ordinarily known as 'love' between a man and a woman – when sex disappears a man and a woman no longer 'love' each other. There is love of feeling which evokes the opposite, and makes people suffer. Later we will talk about conscious love."*

> *Beloved Osho, Can you please talk about conscious love, both in a man–woman relationship and in the master–disciple relationship?*

> *OSHO: The question is not only about conscious love. The basic thing is consciousness. A conscious person does everything consciously – love or painting or dancing or making a cup of tea; it does not matter what. Consciousness prevails over all your actions, just as unconsciousness prevails over all your actions. You love unconsciously, you hate unconsciously, you do everything unconsciously.*

So the whole problem has to be reduced to these two words: consciousness and unconsciousness.

Love can be a good example.

People say they love, but they don't know what they are saying, they don't know what they mean – because they love a car, they love a woman, they love a certain brand of cigarette, they love the football matches. It is difficult to figure out what they mean by love.

Unconscious love is simply a magnetic pull towards something irresistible. But you are not going towards the object of love, you are being pulled. You are fast asleep. In your whole life, you are behaving like a somnambulist. There are many people who wake up in the night without waking up; they get up, not wake up. And they will do things, and they will go back to sleep and in the morning, they will not remember what they have done.

Many times, when people think that their houses are haunted by ghosts, it is just that somebody in the house is a somnambulist who does things in sleep – sets fire to things, throws things away, and goes to his bed and sleeps well. And in the morning he is as puzzled as everybody else – not that he is deceiving anybody, just he has no remembrance.

Your life – which one minute is full of love, the next minute all the love disappears. Not only that, it can turn into hate; you were ready to die for the person and the next moment you can kill the person.

According to Gurdjieff, and according to those who have awakened, the whole of humanity is asleep, sleepwalkers. Things are happening but you are not the doers because you are not conscious enough to do something.

You fall in love. You cannot say that it is a conscious decision on your part. Perhaps the woman's hair is just like your mother's hair. Every male child loves his mother, that is his first love object. The baby girl loves her father, that is her first love object. And slowly, slowly, the father and mother become imprints – the mother becomes imprinted on the boy, the father becomes imprinted on the girl.

And when the boy falls in love with a woman, his unconscious mind signals him, "Here comes your mother." You don't hear it. And nobody can be exactly like your mother, so there can only be some similarity – the way the woman walks, the face of the woman, the eyes of the woman, the way she talks. Anything can trigger in you the old figure waiting in your unconscious, deep in the well inside you.

The same happens to the girl – the way the man walks, just the sound of his boots, may remind her of her father. Any small thing can trigger it, and suddenly you feel a great love arising in you. But because it is only fragmentary, it cannot be very lasting. It can be lasting if you don't ever meet – then you will never come to know about the other fragments. So the most successful lovers in the world are those who never meet. They make the most romantic, beautiful stories – no quarrel, no nagging, no fighting. And they never come to find out that this is not the woman made for me and I am not the man made for this woman – they never come close enough to know this.

But unfortunately, most lovers get married. That is the most unfortunate accident in life. That destroys the whole beauty; otherwise they would have been Laila and Majnu, Shiri and Farhad, Soni and Mahival – great lovers of history. But all those great lovers never met, never lived in a one-room apartment in Bombay.

Once two persons are together, then other fragments of their lives are bound to surface. You have to become acquainted with the whole woman you have got; the woman has to become acquainted with the whole man she has got – and there is the trouble, because then slowly you find that the fragment you love is very small in comparison to the fragments you hate. Now just the colour of the hair does not help, nor the face nor the eyes nor the nose – nothing helps.

In the West, women have been asking me – because their love lives are not going well. ... Nobody's love life is going well; it simply does not happen. So those poor women were asking if they should get their nose fixed by a plastic surgeon – because the husband is continuously talking about her nose, that it looks Jewish. And he is so much against Jews – the moment he sees the nose, all love disappears. The poor woman is ready to fix her nose.

I said, "Don't unnecessarily torture your nose. He will find something else; this is just an excuse. Right now he may become accustomed to this nose, but if you fix it then every time he sees you he will see that this woman has a fixed nose, she is really a Jew behind the nose. It will be very difficult for him to forget this. And the money is going from his pocket to fix your nose. You just leave it as it is."

In fact, for centuries a wrong concept has been prevalent: that lovers should like each other in every possible way. That is absurd. Lovers should make it clear – "These are the things I don't like." Both should make it clear, that these are the things I don't like, and these are the things I love. And there is no need to quarrel about it every day because that quarrel is not going to change anything. They have to learn to accept that which they don't like – a kind of co-existence, a tolerance. This is for the lovers who are not awake.

A conscious love is a totally different thing. It has nothing to do with love as such, it has something to do with meditation, which makes you conscious. And as you become more and more conscious, you become aware of many things. One: that it is not the object of love that is important. It is your loving quality, your lovingness that is important, because you are so full of love you would like to share it. And the sharing has to be unconditional. You cannot say, "I will not share if your nose is Jewish" – what has sharing to do with noses?

Conscious love changes the whole situation.

Unconscious love is centred on the object of love. Conscious love is centred in oneself, it is your lovingness.

Unconscious love is always addressed to one person; hence there is always jealousy – because the other person also knows that unconscious love is always centred on one person, that it cannot be shared. If you start loving somebody else, that means you have stopped loving the first person. That's the jealousy, the continuous fear that your lover may start loving somebody else – as if love is a quantity.

Conscious love is a quality, not quantity.

It is more like friendliness – deeper, higher, with more fragrance, but similar to friendliness. You can be friendly to many people; there is no question of jealousy. It does not matter that you are friendly to five persons or ten persons or ten thousand persons; nobody will feel deprived because you love so many people and his share is going to be less and less. On the contrary, as you are able to love more people, your quality of love becomes mountainous. So whoever you love gets more love if your love is shared by many people. It dies if it is narrowed. It becomes livelier if it is spread over a vast area – the bigger the area, the deeper are its roots.

Consciousness gives everything a transformation. Your love is no more addressed to anybody in particular. It does not mean that you stop loving. It simply means you become love, you are love, your very being is love, your breathing is love, your heartbeats are love. Awake you are love, asleep you are love.

And the same is true about everything else – your understanding, your intelligence, everything goes through the same change. You become the centre of the whole existence, the centre of the cyclone, and everything radiates from you and reaches anybody who is able to receive it.

It is not a question of loving someone for certain reasons; it is love simply out of abundance – you have so much that you have to share it, you have to radiate it. And whoever receives it, you are grateful to the person.

Conscious love makes you a doer, a being, a soul.

In unconscious love, you are just an emptiness – dark and dismal, hungry and thirsty. In unconscious love, you are a beggar. You are begging for love, because love is nourishment.

And this is hilarious: You are begging for love, the other person, whom you are in love with is also begging for love – two beggars holding their begging bowls before each other, and both begging bowls are empty.

Conscious love makes you an emperor. You don't beg; you simply give. And you give because now you see that the more you give the more you have. So whoever accepts your love, you are grateful to the person.

The whole earth can become full of love, an ocean of love, but only with conscious people.

With unconscious people, it is just a disgusting place, nauseating ... everybody pretending to be loving, and nobody is loving. Everybody is trying to exploit the other, and the other is doing the same to him. And because both are empty, sooner or later they are going to start quarrelling: that "You deceived me," that "You cheated me," that "You betrayed me," that "You are not the woman you pretended to be," that "You are not the man you were showing yourself to be." But with beggars this is going to be the problem.

I have heard about a beggar who used to sit by the side of a bridge. One day he asked, "Give something to a blind, helpless old man."

So they gave him one rupee. He looked at the rupee and he said; "This is not real, this is false." They said, "But you are blind!"

He said, "I am not blind. The man who usually sits here is blind. Today he has gone to see a film. He is my friend; I am just sitting in his place."

But everybody in this whole world of unconsciousness is a beggar, trying in every possible way to snatch some love, some attention, some sympathy – because love is a necessary nourishment. Without love, you cannot live; just as food is necessary for the body, love is necessary for the soul. And everybody is suffering without *love, because without love your souls are dying.*

But what we are doing is not right.

The right way is to bring consciousness to yourself.

And from consciousness there will be many revolutions in many dimensions. Love will be one of the most important dimensions, and you will find the golden key of how to get love from the whole existence.

The secret is: whatever you have, give it, share it.

Once the universe knows that you have become a sharer, then all the sources of the universe become available to you. They are inexhaustible".

This man was definitely a genius and a madman combined. At times he spoke great teachings. Sometimes he spoke like a madman.

In regard to "enlightenment", the nature of existence is that it is always in a flux of change, and humans are part of existence. Yet, as I had come to see in Osho, there is perhaps some form of occasional enlightened activity and then great moments, accompanied by great words.

Look to the teaching, for one must eventually forget the teacher. There is an old saying that explains this: "Look at the moon, not the finger pointing at the moon."

This means if you just look at the finger (the master) then you will never see the moon – the moon being what the sage is pointing out to you. What happens to each individual then has the potential to be unique, just like a fingerprint.

Chapter 26 - Home Again

When the long journey in the plane was finally over, the joy of being with Cheyanne again was a feeling that began in my stomach and exploded up my chest. Unselfish love is a wondrous thing.

Her mother had bought a house on the Tablelands with the proceeds from the sale of my home. It was a long way from where I lived so I purchased two horses and rented a caravan on a farm on the Tablelands, to be near my daughter.

Cheyanne loved horses so for the next six months we would compete in horse events and often ride in the thick rainforests that grow in the area.

What I found somewhat strange here was the fact that I still wanted to be with her mother, my old partner, even after everything that had occurred. I seemed to want to be with her more than ever, and imagined that I loved her in a deeper way after all my experience in India. Why did I want what I could not have? Probably because the dream is safer than the reality. It obviously takes a lot of pain and effort to wake up from the great romantic dream. It would take me many years to finally realise that love has no future because the future is created by the mind while loving is not. Love is found in feeling which creates presence. The mind interrupts exalted feeling.

One day when I went to their home to pick up Cheyanne it hurt me to see her current boyfriend with her and my darling daughter. What was he teaching my daughter, I wondered sadly? God, this life can be complicated. Why must I suffer?

I never said anything about any of this until one day when Cheyanne was at school and her mother was home alone. Her current boyfriend had been dumped. I begged her to come back to me. I suppose I was, if not crazy, then definitely unconscious as I tried to get my own way. I knew nothing of surrender.

She wisely refused.

I once again wondered why I could not fit into this normal world of relationships where it seemed that I did not belong. I had no direction here apart from being with Cheyanne. Fortunately, destiny takes over when surrender occurs and signs, when presented, are followed.

After six months on the farm while spending all my spare time with Cheyanne, a new quest was revealed. Osho was already in America. The first wave of people were already on their new ranch while the therapists and others were living at a place that they had named "Geetam". It was a hastily erected tent town with a few basic buildings, situated in the desert close to Las Vegas.

I told Cheyanne that I must go away again but I promised that I would soon send for her in the school holidays. This excited her and made her happy as she has always loved to travel to far-off places.

I left the horses with her. One of the horse's names was "White Eagle". My Aboriginal friends on the station where I once worked had named him and given him to me. He was a mighty horse and I always won the figure of eight race with him.

One of my old teachers, Krishnamurti, once said that only a happy man can bring about a new social order; but he who is identified with an ideology or a belief, or who is lost in any social or individual activity, is not happy.

As I randomly read these words, I realised that it always appeared that I received support and assistance in this quest when I most needed it.

I wondered whether there was an intelligent energy field that was either guiding me or was simply reflecting back my longing?

Chapter 27 - The United States

On arrival at Geetam, near Las Vegas, one of the women who ran the place led me to a shed in which there were single cots. Each bed was placed one foot apart, running from wall to wall. There were probably about a hundred beds in a small, tight area. She led me to a cot in the middle of this shed and told me that this would be my new home. As I stared in shocked disbelief, I wondered why I kept coming back to Osho's places?

I turned to look around and discovered that no more than 4 feet away from me, on a single cot, were a man and woman madly engaging in wild sex in the middle of the day with not a care in the world that anyone was watching.

The woman who escorted me here said, as she pointed, "Don't worry about them". Unfortunately, I looked to where she was pointing and was shocked to witness a perfect vision of his penis sliding in and out of her vagina.

I decided to quickly get out of this shed and into a nearby tent as soon as possible. This I managed to do in two days with a little bribe placed in the right hand.

These tents were erected on wooden floors and each tent had more than one person living inside. My tent had only one other person with whom I had to share, not three, and my tent mate was a male.

At night I needed to sleep after the long hot days, but sleep did not come easily here because as late as midnight the sexual noises filled the air. It was soon obvious to me that this sex fetish had become some sort of weird competition to see who could produce the loudest moans and screams, so showing everyone else that this person was getting the most enjoyment and best sex from the best partner.

After a few nights of this madness, on the fourth night at

11.30 pm the loud competition was fierce. I waited for a lull and then screamed at the top of my voice, "For God's sake, you are all supposed to be enjoying this, not suffering."

There was complete and utter silence for ten seconds. Someone laughed out loud and then everyone joined in the laughter. After that the nights were much quieter and we could sleep in peace.

One day when we were playing volleyball, dressed in our maroon robes with our malas around our necks, a large crowd of onlookers were watching. A television crew of six people wandered past, dressed in Western clothes. I pointed and yelled out, "Look, white people."

The same shocked silence followed for ten seconds and then everyone laughed loud and long. The truth is a funny elixir. We were all known as the "Orange People".

I had been there a couple of weeks when I was informed that the therapists – and they were all still there as the ranch in Oregon where Osho was to move to had not yet opened – had joined together and decided that I was angry and needed to be taught a lesson. They obviously couldn't decipher the fact that I was not angry, just bored.

Yet not one of these famous black-robed shrinks had discovered that I was a Vietnam veteran. I had learned the hard way not to tell anyone – especially therapists – this fact.

The therapists had decided at a closed meeting to summon everyone at Geetam to meet in the large group hall where I was to be taught a lesson in front of the entire Geetam population.

Unaware of what was supposed to happen, I was sitting quietly in the middle of the crowded hall.

The man of the moment, second only to Osho, was a therapist named Teertha who would become world famous in his own right after Osho kicked him out of the ashram.

Teertha always sat on a raised cushion in front of everyone with his two lovers, come handmaidens, a blonde and a brunette, both of course of notable beauty, sitting on each side of him.

After the preliminary openings, he pointed at me and told

me to come and sit by his feet. Of course, as usual I had no idea who he was pointing at and looked behind me. After a bit of this back and forth I realised he was actually summoning me so I stood up and went and sat cross-legged in front of him, having no idea what to expect but fearing the worst because of my loud unholy jokes here. Would I be sent to another nuthouse, I wondered, or was I already in one?

He sat and stared at me for what seemed like a long time but then he responded in the strangest way.

As he sat looking at me, tears began rolling down his face. He then said, "Everyone told me that you were angry. It is not true. I do not quite know how to tell you this but I will try. I see a great light in your chest. The only way I know how to describe it is to say it is like a great love. You are standing next to a car holding a bunch of flowers and a beautiful woman, who you love with all your heart, comes towards you. You give her the flowers and open the car door for her. I hope you understand that this is the best way I can describe the light I see in you."

What a surprise! I think he was talking about an all-encompassing awareness of conscious love and used the flowers and car as a symbol of this type of love he was seeing.

He became a good friend and I taught him the easy art of running. He was quite sick at the time and I thought some exercise would help him. He did get better.

One day the woman in charge told me that I was now wanted on the ranch. The woman (they seemed to run everything for Osho) still thought I owned a house. She said I should sell my house and give all my money to Osho and then the ranch would be my home for life.

Honestly, it took amazing willpower to stop myself from laughing out loud. I was not like the other devotees who donated all their money to Osho. Are these people so desperate to be at the master's feet, or just momentarily insane? Were they this desperate for approval and attention?

I told her I would sell my home when I returned to Australia so she told me to leave for Oregon as soon as possible.

I packed my bag immediately and happily left alone for the

long drive from Southern California to Oregon in the far north of America.

On the first night of my long drive, I stayed in a motel in a place called Big Sur. Watching the news that night I was surprised to see that there was a story about the ranch and the "Orange People", as Americans now called us. An Indian woman named Sheela, who was the big boss at the ranch in Oregon, gave an interview with a huge 38 pistol strapped to her side. When told that some people may take offence at this pistol she replied, "Tough titties." They then showed her getting into a helicopter with a sannyasin in his orange robe hanging out the side of this helicopter and pointing an M16 automatic Armalite rifle just like the one I had carried in Vietnam.

Right then and there I decided that I was not going to the ranch. Vietnam was enough of guns and madmen for me in this lifetime.

I was not sure where I would go. America is a big place with a lot happening and when I found somewhere interesting and safe, I would send for Cheyanne.

The next night I met up with two friends, a South African and an Australian, who had already been to the ranch and were heading to Los Angeles. They told me that they could not tolerate all the guns and the fools who did not know how to use them. Even worse, the ranch had also set up video cameras in all the rooms. The rule was that if you made love, you were not allowed to kiss and you all had to wear gloves on your hands and the men wear condoms. One may as well have a bath in a raincoat, I thought. I decided that I definitely was never going to this ranch with all this madness going on.

Some things that happen are unexplainable but in retrospect it makes me realise that there is something – call it evolution, destiny or grace – leading me on through this sometimes minefield of life. I am being led from the inside and the outside and know that I will be shown my next destination.

Wondering where I could go in the USA, I opened the drawer next to my bed in the motel, and lying there was a magazine about a place called "Esalen". Someone must have left it there. I read that Esalen was a sacred Indian site on the coast between Los Angeles and San Francisco. Hot springs ran out of the side of the mountains and had been piped into large man-made hot tubs. These springs supposedly had healing powers.

The book also stated that many world-famous teachers also lived there and these teachers taught groups of visitors from all corners of the earth.

I immediately realised that this place was the Poona of the West but without the lone guru. I was excited to see that this was a place of learning and teaching unfettered by a single religion or guru. Yes, this was the place I was searching for. Thank you!

I looked through the magazine for one of the many introductory groups that I needed to sign up for in order to get a bed there. Amazing teachers were listed such as Joseph Campbell, Gabriele Roth, Fritz Pearls, Feldenkrais, Terence McKenna, Dick Price, the LSD teacher (Stanislav Grof), Ram Dass (Richard Alpert), Timothy Leary, Lily the Ketamine man, and many other famous facilitators who were often in residence at Esalen teaching different courses. All these famous teachers/masters of their chosen field were welcome and gathered side by side here.

Life at Esalen became a fruit bowl of wisdom offered freely. I had a special experience with Joseph Campbell, the author of The Hero with a Thousand Faces and Myths to Live By. One day while having lunch at a large communal table in Esalen's mess hall I asked him about living the true life.

He said: "We must be willing to get rid of the life we've planned, so as to have the life that is waiting for us." In his Western wisdom he was alluding to the power of surrender.

And Stanislav Grof, another amazing pioneer of human potential, was also there. In a book titled Cultural Crisis and Transformation the following was written about him:

From Grof comes the recognition of the titanic destructive and transformative energies of the perinatal unconscious that are intrinsic to the transpersonal death-rebirth process, tied to the necessity of undergoing a fundamental collapse of the old matrix, a death of the old order – the necessary loss of the womb, the expulsion from the garden, the destruction of the old identity, ego death – for the possibility of both biological birth and spiritual rebirth.

Carl Jung, on this same track, wrote a personal letter to a friend in 1945:

"There can be no resolution, only patient endurance of the opposites which ultimately spring from your own nature. You yourself are a conflict that rages in itself and against itself, in order to melt its incompatible substances, the male and the female, in the fire of suffering, and thus create that fixed and unalterable form which is the goal of life. Everyone goes through this mill, consciously or unconsciously, voluntarily or forcibly. We are crucified between the opposites and delivered up to the torture until the 'reconciling third' takes shape. Do not doubt the rightness of the two sides within you, and let whatever may happen, happen. The apparently unendurable conflict is proof of the rightness of your life. A life without inner contradiction is either only half a life or else a life in the Beyond, which is destined only for Angels. But God loves human beings more than Angels."

Their wisdom was a support anchor in my mind as my own time of this intense, almost unbearable suffering was yet again slowly approaching. I would desperately need to pass through the personal crucifixion and intense burning that herald deep transformative change. It is no wonder that it is said that people need a master to pass through this "dark night of the soul".

I decided to pay for a five-day group in Esalen that offered "Nine teachers, some from as far away as Sirius B".

Wow, nine teachers all there to help me. Great odds, I thought.

I do not know where Sirius B is but if there are nine teachers in the room, I felt sure that at least one would have something to support my inner journey.

Naive as usual, if not a little stupid, would probably be a good way to describe me here in my complete misunderstanding of the "Nine from Sirius B".

When I entered Esalen, at the front office I was given a written description of this mighty healing centre. I would soon discover for myself that Esalen is a magic place and even this description falls far short:

Esalen is a state of consciousness as much as it is a physical place. It is a pagan monastery, where seekers of every description come to find light. Breaking out of the crumbling structures of their past, they come to find themselves. At this poignant moment in their lives, Esalen stands like the Temple at Delphi, where paths inward are offered, where they come to discover again, their bodies, their feelings, their pain, their knowledge, their happiness at being alive. For many, Esalen is where the tide turns in their private revolution against the inner tyrants of their past.

But Esalen is Hell as much as Paradise. It is a climate only for those who are both vigorous and capable of total defeat. The air is rarefied, the energy from the mountains, the creek, the ocean, is powerful and prone to dramatic shifts. Underneath Esalen's sign at the top of the hill is an unwritten dictum: Give up hope all ye who enter here. For here your nightmares must come true in order to fulfil your dreams. Here you are forced to fall flat on your face before you can drink the cool, sweet waters of joy. Many do not enjoy too long a stay, for here is where the mirror is ruthlessly turned around to face inward: the demons, flushed to the surface, are no longer "out there". The pace of karma quickens and comes home.

Some leave gladly. Others with anger. Some depart with tears, like the crippled boy who cannot enter the mountain with the other children following the Pied Piper. Some never leave, and they become Esalen.

The ocean at its feet, extending immensely into the distance, lets us breathe, expands our inspiration, absorbs our poisons, promises the infinite. The mountains and cliffs converge to form a tightrope for our existence – it is a land of the warrior. The occult energies of the creek flow from a canyon enchanting beyond words. And at the heart of this temple of nature, springs pour forth warmth from the womb of the earth; in this sanctuary, our bodies and souls are nourished, made supple and young, healed after the armouring of the cities. The baths are literally a Fountain of Youth.

Esalen is a renaissance court, with its geniuses and fools, its royalty and peasants, its knights and ladies, its musicians and scholars, its astrologers and ministers, its rogues and lovers. Here, at this Westernmost frontier of the continent (and therefore both outlaw country and yet closest to the east) lies a centre of learning, of culture, of energy. A threshold for change, it is both an incarnation of the ancient and a gateway for the unsung.

Esalen is where the archetypal level of reality breathes itself more visibly – somehow the magnificence of its beauty draws out the full power of the human spirit. Esalen is a Kingdom of Death and Rebirth. It is a place inside each of us.

This fascinating description of this beautiful place inspired me. That night as I sat on the side of the mountain looking down, I saw that Esalen appeared to be like a mystical little fairy village. I was excited and inspired, looking forward to my meeting with Jenny and the nine teachers the next morning.

Next day, after a lovely nude swim with a crowd of other seekers in the famous Esalen hot baths, we gathered together in a large room overlooking the pristine Big Sur Pacific Ocean. There were twenty-four of us and I immediately took note of the fact that twenty-two were women. The only other man was quite young, about seventeen years old.

After we were all seated Jenny and the nine teachers from the city of Sirius made their grand entry.

Jenny was a beautiful dark-haired, large-breasted English woman a few years younger than me. Next to her was Russell, her New Zealand husband, who was ten years younger than me. I was nearly forty years old at the time.

Where were the other nine teachers, I wondered? Maybe they would join us over the next five days?

I soon realised how naive and uninformed I was. I felt like a dumb outback cowboy just in from the bush. I didn't realise that the "Nine" teachers from as far away as Sirius B were all in this pretty woman's head and Sirius B is some mystical planet in outer space.

The women in the group asked her questions. While looking at the questioner, Jenny's right hand, holding a pen, would automatically write and answer any and all questions.

The only good thing about this strange automatic writing, which I immediately considered weird, was that Jenny's very large breasts, unconstrained by a brassiere, would bounce around as if they were in some sort of wild dance while she was writing furiously.

Yet I did not notice this bouncing until I relaxed the next day. On the first day I was just angry. The therapists at Osho's Geetam would probably have been pleased to note this. However, anger can be a protection and not something to be healed by rabid therapists who think they can make an un-perfect world perfect by pointing out everything negative in people who are struggling to find themselves in a new light. Perhaps unconscious hangovers from the Garden of Eden fable drives these therapists to seek perfection in others.

I sat huddled in a corner and tried to be invisible, never saying a word. I didn't want to be in this room which I considered, at the time, was full of mainly stupid witch-like believers. They all believed that "The Nine" from a planet called Sirius B were offering truth to their many and varied questions. Crazy California!

Soon, however, this lovely woman and her unreal abilities would impress me and a lifelong friendship would develop. Like most of these amazing people whom I would meet at this magical place of Esalen, Jenny had a special talent that made her different to normal people. Extremely different.

It soon became obvious that these wonderful people, all at the forefront of time, were not able to fit into a normal life due to their unusual and unique talents. How sad for them, I often thought.

In order to soothe her nerves from this intense channelling, Jenny drank Coca-Cola all day while she chain-smoked the strong French cigarettes called Gauloises. She also did other unhealthy things. Yet at this time she was the undisputed Queen of Esalen. When the directors of Esalen wanted to make an important decision, they would ask Jenny and the Nine.

I remember, years later, being in a room in San Francisco with Jenny and Russell. There were about 300 people gathered to see her. Among them were politicians, army officers, musicians, businessmen, plus a NASA scientist. This scientist eventually stood up and said to her, "I have been working on a formula for ten years to develop a fuel for the space rockets but I cannot finish the formula I am looking for."

Now Jenny, who did not even go to high school in England, said to him, "Tell me what you have so far."

The scientist rattled off a long formula full of algebraic symbols. When he had finished, I watched in complete amazement as Jenny rattled off a similar algebraic sentence and the man screamed out at the top of his voice, "Eureka, that's it." He then ran from the room.

Things like this happened all the time while I was travelling with her and Russell. She really had an amazing ability. She told me that she entered into the minds of her questioners and sometimes found it difficult to re-enter her own body.

Other great lights who came through Esalen over the years while I was there also suffered in their uniqueness.

Werner Erhard, who offered to the world a teaching called "EST", and with whom I spent three days, constantly drank strong black coffee while he smoked huge cigars. This magnificent physical specimen, like other great lights at Esalen, was quickly destroying his physical body and did not seem to be conscious of what he was doing to his health.

Also, the amazing Jay Zee (Janet Zuckerman), one of Esalen's most celebrated group leaders, could not stop eating and weighed about 130 kilos. Years later, on her deathbed here at Esalen, I sat with her, alone and holding her hand, just an hour before she died. I asked her this question, "Jay Zee, what is prayer?"

She gazed far off into the sky and said, "Prayer is listening to the silence." What a wonderful description of prayer.

Osho used to say that prayer was not begging.

Most of the talented teachers in residence at Esalen also suffered physically. Being different, or before your time, can be a burden that burns, seemingly, sometimes worse than death itself, and so they slowly kill themselves in one way or another.

Focusing on the inner spiritual world, as is also the case in India, appears to allow people to ignore their physical body. The awakening into multidimensional awareness and then maintaining integrity on all levels is quite a challenge it seems.

Back in the room with Jenny and the Nine, on the second day of my arrival at Esalen, I was again sitting in the corner, angry that I had wasted so much money hanging out with what I still considered to be a mob of freaks listening to a mad woman. How can they believe this nonsense, I thought?

Jenny suddenly turned to me for the first time and said: "Why are you sitting in the corner all dressed in black like a Vietcong? Haven't you had enough of Vietnam?"

As I have said before, I tell nobody that I have been in Vietnam and it was impossible for anyone here to know this. Never tell a gaggle of therapists that you have been in Vietnam as their minds go immediately to thinking that you are damaged in some way, and they attack. Maybe I am damaged but I do not think about it and I will deal with it myself. It was not in my resume' to trust therapists and psychiatrists anymore. I had already done enough of that.

She continued speaking as she turned back to the group of women, shocking me again, as she asked them, "Did you know that all you women here are in love with this man? Well don't waste your time on him as he is only interested in God."

I was again shocked by what she said. I have ignored God or whatever you wish to call "it" since my father took me to a large Catholic church when I was a small boy and said to me in an angry voice, "If I ever catch you in a place like this, I will flog you to within an inch of your life." Subtle, my dad. And Vietnam had completely cured me of belief in anything or anyone.

I would not become aware of "existence", "it", the "unnamed", "evolution" until much later in my life, and in a totally new way. Yet, I knew somehow, without a shadow of a doubt, it could only reveal itself in "its" mystery. Catch 22. I had to continue my searching, whatever the cost to this body and my reality. Perhaps I was a bit like Jenny and the others here. Like finds like then unites.

Back in the room, Jenny then totally ignored me, but I finally sat up and really studied this freak of nature.

Here are some of the questions and answers that took place over the next few days. And now, in this room, relaxed and attentive while looking closely at her, I noticed the amazing dance of her breasts, unfettered by a bra, as she busily wrote what the Nine were supposedly channelling.

And, even years later, though I loved Jenny, I still did not believe in the Nine. What she does is simply the almost unbelievable talent of herself, Jenny, this amazing person, unique and different to any human I have ever met. How I wish she could have just owned it but obviously it is extremely difficult to be so different to others and live a normal life while facing the real truth about one's self. It seems obvious, however, that surrendering to an imagined higher power of some sort allows a human, unfettered by ego, to speak some deeper truth. This is what Christians and other religious followers do. They surrender to an imagined god. Perhaps I may one day be able to surrender to the mystery. The limitless that cannot be limited.

"THE NINE"

The following is an example of what the Nine were communicating through Jenny during the question and answer sessions. It seemed obvious that the quality of the wisdom was determined by the receiver.

> Question: Eastern religions such as Tantra value the importance of not losing the sexual fluid, which is the spiritual fluid. I tend to disagree. What do you say?
> Answer: We see you discussing sperm. Sperm is transient, and is the physical result of sex. The physical result of the sexual act is union or the hole, one and the same.
>
> Q: My interpretation is that you are talking about the physical versus the spiritual in terms of sex.
> A: There is no difference, for sex transcends both. It is a

union of both – a union of spiritual.

Q: What is the distinction between beliefs and reality?
A: One is pre-thought, one is actualised.

Q: Give me a definition of creation?
A: Creation is.

Q: What is truth?
A: Truth is a statement that within a moment called time, you believe.

Q: What is emotion?
A: Emotion is the apron of time and the guardian of death.

Q: Is belief an emotion?
A: Nearly.

Q: What is love?
A: Love is the expression of hope in the emotion, the physical reality. The energy from love is a divine power.

Q: Why is sex at the base of all human relationship?
A: Because it is at the base of your physical body, just before your legs split.

These questions and answers made my head spin, yet her ability to immediately answer the many and varied questions was impressive. Some of her answers just left the mind floating in nowhere land. The questions kept coming.

Q: Please comment on monogamy versus polygamy in terms of mankind finding sexual fulfilment.
A: It does not have a "versus". During sex, the ultimate spiritual involvement is felt. If you want to share, then God be with you, for you sure aren't with yourself.

Q: What is the drive in so many individuals to transcend?
A: Drive to transcend is called "Please don't leave me in this place".

Q: What is music?
A: Music is the state of the soul that needs expression.

Q: Is man on the road to destruction?
A: Or perfection, yes.

Q: So God isn't what we think it is?
A: God is. The whole of creation is God. There I am, and we are, and God is.

Q: What is God as portrayed in the Bible?
A: God is. God has no why, what, how, you, they, I, etc, etc. God is. Within this statement is the truth. You see, the concept of the supreme being is your own, and therefore not liveable by God. God has no subdivision or comparison, so again we come back to – God is.

Q: The Bible is presented in two books, the New and Old Testaments, with Christ in between, and as a result of that story the Jews have been persecuted. Why is that, and what does it mean? Why is the Bible divided into two?
A: Before Christ/after Christ has not so much to do with persecution but rather a statement of the end of one era, the beginning of another. A comma within a millennium. The Jewish race pay a price still for being the richest tribe. The price is persecution. The manifestation is envy. Tribes of wealth in any form are persecuted for the "You have more than I" syndrome. Then blaming the death of Christ, this seems merely to be a convenience.

Q: How do you see the Bible in relationship to other holy books?

A: Mainly that of a different cover.

Q: When the Bible talks about God, it also speaks of gods. Please comment on the multiplicity of gods.
A: Gods is the simple need of man to see the same phenomenon within different shapes. You all would furnish your houses differently. The same with gods. Some want a thunderous one, others a love child, others a green piper. Same chair, different cover.

Q: Are we all gods?
A: Of course.

Q: Where did Jesus get his knowledge from?
A: From himself.

Q: Are his miracles, such as feeding the multitudes, actual events, and how were they brought about? How was that manifested?
A: Manifestation is being able to pull the resources of every shape and size ever designed into being. Jesus was able to transcend the boundary of skin we spoke of earlier. Unlike most who experience out-of-body, he knew that the materialisation was the space before he re-entered the body. If you try to manifest after re-entry, you have the star-studded phenomenon called "wish". So, in a word or two, yes, he did perform said miracles.

Q: Do you think any particular spiritual path is worth following in the next decade that will become useful for creating man's spiritual transcendence?
A: The only form is the being. The form is a part of the whole. Some reach this In-Blink stage by horse-riding, others play football, and others meditate. Some even talk to the Nine. All are valid ways; therefore we say again, if it feels good, do it.

Q: Is there one true religion?
A: Yes, Life.

Q: What is wisdom?
A: Wisdom is the state of consciousness much like an empty room. It will use any furniture it needs.

Q: How can I like the bits of myself that I don't like?
A: Stop seeing bits.

Q: Would I be able to find contentment in a committed relationship?
A: Contentment and commitment have no union, so no.

That is a small taste of "Jenny and the Nine".

After five days the group ended and everybody hugged and said their goodbyes. As I was walking out the door, Jenny and Russell asked me to remain.

Russ surprised me when he said, "We want you to be with us as we see you in our future." I did not know what he meant but I had grown to really like this strange pair and thought, why not?

From then on, when Jenny led groups of people, Russell would sit on one side of her and I would sit on her other. I soon discovered that I was there to protect her because she made some people angry with some of her truths. I played this role during all Jenny's groups at Esalen, plus the many groups she led in cities around the USA. Usually, I only had to quickly stand in front of anyone angrily approaching her because she had questioned their primitive beliefs. Once I had to physically stop a man who was running at her, but I never hurt him, just lay him gently on his back.

When a particular belief or ideology of a person, which they may have desperately clung to for their entire life, was exposed through logic, some people appeared to suffer great fear and then act in a crazy and angry manner. These people were lucky enough to experience emptiness for a moment but their emotional mind quickly returned to the comfort of their particular belief system which then chained them to the intellect and not their hearts.

Jenny also seemed to court danger when we were working in San Francisco or other large cities. One day in San Francisco, as we were walking down the street, she dragged Russell and I into a bar full of tough-looking people, a mixture of both black and white. After a few bourbons she began telling these people what was obviously amiss about them. I became nervous and tried to get her out of there but she was having none of that. Finally, two men came over and began threatening her with closed fists. I moved quickly in front of them and blocked their path. One immediately threw a punch at me which I easily caught with my hand, and said: "I'm her bodyguard and I really don't want to hurt anyone, though I easily can."

Fortunately, they believed me and I quickly got Jenny out of that bar. Outside I was a little angry and asked her why she confronted those men as she could have easily gotten us killed. Jenny just laughed and said that those people needed the lessons.

I continued to accept Jenny as she was, in all her duality. My time with Jenny and Russell offered moments of mystery far outside anything that could be considered normal.

Jenny was not really a sexual being and was not interested in her body at all. But she had a heart as big as Christmas and an honesty that was refreshing. But, like Werner Erhard, and the many others endowed with gifts that made them extraordinary, Jenny never could truly relax. Deep inside she was a scared little girl. She was here on earth, it seemed, before her time, and in a hurry to leave.

Just one of the many phenomena that occurred around her was that Russell and I merged in a unique way. For one thing, we began to look like each other. In Esalen, people who had known Russell really well for many years would approach me and begin talking to me as if I were him. When these people finally realised, sometimes in mid-sentence, that it was me, they would do a double-take and pull back their heads in amazement.

One day, for instance, I was in the line at the Esalen dining hall waiting to be served lunch. I was standing behind two people who had been working at Esalen for many years – Benji, a therapist, and his beautiful wife who worked on the famous Esalen massage team. Massage was just one of the many things that Esalen was world famous for.

Benji's wife said to him: "It is absolutely amazing how those two look like one another."

Benji turned to her and said: "They do not look anything like each other at all." He glanced up and saw me and said: "Isn't that right, Russell?"

I laughed, and when he saw his error, he got such a shock that he staggered backwards and nearly fell.

Russell and I had similar mouths and faces, but he had a large broken nose from playing Rugby Union, whereas mine had been straightened many times, and eventually shaved by a plastic surgeon, thanks to Vietnam.

The only reason I can think of for this peculiar phenomenon occurring was the great love we both felt being around Jenny.

Chapter 28 - Linda

Over the years I would come and go to Esalen, sometimes bringing Cheyanne. It was a magical time as I met many amazing people who took me in and shared their lives with me.

One person I met, who deserves a chapter of her own, was Linda Tellington-Jones – a truly remarkable woman who could definitely communicate with animals. I travelled around America with her for a while, taking care of her and watching her do things that seemed totally supernatural. To mention just a few, for example, one day we were called to a nearby ranch because there was a huge rattlesnake living under the ranch house and the people could not remove it. Linda knelt down in front of this totally frightening creature and began talking to it. The snake came towards her and I thought, oh my God, this woman is going to die. She stood up when the snake was by her side and walked it off the property, talking to it all the way as it wriggled along by her side. I just stood there shaking my head.

The power of human love and her compassionate connection to all living creatures was amazing to witness.

Another time she was called to a zoo that had a big old gorilla imprisoned in a steel cage. This once docile creature had turned into an angry, raging nightmare that no one could get near or do anything with. She asked for the cage to be opened and stepped inside. Once again, I thought that she was about to die, so I asked the keeper for a large gun, but within moments she was actually sitting on the gorilla's lap. She then turned to us with a big smile and said, "He just wants to go for a walk."

She then held his big hairy black hand, stepped out of the cage and took him for a long walk. When they returned the gorilla seemed happy and content. Linda confirmed this.

Once we were called to a place where a magnificent black

horse was being held in a small round yard. The owner said that they would have to kill this beautiful horse because no one could get near him. If you stepped into the yard he would rear up and try and kill you immediately.

As we stood looking at him, his ears came back and he reared up and screamed loudly at us.

I turned to Linda and told her not to go in there with him.

I then turned to the owner and asked him to get me a rope and told him I was a horse breaker in Australia and in five days I could be riding his horse.

Linda just laughed and said, "Open the gate." Then, to my wide-eyed horror, she stepped inside with this rearing, black, angry, powerful horse.

To my head-shaking amazement, within minutes she was rubbing the horse behind the ears and it had its head almost on the ground with its tongue hanging out, its eyes half closed. It looked as if it was in pure ecstasy. It never gave the owner trouble again, so we were told.

On that very same day we saddled up that same horse for Linda, plus a magnificent Arab for me, and we went for a ride into a huge forest. Here, Linda convinced sceptical me that fairies were real. She offered to show me some fairies.

When we came to a certain place in this forest she laughed and said that they, the fairies, were there. I looked around and saw nothing, and asked, "Where?"

She said: "No, no, you cannot look directly at fairies as they are very shy. You must look off to the side and you will see them from the corner of your vision."

She pointed to where they were and showed me how and where to look at them. Sure enough, I saw movement at the edge of my vision, but when I quickly swung my head to look directly there was nothing there. Linda laughed at me with her tinkling laughter and I thought that this woman might well be the Queen of the Fairies.

She also massaged sick animals and they would immediately improve. Horses, for instance, limped up to her and then they soon cantered away in a different energy with strong legs and no limp. I questioned her on how she did all these miracles as I wanted to do them too. By now, I thought of myself as her apprentice not her protector.

She showed me what she did and tried to teach me. I did all the moves she taught me but soon realised that it was this woman's amazing love, admiration and fascination of animals and creatures that was the secret to her powers. She simply loved creatures in a mouth-opening agape fashion, and the creatures could feel it. I never did get the hang of this while I was with her, because I did not have this depth of love and amazement, free from the constrictions of thought, that she seemed to naturally possess.

I realised, in the company of beautiful Linda, that intense feeling simply stops the mind which then allows agape connection.

Feeling the mystery simply negates the mind so how can anyone tell you what god is, or even what some god says or does. It is not possible, for god is not, and indeed cannot be explained by the mind though many have tried.Eventually, I finally gave up trying to do what she did, and went back to Esalen. However, I did have the experience that agape love has a lot to do with healing and this could be done with humans.

Chapter 29 - Other Amazing Teachers

At one point, back in the village, the powers that be embraced a particular health experiment and made it a rule (making rules was unusual for Esalen) that salt and sugar were now banned in Esalen. In place of sugar, they supplied something called "Stevia" which was some sort of South American vine sweetener that was supposedly a healthy sugar. I tried it once and it tasted bitter and terrible, like some sort of poison, so I trusted my taste and never touched it again.

The salt they supplied was some type of vegetable salt that smelt terrible and tasted even worse. I immediately went to Monterey and bought my own real salt and sugar. Then, at each meal sitting, I smuggled my salt and sugar into the dining room.

I am not a person to blindly follow rules. And Californians, believe me, can definitely be an intense bunch of do-gooders in their quest for health and hidden secrets.

Being the only rebel to have this salt and sugar made me very popular and the large table I would sit at always quickly became full.

Joseph Campbell, Gabrielle Roth, Dick Price and other awe-inspiring people would borrow my salt and sugar and chat away with me. I loved asking them questions.

I once participated in a dance weekend with Gabrielle Roth. What an amazing shaman of a woman. Her groups were always packed and fully booked out months before they began.

At the beginning of her session our group of about fifty people were dancing freestyle while her personal band played. Gabrielle suddenly swept into the room dressed in black with a large black cape draped over her shoulders. She immediately challenged everyone to be brave and be real, as she led us on the dance journey, she taught using a system she had devised called

"Five Rhythms". Five Rhythms would soon be practised all over the world.

At the end, after five days and nights of that intense group, and feeling quite refreshed, I quietly left the room while all the participants were hugging and exchanging phone numbers. As I was walking away, down a path, Gabrielle came out and called my name. I turned and she came up very close, face to face, and looked deep into my eyes. After a minute of just looking, this amazing woman said, "I loved you from the moment I saw you."

This shocked me and I wondered what she was seeing. Was it some sort of deep pain left over from Vietnam (even though she did not know I had been there) that was reflected in my eyes? Could it possibly be that Gabrielle was feeling this?

Danny, one of Esalen's leading psychics who ran the kitchen, had recently told me that I had the kindest eyes that he had ever seen. Maybe she was seeing that?

Gabrielle and I became close friends. We would walk together, have hot tubs together – naked of course as this was Esalen – and we finally travelled to New York together where she showed me her city.

Gabrielle was married to the leader of her band and it amused her that people were saying that she had an Australian lover. This was not true. We were friends, never lovers, and I never approached her in any sexual way, though I loved her dearly, but in a heroine worshipping way because she seemed so accomplished in who she was and what she did. This breathless type of love that comes about when I worship a woman leaves me powerless to approach that person as a lover. Being lovers may have taken away that sacred feeling anyway.

When I finally left New York, after we had walked half that city, we hugged for a long time and then exchanged presents. I had been reading the "Mother Peace" tarot deck and when I read for her, she was fascinated so I gave her my deck. She, in return, gave me her treasured "I Ching" book and the coins that go with it.

We wrote to each other for a long time but I never saw her again and she died quite young. Only the good die young?

On my return to Esalen I had another magical experience. A particular group of scientists and doctors had been coming to Esalen for seminars for quite some time. This particular time they had a "Eureka" moment where, using only nutmeg and sassafras, they invented a drug soon to be called "Ecstasy". This drug was legal in the USA for the next two years. Psychiatrists used it in sessions with their patients. I read some reports by these psychiatrists, and one of the many things they wrote was that "one session with a patient on this drug was worth twelve months of therapy". Perhaps the therapists were taking it too. It definitely brought out the unhindered, open loving truth in people.

Ecstasy was eventually banned and put on the dangerous drugs list, worldwide, simply because street people began misusing it. They began taking it every day searching for endless pleasure and that elusive high.

Pure ecstasy by itself cannot kill you because you cannot overdose on it. Once you have taken 125 ml of this particular concoction (I don't like calling it a drug), anything more you take has no effect. However, some people become so happy on it that they scull a bottle of alcohol or overdose on some other drug. Ecstasy acquired a bad reputation even though it was not the drug causing any fatalities. This drug then became a huge money earner for criminal gangs and drug dealers who then laced it with fillers. Fillers such as speed, heroin and others , which could cause death, were added to the nutmeg and sassafras.

Anyway, one late afternoon I was still suffering a terrible hangover after my birthday party which had been held for me by the Esalen staff the night before. A woman, whose name I will not mention, but who was one of the leading lights and leaders of Esalen, invited me to her home on the cliffs overlooking the glorious Pacific Ocean.

When I told her of my hangover, she handed me one of these newly invented ecstasy tablets. I refused to take it and told her that if I took another drug I would most likely die.

She laughed, and promised me that this tablet would totally

cure my hangover.

I eventually trusted her and so I took it in the hope of getting rid of that whopper of a hangover.

Within the hour I was sitting on a deck chair overlooking the ocean, watching the otters play, with a smile on my face and the hangover just a distant memory. I was totally cured by this magic tablet.

We later made love like I have never loved before and during the lovemaking I asked her to marry me. She said that her answer would be yes if I asked her again in the morning.

Next morning, I was embarrassed because marriage was suddenly the furthest thing from my mind. It had been a long time since I had trusted in marriage and I thought of it as a Christian plot cunningly combined with romantic love to trap innocent people in unconscious behaviour that led to unhappiness and sometimes murder.

As I grew older and wiser, I came to have a better understanding of a different way of loving that did not create jealousy and violence. But that understanding only came after much deep suffering.

I apologised to this beautiful, big-hearted woman and a tear fell from her eye as I took my leave. Her last words were that she thought I came from a different world than hers. She was correct about that.

Ecstasy had shown me the incredible power and the hidden depths of love which could be possible in myself. I knew that I did not need to take it again as it had already given me the strength to continue on this journey of discovery where I would eventually feel much more than I had ever felt, even on this magical pill. Ecstasy gave me a glimpse of the human potential that must eventually reveal itself in its own time through evolution and understanding. I had just stepped out of time and the normal world for an evening.

One commonality I had noticed about all these amazing Facilitators, at Esalen, running classes and transforming people, was that their work had a foundation of love and support for struggling humans of all colours and beliefs who flocked here.

Love is an important key that I would have to keep in mind. What did I need to learn and perhaps be rid of in myself to feel this great love? I did not yet know. Life is about learning, especially learning what not to do. Then Existence becomes my mirror.

I returned to Australia with Cheyanne as my six months American visa had expired. I would return to Esalen again and again to witness, learn, and partake in more of its hidden mysteries, surrounded by its unusual people and leaders.

During her time with me at Esalen my daughter, who was now eleven years old, became friends with Dick and Jenny Price's daughter, plus children of other famous teachers in this magical place.

Chapter 30 - Australia Again

When I returned to Australia, I was no longer able to return to my old ways. I was sick of working in nightclubs and hotels as a crowd controller, or as a bouncer as people called us back then, so I decided to earn my living in a more normal way where I did not have to deal with drunks or drugged people who were liable to erupt into violence.

For a short while I sold boats for a friend but there was not a lot of money to be made as I did not know much about the type of boats he sold. I successfully applied for a job with a firm called Right Home Improvements, which marketed a plastic wall cladding made by Iplex Industries of Australia Ltd. While selling this product I also sold a roofing tile for another overseas company.

Within a few months I was promoted and given the title of Product Manager because I had sold more wall cladding and tiles than all the other workers combined. The money was rolling in and soon I was training the other sales people and receiving a percentage of their sales as well as mine.

One day I sold the cladding to a very nice working man who owned a home that had been left to him by his recently deceased mother. Having watched me play football years ago, he trusted what I told him.

I assured him that this cladding would definitely increase the value of his somewhat run-down home and would protect the timber, which was rotting in places. What I told him was all true.

The cost of the cladding was only $3980 and he was able to afford this amount as I had worked out his earnings and expenses.

Out of this $3980 I was to receive $750 commission while my boss got $400 for doing virtually nothing. The company, Right Homes, received $1200 so the cladding was certainly a lot cheaper to produce than the initial $3980 that I would charge him.

I thought about this and concluded that everything in this society operates only for profit. Lawyers, doctors, bankers, insurance companies, etc, all have large profit write-ups, while the common worker pays for the hefty write-ups to make the wealthy even richer.

Some doubt about what I was doing entered my mind but I dismissed this because at least his home would be worth more money and last longer.

A few days later the manager, a big burly man from New Zealand, showed me the final finance papers he received from AGC for the $3980 loan which I had arranged for Ian to borrow.

AGC is a large company that advertises itself as being "friends with money".

I was ignorant of the fact that my boss had a little side deal going with someone at AGC. The final client contract shocked me:

PRINCIPAL SUM TO BE ADVANCED BY PAYMENTS AS FOLLOWS:

    To AGC Limited: $288.50
    To AGC Limited Townsville on a/c of Deputy
    Register of Titles: $26.00
    To Stamp Duty: $70.50
    To Right Home Improvements: $3980.00
    To AGC Limited (Cairns) A/C 50401010-26311-094: $272.00
    To Deputy Commissioner of Corporate Affairs: $4.00

    Principal Sum: $4591.00.
    Plus Interest [this is what really shocked me]: $3213.00

$7804.80 TOTAL

    Total payable by 60 monthly instalments of $130.08.

I was now shocked and ashamed because of this contract that I had instigated. I had already quoted my client a much lower figure, at a much lower interest rate, which was nowhere near as high as the loan payments costing $7804.80. And this same company was already into him for $90 a month on his car loan. So now they would have him working like a slave for $220.08 a month, for years.

This honest hard-working man only earned a wage of $448.00 a month in the steel factory where he was employed. So that would leave him $228 a month to buy food, pay rates, electricity, have a drink maybe, and a meal out, if he was lucky.

This man was going to be working like a dog for years to pay off blatant, apparently legal, thieves.

Must we prey on one another like this? Is this the stage we have reached where money negates our morals and we make slaves of hard workers? It certainly appeared as if this was true.

What totally fucked-up law gives this insurance company, and my boss, the disgusting power to charge 80% interest on a loan? And what law even lets us sell a product worth $800 for $3980.00?

The same law that is passed by money hungry government men, who are feathering their own nests in the four years that they are voted into power for. The same law that puts young people in jail for years just for smoking marijuana, yet encourages them to become alcoholics or tobacco addicts, which are much worse drugs and kill many more people than marijuana ever would or could. This society is eating away at its very foundations because it is based on immorality and greed. This simply means that this society will eventually change through total collapse.

Why aren't our religious-based moral laws working? Because religions are not working? Where is our joy and elation? Is it only in drugs, plus tobacco and alcohol, because of the unreasonable working demands forced on most workers in this society? One of the major reasons why it is unbalanced is because five per cent of the population possess about ninety per cent of all the money.

Anyway, I could not allow this rip-off so I tore up the contract in front of my boss. He lost his temper at losing money and like all big men thought he could play the bully with me simply because I worked for him and was smaller.

He screamed in rage and quickly advanced towards me with fists clenched. I merely hit him with a straight open palm that landed on his advancing, unbalanced chest. His own huge weight threw him spectacularly backwards over a table and he landed on his back on the floor.

"You're fired," he screamed from the floor.

I walked around the table, trying to control my temper, and standing over him I said, "You cannot fire me because I already quit when I read your dirty contract."

With no job, the city and its social scene that I began living in, revealed yet again the seeming hopelessness of human relationships where divorce had reached about 60 per cent, and people were living drunk and drugged on the weekends, merely to cope with their slave-like lives.

I felt very sad. Change can happen. We cling to our current beliefs and feel lost and lonely without them, yet it is beliefs that are keeping us unconscious.

A few days after losing my job I went to a large party. A hat party. Everyone in town seemed to be there and the party was packed with hat-wearing people. It was being held in the grounds of the city swimming pool. The food was excellent. The people looked beautiful, dressed up to the nines in groups of up to seven but mainly in twos or fours. A select few tried to circulate and communicate with strangers and some succeeded. The drink was being thrown down quickly and cigarettes sucked hungrily to help ease the social tension and shyness. By 11 pm everyone finally began dancing, throwing off their shyness and paranoia and the heavy social mantle that somehow keeps us trapped.

I studied the crowd as some smoked pot while others swallowed Mandrax and Quaaludes, or pills of some sort.
They had perhaps experienced and felt, on other nights, that elusive something when happiness prevailed, when dancing

came easily, and life was good for an all too fleeting time.

It was sad that this never-ending, albeit unconscious search for freedom was not only happening here in Australia, but all over the world.

A conscious, free humanity where people did not need this over-indulgence to relax and feel normal must be possible.

One day this elusive high will come without an excess of tobacco, alcohol, drugs, or prescribed tablets. We shall actually feel the ecstatic wonder of our breath and movement and we shall move slower. This global transformation can only arise naturally because of a new understanding of what man has continually searched for. Then gods will no longer be searched for or, indeed, worshipped, and will no longer need to be named. No such separation from gods, in human thought, shall exist. The religious belief based on duality and separation from nature taught us by many different religions shall no longer rule the human mind. We are one with all and everything and in this we shall also find true responsibility for all our actions on earth in this fleeting body.

I finally realised once again, fortunately or unfortunately, that I did not fit in to this world I found myself in. This was not the home I could return to, as life is like a mirror and I needed to change my reflection before I got lost in it.

I was forced to search even further and find the information needed to make my vision possible. If I had a choice, I would be a settled father and content husband, living a happy life like I saw in the movies and read about in books.

But Western society had become a feeding frenzy and I needed to fast.

> Man eats life
> Or life eats man.
> Something unknown
> Arises to lend a helping hand.

Where must I go next? The answer, as usual, soon revealed itself when I randomly read these words:

WHAT COMES, LET IT COME.
WHAT STAYS, LET IT STAY.
WHAT GOES, LET IT GO.

Poonjaji wrote this. He is an Indian guru from Lucknow in northern India.

Papaji giving Satsang in Northern India.

Chapter 31 - India

Cheyanne and I soon set off for India yet again. Enlightened masters are often reported to live in India and one such man was named Poonjaji who was a disciple of the late Ramana Maharshi, the famous saint of Arunachala.

Arunachala is a sacred mountain to Indians and is found in the south of India.

Poonjaji, or Papaji as I would soon call him, lived in the north Indian industrial city of Lucknow.

On our way to Papaji's ashram Cheyanne and I returned to Poona to pick up my Indian Enfield motorcycle which I had kept in storage.

Because we did not wish to visit Osho's ashram, I quickly booked a sleeping compartment for the long train journey north. I loaded my large 500cc Enfield on this train and we were ready to go.

While waiting to board the old Indian steam train, a man named Annant from Byron Bay, who was also travelling with his young daughter, came up and excitedly asked me if I knew that there was an enlightened man travelling on the train.

No, I politely answered, I did not.

Annant excitedly replied, "Yes, and we are going north to live with him on his ashram."

Obviously, they were his followers or disciples. With those words he hurried off to look for this so-called enlightened Indian man.

Cheyanne and I found our allotted sleeping compartment; it had three beds so we had to share it with a sweet little Indian man named Samdarshi. After introducing myself and my daughter, my curiosity got the better of me so I set off through the many carriages of this train searching for the enlightened man.

I finally found him, or so I thought. A large man with a beard was sitting cross-legged in a carriage surrounded by other Indians, and he was talking to them in Hindi as they listened with seemingly rapt attention.

I had seen this man in the past at Osho's ashram. Fair enough, I thought, he must have become realised in some way, but I was not drawn to him at all, so I went back to our carriage.

I still did not get this enlightenment business. It seems like a word, and thought, that has been substituted for the word heaven or paradise by avid seekers. Just another thought to relentlessly pursue, though at least the followers think they may find it in this lifetime, not after they die, as in heaven. This enlightenment then is a small step away from the pursuit of heaven or paradise in an after-life.

The train finally pulled out after a long Indian delay (trains always seem to be late here) and I soon got into a conversation with the little Indian man in my carriage named Samdarshi. We discussed life in general. He seemed quite intelligent and had a sweet nature.

At the first stop someone knocked at our door and there stood Annant, the man from Byron Bay. He brought his hands together in front of his heart and bowed to Samdarshi who then invited him into our carriage. Samdarshi then excused himself and went to the bathroom.

There are more seekers per head of population in Byron Bay than in any other town in Australia, if not the world. Byron has always been an unusual place.

Annant turned to me and said: "Do you know how blessed you are to be travelling accidentally in the same carriage as the Master?"

To my amazement he was referring to Samdarshi. This was the enlightened Master sitting here in the carriage with my daughter and I.

By the worshipful look on Annant's face I thought he would be willing to pay me a huge sum of money to take my seat and travel with his master. I was sorely tempted to begin negotiations, but I did not and Annant reluctantly left our carriage when Samdarshi returned.

During our journey we spoke of many things and Samdarshi finally insisted that we get off at his station and come to his ashram with him.

Why not, I thought, one guru is as good as the next.

At a large town near his ashram, we got off the train, together with Annant, his daughter, and many other Westerners.

The large Indian man who I originally thought was possibly the "enlightened one" was just another follower and together with a large group of Indian disciples he also alighted from the train. We booked into a hotel because it was too late to set off to his distant ashram.

That evening he invited us to have dinner with him and I questioned him extensively on philosophical subjects and his road to enlightenment. He said that he had used years of meditation as a tool for his awakening and finally one day realised that it was not the mind and its simple answers that he had been looking for.

For me, though, I was aware that facing my greatest fears, together with any other painful lessons in life, held the keys to a different consciousness. It was not my path to sit alone for years meditating. I needed to learn from experience. I quickly realised that I was on a different path to Samdarshi and his many followers. He had told me that they meditated for hours every day.

I decided to continue riding north the next day. When I told Samdarshi of my change of plans he was disappointed and tried to get me to change my mind, but I was determined to see Papaji as soon as possible. Cheyanne and I rode off on the motorcycle into wild and vast India early the next day.

After a long ride we stopped at a town called Khajuraho, where are the famous Tantric temples with many statues that show people in different positions of sexual coupling.

A lot of the intricate and beautiful carvings had been destroyed or damaged by the conquering Muslims who, like the Christians, seemed to have a terrible aversion to the art of sexual freedom. Controlling everybody's sexuality is high on the agenda of religions. This seems to me like sexual fear and the desire of these men to control women, which by the way is eventually impossible.

We stopped in an old motel in Khajuraho for another night and then after a very long ride we found an ancient, isolated, magnificent palace where we could stay for very little money. We were surrounded by luxury and lots of servants anxious to fulfil our every wish. There were no other tourists there.

Our room was huge and fit for royalty, complete with lush carpets, beautiful wall hangings and expensive paintings.

I was warned by the staff at this palace that on our last day's ride we would be passing through a part of India that was notorious for robbers who waylaid travellers.

I had also been warned that in India at this time in the 70s and 80s, you never stopped for anyone, even people dressed as police. Even if they were really police, they would demand a payment just to allow you to continue your journey even if you had not broken any laws.

We rode past two men on a motorbike who were heading in the opposite direction. They were carrying a rifle, and they immediately stopped, turned, and gave chase.

Most motorbikes in India have a 250cc engine. Mine had a 500cc motor and a German carburettor that I had fitted. It had been rebuilt to go faster than any normal Indian bike. Opening the throttle, we soon left them far behind in the dust.

When we passed through many different towns and cities, we had often been ordered to stop by the poorly paid police but I just sped past. They never gave chase because they had no vehicles.

I had not broken any laws and they would have given me grief until I had paid them enough "baksheesh" to let us continue our journey.

India runs on baksheesh. It is an accepted practice. Basically, it is bribery. The same thing happens in the West but it is just more sophisticated and hidden, such as in "political donations" by rich individuals and companies who want laws changed in their favour so they may make more money.

"Lobbying" to politicians in the halls of parliament in the West is also another form of bribery to get laws changed to suit large companies or rich individuals.

Backsheesh in India is probably more honest and straightforward.

Finally, we arrived in Lucknow. By the time we rode through the city and booked into a posh hotel near Papaji's satsang house, our faces were black from the smog and dust that settles over this heavily populated and thickly polluted city. Luckily, we had been able to breathe freely because I had purchased expensive gas masks in the West to be used for this purpose in India.

Satsang was held every day in a large hall that could hold a couple of hundred people. Inside, everyone sat on the floor on only one cushion so the people behind could see Papaji who sat at the front on a simple wooden chair, not on a raised stage.

The first day that I went there Cheyanne came with me but satsang was boring for most young people. After that introduction she spent her time with other Westerners of her age who gathered together in the back of the hall, playing games, while I remained in satsang.

People came from all over the world to listen to this eighty-year-old man. When we were all quietly seated, he entered with his entourage which consisted of two beautiful women, plus a rich Israeli man and a European man who organised everything for Papaji. This European man was also wealthy, I later found out.

The women escorted this large, strong-looking Indian man to his chair. He sat and closed his eyes and we all meditated with him for ten or fifteen minutes. He then said a blessing for the world and began speaking.

He spoke eloquently for a while about life and how to live it and then members of the audience handed him their written questions. This was the usual form of satsang with other masters I had sat with. His answers soon showed me that this old man had indeed reached a place in his life that was different to mankind's normal awareness.

I spent several days at the satsangs and eventually I was noticed by Papaji. When he learned that I was Australian he asked me if I liked cricket and if I knew anything about it. I rattled off the names of not only our best cricketers but also some of India's.

He immediately invited me to his home to watch the coming cricket test between our two countries.

The next day when I knocked on the door of his little home, a gracious woman named Garimo let me in, bowed to me, and said, "Welcome to the home of my master."

Garimo was a favourite of Papaji's and he was always happy when she arrived from her home in the USA to visit him. She would soon be ordained by Papaji and become world famous in her own right and have many followers. She would eventually ask me to be her bodyguard and driver.

I walked through a small room and into the kitchen where Papaji was seated reading the Times of India newspaper. He welcomed me and said the test was about to start. We discussed the cricketers and their merits and then he passed me a newspaper and we sat together reading.

It is worth mentioning here that the floor of his home was made from hardened buffalo and cow shit.

There was a knock on the door and Garimo announced that there were two women, who I knew from Byron Bay, here to see Papaji. When they entered the room, they immediately fell to the floor and wriggled like snakes to Papaji and then began kissing his feet.

I had watched their amusing entry in amazement, but Papaji had the paper in front of his face and did not see them until they began kissing his feet. He looked at them wide-eyed, bent over and helped them to their feet all the while saying, "No, no, no."

They gushed platitudes at him for a few minutes but then the cricket started and they were quickly shown the door by Garimo.

People seem anxious to quickly make themselves less than both gods and gurus.

Over the weeks I spent many hours watching cricket with this interesting man and it felt just like I was watching cricket with a mate in Australia. He knew everything about cricket and the players and we had fun commenting on the play, often laughing together.

It seemed as if he was happy to be treated as a normal man for a short time instead of always being worshipped like a guru.

One day when the cricket tests were over, he told me that he was planning satsangs in the forest outside Lucknow. He asked me if I would come with him and decide on a place to hold these gatherings.

We climbed into his Indian car, together with his driver and a good-looking blonde German woman. It became obvious, as Papaji spoke to her, that she did not like men.

Papaji asked me if I would drive, then he had his driver get in the back seat next to her. As I was driving, he then asked her to kiss his driver on the cheek. She seemed horrified, but after some good-natured urging by her "master", she eventually kissed him. This brought much laughter to Papaji, his driver, and myself. I think this intelligent man was just kindly showing her that disliking men, because of a past relationship, or something else, was not a smart option and did not serve her. There is only one self after all.

After driving to a few places, we eventually found a clearing in a forest that I thought was suitable and safe. I had searched the forest and saw how I could defend this man from any attacks by mad religious believers who hated anyone who did not share their beliefs or their chosen god. In India there are many gods but the believers in the Muslim god were the most dangerous.

Many wonderful and interesting satsangs were held in this forest clearing.

Each day I would ride out before Papaji arrived to make sure no one was hiding in the forest. Even some Hindus took exception to these new Indian spiritual masters. These religious zealots had been known to plant hidden bombs or simply shoot people because they were not of their exclusive sect or religion.

I had picked out four of the strongest looking men from Papaji's congregation and spread them out around the perimeters to keep watch.

Papaji revealed many living spiritual truths, including the one self and the eventual evolution of human consciousness, or, in other words, an awakened state of being.

This seemed to be just what I had been writing about during my search over these past years. I wondered if this was why I was drawn to this lovely man.

After we completed the series of forest satsangs and then held them once again in the hall, a couple of interesting events occurred. One morning someone had told Papaji that one of his closest helpers had been selling the drug ecstasy and this was why he was rich. When this man entered the hall Papaji angrily grabbed hold of him and physically threw him out, telling him in a loud voice to never return.

During satsang a woman stood up sobbing and said to Papaji, "I thought you were enlightened. How could you get angry like that? An enlightened man does not get angry."

This was a prime example of a follower's mind and one who thinks that they know what enlightenment is and how an enlightened person should act.

Papaji calmly spoke about anger saying that there is nothing wrong with anger. Anger can protect you in an emergency or in times of trouble. He said, "I was angry, now I am not."

Then he told a parable to make the point. Two monks were on a journey and they came to a river where a woman was crying hysterically. Her baby was on the other side with no food or protection and she could not cross the river. The older monk told her to climb on his back and he carried her across the swollen river.

An hour later, while the monks were continuing on their journey without the woman, the young monk became upset and said to the old monk, "How could you let that woman climb on your back, pressing against you, when we have taken a vow to never touch a woman?"

The old monk replied, "I carried the poor woman across the

river and left her on the other side. You, on the other hand, are still carrying her."

Papaji then said to the woman, "My anger is passed and forgotten, while you are still carrying it."

Another interesting event that occurred was when Garimo, who Papaji loved, returned from America with her husband.

Now in satsang, as Garimo and her husband knew, we were very careful not to come into the hall when we had the flu or if we were coughing. Papaji was old and we had to be very cautious with hygiene and visiting Western diseases in order to protect him.

The other rule was for each person to sit on only one small cushion so others behind us in the packed hall could also see Papaji. We happily adhered to these two rules and made sure everyone in the hall could see him.

I was surprised to hear that, after arriving, Garimo's husband had informed some of the devotees that he was also enlightened.

This supposedly enlightened man entered the hall with three large cushions and plonked himself on top of his stack of pillows at the front of the hall, as near as possible to Papaji. He was obviously infected with the flu as he kept coughing and blowing his nose during satsang. Everyone sitting directly behind this fat, thoughtless, "enlightened" man had to move to either side of him so that they could also see Papaji. I watched as he turned around and saw the empty spaces stretching all the way to the back of the hall. He just smiled as if this was his sacred right because he imagined that he was enlightened. He continued to sit in his elevated position coughing and loudly blowing his nose. Does enlightenment in his case mean that you are so special you do not have to take any normal person into consideration?

This extremely self-centred man also had no consideration for his master's delicate age, health or wellbeing. I can say without any shadow of a doubt, from my later meetings with him, that he was the most self-centred individual I have ever met. He actually believed that he was enlightened and therefore above others.

Albert Einstein said: "A human being is part of the whole, called by us 'Universe', a part limited in time and space. He experiences himself, his thoughts and feeling as something separated from the rest – a kind of optical delusion of his consciousness."

Einstein thought we had to free ourselves from this prison.

One thing I have noticed in my meetings with many different gurus and other master teachers is that they are relieved to meet people who do not worship them or expect them to be perfect. When I speak to them as an equal human being they gratefully respond as an ordinary human being.

Is this why Garimo, who seems like an intelligent and nice lady, is with this man? Is she sharing her honoured teaching stage with him in order to save her marriage and to have a companion who does not worship her?

Anyway, what follows is a small taste of satsang with this wonderful old man, Poonjaji.

Satsang with Papaji: The first for 1993

About 200 people from all corners of the world are seated in an Indian house especially prepared for satsang. Papaji, or Poonjaji, a now 81-year-old disciple of Ramana Maharshi (the now-dead enlightened sage of Arunachala), has himself attained mastery, and is seated cross-legged on a couch in front of us. His eyes are closed and he has been sitting silently for the past half hour, as have most of the people in this hall.

He opens his eyes and begins with a short chant of the sacred word Aum. We all join in. He then speaks, beginning with a short prayer.

"Let there be peace and love among all beings in the Universe. Let there be peace, let there be peace. Aum, Shanti, Shanti, Shanti. Namaste! Welcome!

"Happy New Year. We are meeting after New Year, here. May this year bring you peace, love and freedom. This is my wishes for the New Year blessing." (As he often does, he then giggles.

A wonderful sound.)

"There are many people who I have not met so I'll be seeing them after satsang if I have time because there are many questions. So I will straight away begin with the questions because some people are leaving tonight. Therefore I will attend to their questionnaires." (He giggles as if this is the greatest joke in the world.)

(These questions are written by seekers in the audience and are placed on Papaji's table before satsang begins.)

He picks up a letter and reads it to us:

Q: "What to do when you desire to sleep with the beautiful woman?" (Everyone laughs.)

A: "Many people are guessing. Many are guessing. [Giggles.] Nobody will relish this answer, I am sure. As this letter is maybe the German version so this is the German answer. Avoid it." (Laughter.)

Papaji turns to a woman and asks her: "You are German. You don't like this answer?"

Lola: "It is too holy."

Papaji laughs.

The next question is also from a German woman who asked about love.

A. "This happens when we feel oneness with everybody. All the beings. We can do it. Everyone. Everybody can do it. [Looking at us.] How? Because the dress that we wear is not me, is not we. The person who wears the dress is more important. Likewise, the bodies are also the dresses. So who is wearing the body? And the answer to that is 'One'.

"So if you come to this conclusion, bodies may live and bodies may go and there is no difference at all. Because That is one essence. And at the same time That can wear millions of bodies and has millions of tongues to speak. If you identify yourself with that One. From where it rises. Time, mind and consciousness. So from there you can recognise that It is speaking all the languages. The birds are chirping. The tigers and the waterfalls are roaring. And the breeze that whispers into the eyes, into the ears, into the plants. He finds himself so identical with It that he doesn't find any separation. So if we go to that essence of stratum from where everything arises, then you can speak, you can do whatever you want. All the activities rise from there and you feel it is you who is speaking through all the mouths. And it is you who is active through all the activities of the world. And it is not difficult. You can do it here and now. How? Just searching, where does the physicality arise. So go on going towards It. Go on going where this thought takes you too. Where the thought arises. The first thought. Where does the wave arise from? Follow the wave. Where does it dissolve? And you will discover something vast from where the waves come. So all these manifestations are just the waves which are rising from within. Within. A very secret, sacred cave of your own heart. And you are this. When you hold something else, your mind takes you to hold any object, any name, any form; then you forget who you have been and who you are. Therefore once again return back to your original status, your own kingdom. And don't beg. Throw away the begging bowl. The begging bowl is desire, intention, hope. This is the begging bowl. Desire and hope. So if you get rid of desire and hope you are back home. Instantly you are back home. Instantly, you see, we become one. All is one and it is all. There is no doubt about it.

"And you have to identify yourself, from wherever you are. Return to this centre. Wherever you may be situated on the circumference doesn't matter. You are to rush towards the centre. And the centre is common for anyone who's in the circumference.

"So you can do it. In this instant. Don't postpone it. So postponement is suffering. Postponement is called mind and postponement is called time also, you see. So to get rid of this and arrive at eternity, happiness, bliss, beauty and love, you have got nothing to do, only give up your old habits of desire and hope. That's all.

"Desire is taking you out of the kingdom. [He giggles.] And this desire itself is a begging bowl which cannot be fulfilled. Which is never full. Can never be full at all you see. So once you throw away the begging bowl, instantly you are emperor." (Giggles.)

So this was just a little of what occurs in satsangs and is a hint at what many of the masters I had met, and would meet, were trying to tell me.

Listening to these conversations and Papaji's wisdom forced me to wonder if Westerners were missing some simple basic facts about meditation? Is it meant to be easy and always enjoyable as a serious meditator once told me? Doesn't struggle make us more humane, more compassionate? Will suffering, disease, or chaos, speed up the process of destroying this egotistical separation of ourselves from animals, nature and other people?

And do disciples or blind followers who struggle for bliss and enlightenment through seemingly endless meditation, chain themselves to a useless cause through grasping onto a belief about eventual outcomes?

When we are in this spiritual atmosphere with a person like Papaji who has found his unique particular way to the source, then it is easy to feel this "no mind" he continually points to – to feel this high, this spiritual ebullience, excitement, gratefulness. Yet when he leaves and we go to the restaurant, or to a party, it is no longer present in us and we long for it again. Is this why people become addicted to being with gurus, holy men or priests, for ten or fifty years and never find their own unique way to their own centre or the light these wise men speak of.

The following quotes come from Papaji and are worth mentioning:

1. Whatever comes, let it come. Whatever stays, let it stay, whatever goes let it go.
2. If there is peace in your mind you will find peace with everybody. If your mind is agitated, you will find agitation everywhere. So first find peace within and you will see this inner peace reflected everywhere else. You are this peace.
3. You are the unchangeable Awareness in which all activity takes place.
4. That ocean of eternal peace is you. What is the difficulty that we suffer from? It is that we seek peace elsewhere and do not experience that we are peace incarnate itself.

5. Look within, there is no difference between yourself, Self and Guru. You are always Free. There is no teacher, there is no student, there is no teaching.

6. When you stop thinking, you are no longer imagining things that are not true.

7. Now is the time to have a direct introduction to this moment. This moment is free of time, of mind, of any notions. Introduce yourself to this moment.

8. Let your will burn in this fire so that it takes you nowhere else. Let your self be burned in this fire of eternity, love and peace. Don't be afraid of this fire, it is love itself. This desire for freedom is the fire of love.

9. Detach yourself from those things, which are not permanent, and rest in that which is always permanent.

10. Surrounded by nothingness you have to do nothing. You have to do nothing to be who you are. Nothing at all.

11. When you know that you are meditating it is not meditation.

12. You have to simply become aware of the movement of the mind, which begins this endless trouble.

13. Nobody wants to work. Not even the breath wants to work.

14. If you see the illusion, you are enlightened, but if you think that you are enlightened, you ARE IN the illusion!

15. One is always free and one is always alone. The mind is only dreaming.

16. If you identify with what comes and goes, you will be unhappy. If you identify with what is permanent and always there, you are happiness itself.

17. Look here and know who you are. Instantly you will find freedom, and this suffering – hitchhiking from womb to womb – will instantly stop. The only way is to look within.

18. Meditation is to effortlessly turn the mind towards that energy which energises mind.

19. Keep quiet. Don't touch the thoughts. Let them be.

20. The way to live a happy beautiful life is to accept whatever comes and not care about what does not come.

21. If you address the mind with "Who are you?" it will disappear.

22. What is the purpose of human life? To return home.

23. Understanding and not understanding are all in the scheme of ignorance, just a realm of the mind. This is not learning. This is your birthright. You cannot study to be who you are. You do not need to understand in order to breathe.

24. Find where the ego rises from and it will disappear.

25. Thoughts are impediments to seeing your own face.

Don't give rise to any thought, and discover who you are.

Matthew Scott Donnelly.

I spent more time with Papaji in satsang, and sometimes watched the cricket with him, together with reading papers and being ordinary. Occasionally he spoke to me about the stillness within and the great peace found there. These things, he told me, are found through awareness.

This would become a somewhat common theme with the spiritual teachers I would soon meet and serve, but not all. Money, sex, worship or ego have defeated a lot of once great teachers.

Then it was time to go, so Cheyenne and I rode all the way back to Poona enjoying old India. From there we flew back to Australia.

After a few days spent getting over jet lag, I began work on the wharf, loading and unloading ships as a wharfie. The pay was good and I soon saved enough money to continue my search and it was back to Esalen Institute in America for me.

I still wished that I could settle down in a nice house with a wife and children, but my body would rebel and become rigid after six months at home and it was obvious that I had no choice but to leave. The one time when I was determined to be "normal" and not leave home again, I lasted for twelve long months. But then the doctors and a surgeon said that I needed an operation on my neck or I would become a cripple. I chose to leave instead. As soon as I left my ordinary life, all the crippling symptoms disappeared. I was becoming aware of the powerful connection between mind and body. Between mind and the physical.

Somehow, it seems, I am not entirely in charge here in this body and yet again I wondered what part I play in creating this life I seem forced to lead?

What is my mind connected to that transforms my body and guides my life?

Esalen Master Teacher and friend John Soper.

Chapter 32 - Esalen

I flew to the USA where Jenny (Jenny and the Nine) and her husband Russell picked me up at the airport in San Francisco and drove me to Esalen Institute in Big Sur, which is about halfway between Los Angeles and San Francisco. Jenny was still the toast of Esalen and was consulted on its everyday running by some staff members and the wonderful Dick Price, who had founded and still ran the place.

Another wonderful soul named John Soper gave me a job as a work scholar 2 in his garden. A work scholar 2 is not paid but is given free board and food at Esalen in return for work.

I continued to meet famous people such as John Lilly, the dolphin man. He was a strange, ethereal type, dressed in a flowing robe, who took strong drugs so that he could communicate with dolphins.

I worked in the garden during the days and attended workshops at night, and on my days off.

One session that I shall never forget happened with Dick Price who was also a gestalt therapist. To put it very briefly, gestalt is a therapy where the therapist does not interfere with the patient and lets the patient create his own reality. Fritz Perls invented this therapy while living at Esalen.

Though I had tried gestalt with other Esalen therapists, it always bored me because the therapists would interfere with any process I was approaching. These practising therapists would surrender to their own overpowering longing to help me and be my saviour.

Dick, however, did not interfere and let me create my own reality. My body seemed to float up into a huge silent place which had no ending or beginning. I experienced the feeling of vastness throughout my entire body. Dick had allowed my mind to dissolve and merge into the infinite space that is outside and also inside my body. This still, mindless place has been spoken about by gurus such as Papaji, Kiran and Osho. This place is beyond any idea of any gods or thoughts. It is beyond thoughts of great love or any other thought either positive or negative. Judgement, interpretation or creating, no longer existed.

After the session I felt light and empowered and I experienced, totally, my every little movement as I took my leave. This state lasted for days. What an amazing human being this Dick Price was.

After this session both Russell and I began hiking into the magnificent Big Sur Mountains with Dick whenever we could. He loved to hike and he enjoyed our company. Though we were both fit from playing Rugby Union football both in Australia and America, we still found it difficult to keep up with this super-fit older man.

The last day we hiked with Dick and his dog, he led us to the water source of Esalen, which he had discovered and harnessed in the beginning when he was establishing this mystical retreat.

The source was a deep waterhole in a fast-flowing creek that was high up on the mountainside. From this place he had channelled the endlessly flowing sweet water to Esalen. Along the banks of the stream were huge cliffs made of pure white rock. We often sat on the bank of this source and meditated with Dick. A great quietness came over me as I sat with this amazing man who could still my monkey mind just by his quiet presence.

A couple of days after our last visit to this sacred waterhole I saw Dick walking away from me and thought it was Russell. I raced up and grabbed his shoulder and said, "Hey, Russ." I got a head-wrenching shock when he turned and it was Dick.

He answered with a smile, and said, "Nearly." I wondered what he meant by that?

The next day Dick hiked into the mountains without his dog and without us. He did not return that night and his beautiful wife, Chris Price, became very worried. She sent for me and told me that Dick had been finalising things for the past weeks and handing over control of Esalen to other people, as if he were about to leave.

No one had been able to find him but I felt certain where he could be found. We hiked up to the source of Esalen and there lay dear Dick. A white rock as big as a VW car had broken off the side of the mountain and had hit Dick on the head. The only wound he had was over his third eye and he was dead. Only his head was submerged in the water where it had rolled into the source of Esalen. It seemed to me that something as profound and unusual as this was simply meant to be and was no accident.

He was carried lovingly, and sadly, back to Esalen.

A huge wake and funeral soon followed and people came from far and wide. Everyone loved this man dearly and respected what he had achieved by establishing Esalen, this place of magic where huge leaps of consciousness happened.

Garimo was there and Gabrielle Roth, just two of the many famous people.

For the next three days, a strange and eerie phenomenon occurred. A white fog descended over Esalen that was so thick I could not see more than a couple of feet in front of me. I was working alone in the garden and felt that Dick was right there in this eerie fog, and a part of it. I realised that he was part of existence now and forever. Is that why, when I thought he was Russell, he had said "nearly"?

That unique and unusually thick fog often sent goose bumps and energy racing through my body.

His wife Chris believed that he somehow felt that he was leaving his body and that was why he had finalised everything.

The following is an excerpt from an interview with Dick.

WH: *Try to picture someone who knows nothing about gestalt. How would you describe what you do?*

Dick: *I would have to say what I don't do because Gestalt is not a doing. What Fritz called doctor/patient, that dyad, I refer to as reflector and initiator. The initiator is the person who formerly was in the "patient" role. My function as reflector is simply to be available to reflect and clarify whatever comes up in that person's process. So I'm never defining how a person should be. I'm available in a particular way – a mirror is a good analogy. The initiator remains responsible for his or her own experience.*

This is very unlike standard psychiatry, where you're put, if not in a jail cell, certainly in diagnostic pigeonholes.

Dick was also very close to Jenny O'Connor who channelled the nine at Esalen.

The following is a small part of a session she had with Dick Price and Russ Lee.

Russ: The first I want to talk about is schizophrenia. I would like the Nine to define "in-blink" and "out-blink".

Nine: Definition of something outside viewerage is difficult. However, total in-blink is a part of whole consciousness. Out-blink, a part of exterior world and/or so-called reality.

Russ: What are the various causes of the schizophrenic response?

Nine: "Causes" implies it does not happen to all, and it does in varying degrees. For the schizophrenic who cannot handle it, it is generally family situations and society's restrictions causing mistrust of his own process, resulting in the painful schizophrenia.

Russ: Why are schizophrenics often paranoid?

Nine: Paranoids? There are many of them psychic. They trust energies, not smiles.

Russ: So their paranoia is accuracy?

Nine: Absolutely.

Dick: Always trust your paranoia!

Dick did not say much in this session but he was a man of few words in my experience.

As for Jenny, I believe she used this concept of the Nine merely to get out of her own way so information could come through her. Worshippers of gods sometimes manage to do this as well.

However, the mere fact that so many rich and wise people, from all over the world, believed that the Nine were real left me shaking my head in shock and wonder. California produces some strange things and stranger people.

Basically, Jenny was an honest, good-hearted woman with little education. She may have known unconsciously, deep inside, that there was something not quite true about what she was doing, and this may well have caused her to take drugs, smoke cigarettes, drink gallons of Coke, and eat all sorts of rubbish, with no concern whatsoever for her body and its functions. She was in a state of nervous fear because of her difference to other people, it seemed to me.

I tried to explain these things to her many times, but nothing seemed to get through and she continued to smoke those awfully strong French cigarettes, and drink Coke, until the day she died.

I would also tell her time and again that eventually she had to pay "The Ferryman", as we all must for how we live our lives. Which simply meant that when she died, she would have to pay a terrible price for her drinking, smoking and pill popping. A price of pain and suffering, unimaginable.

Jenny and Russell eventually came to live with me in my home in Australia until I found them their own home in Byron Bay, where she raised her two sons.

She died recently in the Lismore hospital. Her arms were almost as skinny as the size of my thumb. Her body was wasted away from cancer and her stomach was so bloated she looked like she was pregnant. She did not die easily and lingered for weeks with an oxygen mask. She could not speak but would move her finger when I asked her questions.

In the last days the hospital staff stopped giving her water, therefore condemning her to eventually die of thirst. Why do they do this? Because they are not allowed – thanks to the religious people – to do the heartfelt humane thing, which would be to give her a tablet and let her die in peace, free from the pain and this horror she was being forced to undergo. Nobody deserves this.

I could not stand to see her die of thirst and so I kept giving her water. Whenever I asked her if she wanted a drink, she would raise her little finger as she could no longer speak and this was our communication for "yes".

She eventually died with her family by her side. What a relief death must have been for her in that poor wasted and diseased body.

It was the same for my mother who smoked her whole life and took tablets for everything and anything, that doctors gave her. At least the hospital staff in Queensland gave my mum overdoses of morphine, and then another drug to stop the involuntary spasms she was having, caused by the morphine. These overdoses, thankfully, finally killed her. Thank goodness, as it was cruel to watch her die so slowly and painfully. We do not let our animals die like this. We put them out of their misery with a bullet, which is much kinder. As humans we should be smart enough to legalise a tablet like they do in Mexico and other more humane countries of the world. But Australia is still ruled by the Christians both in and out of politics. I know of no elected Prime Minister who does not go to church on Sundays and claim to be a Christian.

And, still in this modern time, the "believers", the spiritual know-alls who actually think they know what a soul is, deny us the right to die in peace free from unbearable pain and horror.

Anyway, please consider this. Each individual must "Pay the Ferryman", eventually, if they choose to abuse the body every single day with either drugs, tobacco, alcohol, or even legal tablets doled out by generous doctors who are encouraged to do this with gifts from the huge pharmaceutical companies. These companies mass-produce these often, debilitating drugs which are poisonous when taken for too long. Even common drugs such as Valium and anti-depressants, to name just two, eventually create very bad reactions if taken for long periods.

The companies who produce these drugs eventually find out about the anger and trauma they produce so they change their names. Valium is now called Diazepam for instance.

Is the reason we use these legal and so-called illegal drugs (even though tobacco, the government approved legal drug, kills more people than all illegal drugs combined) because we have been traumatised as children as the psychiatrists tell us, or is it simply because we prefer to remain unconscious so we can use our damaging little sugars and spice to give us a break from the lives we are often forced to live in our current unbalanced societies ruled by religions and money? I do not know quite yet but I suspect it is the latter.

I stayed on and helped out at Esalen in any way that I could after Dick's death.

When my visa expired it was time for me to return to Australia once again.

Original Australians straight out of the bush.

Chapter 33 - Motorbikes & Deserts

Back in Australia I was again working two jobs so I could save enough money to continue this unrelenting search that seemed to keep from me living in one place for too long.

After a few months of saving, I decided to ride my old Goldwing motorbike (often called a "Leadwing" because they are so heavy) into the centre of Australia where I would do a seven-day fast in an isolated part of the deep central desert. Perhaps there I would learn something different and not have to go overseas again, or so I thought.

I rode the thousands of kilometres out to Alice Springs and prepared myself for a seven-day fast. I did this by cutting out all sugar, alcohol, coffee, bread and dairy, eating only salads and fruit for a couple of weeks.

Alice Springs was a lonely place for me because I did not meet or talk to anyone. I set up my camp just outside of town on the dry, isolated riverbank.

Ancient Aborigines, visiting Alice after walking vast distances in from the desert, often strolled past my camp. I ignored them because they were often drunk. Fortunately, they also ignored me.

When I felt ready for my seven-day fast, I rode directly due west of Alice for a couple of hundred kilometres. I then took a small unused dirt track which led to the base of a mountain range where I knew, from my maps, that I could find water, even in this desert country. I needed plenty of water because I could only carry a couple of litres, together with my tent, bedding and all my gear, on the motorbike.

Riding towards this isolated range, even though I had let some air out of my tyres, I nearly fell off the bike in the deep sand. If I had fallen and this 1800cc "Leadwing" fell on top of me, then this is where I may have died. Nobody comes this far into the outback, phones do not work, and I would not be able to get the heavy bike off me. And I did not have an iPhone in those days. I don't know if they even existed.

Also, if I were to get bitten by one of the many venomous snakes such as Taipans and King Browns, which are common here, then this is where I would also most likely die, alone in the vast Australian outback.

I faced each fear as it arose and rode off the track and across the sparse dry land. I finally came to a waterhole on a creek that was unfortunately not flowing at this time of year in this nearly always arid and dry country. I searched along the bank of this isolated waterhole and the water looked clear and free from any dead animals. This is where I decided to do my quest as finding a fresh water spring would take me deep into the mountainside, followed by a lot of digging.

I had brought no books, radio or writing paper. On a quest of this nature all distractions must be left behind, according to the wisdom of the Red Indians of America and Buddhist monks plus Aborigines on a "Walk-a-bout" journey.

Each day I practised the sitting and walking meditations which I had been taught in the Buddhist temple in Thailand many years ago. Sometimes I simply went for long, slow and mindful walks. One day I became disorientated and was totally lost in the mountains. It took all my bush skills to find my camp after four hours. I was able to remember what my Aboriginal and white families taught me when I was young. I took note of the lay of the land and the way the gullies ran, which way the wind blew, etc.

That day lost in the bush the thought of death often assailed me because if I remained lost no one would be coming to save me. I pushed these thoughts aside because I needed to remain calm and centred to find my way back to camp.

Nature in this so-called desert was prolific, especially among the trees growing on these mountain ranges. There were birds, butterflies, insects, wild horses and donkeys, pigs, fish, and all sorts of hidden foods if you knew where to look.

My Aboriginal uncle on the family cattle station had taught me well, when I was a kid, and I still recognised lots of edible foods. Out here, where people sometimes die of starvation, there are bush bananas, edible weeds and flowers, a type of apple nicknamed Chinky, a type of potato, and many other interesting plants that can be eaten. Of course, I did not eat any of them, as I was true to my quest, but it is fascinating how food, which I would never see otherwise, revealed itself when I was starving.

As each day passed, I experienced different emotions exactly like I had in Thailand when I was training to become a monk. Here they arose faster because of the added pressure I was putting myself through, surviving in complete isolation with no food whatsoever.

One day I yelled and screamed at all of existence as I became consumed with anger. The next day I felt calm and loving towards all people and felt the wonder of existence.

The days seemed to take forever to pass with nothing to amuse or distract me.

I drank lots of water from the waterhole after my own water ran out.

I have never been a person who allows myself to see things that are not explainable although I did run away from the imaginary cobra at the temple in India. It did seem real at the time. However, one day I was sitting in my camp and I looked up and saw a small round object hovering in the sky above me. It had no propellers or engines and made no noise. It just hung suspended in the sky for a few minutes and I was mesmerised, with my mouth hanging open, wondering if it was going to land.

Suddenly it shot away at a truly amazing speed and was gone in a flash. I did not know what it was and so I did not dwell on it.

On the seventh and final day I took a walk around my large waterhole to say goodbye and thank it for keeping me alive. On the other side of the river, I was shocked to see a completely submerged cow that must have been there for a long time as it was coming apart and being eaten by fish and water insects. I wondered how I had missed seeing this when I arrived. I hoped that the rot had not contaminated the entire waterhole because I had been feeling slightly ill all day. I realised that this rotting cow could have been the reason why.

On the eighth day I packed up my meagre belongings and rode out. On the way I dropped the bike and quickly stepped off it as it came to rest facing at a 45-degree angle down an embankment. Petrol was pouring out of the tank. This was just too much to bear after all that I had put myself through. I screamed as a white-hot anger consumed me. I then grabbed the bike and lifted it back to a totally upright position on the track. I do not know how I did this as a fully laden Goldwing is impossible to pick up and weighs much more than I can move and especially not from a 45-degree angle facing down an embankment.

I had amazed myself with the hidden, unnatural power we humans can sometimes exhibit.

After a long hot ride in the blazing sun, on the narrow and dusty dirt road, I finally came to a little shop, which was about 128 kilometres west of Alice Springs. A lady came out to the counter of this small shop and asked me what I wanted. I told her in a quiet, sickly voice that I would like an ice cream on a plate with a mashed banana. Now this lady must have thought I was some sort of drugged idiot. She replied in a loud angry voice: "This is not a bloody al la carte restaurant, fella."

Now normally I would have probably told her to stick her shop where it hurt the most and I would have stomped out, but the fasting week, while drinking polluted water, had done something to me and I was feeling totally vulnerable and not in the least reactive or aggressive. I said softly, "Oh, I am so sorry but I have not eaten anything for a week."

This angry lady looked at me aghast. A tear formed in her eye, and she turned and walked out into a room at the back of her shop. As I was slowly getting up to leave, she returned and said, "Wait!" In her hand she had a plate with ice cream and a banana, and she sat me down, mashed the banana into the ice cream and spoon fed me. I was amazed and grateful, even though, as I now know, ice cream is not a smart way to break a fast.

I thanked her profusely and she flatly refused to take my money even though I tried to insist.

Riding back into Alice Springs I had to keep stopping as I was quickly growing sicker and weaker.

After a long time with a lot of stops I finally arrived back in Alice and some unusual meetings immediately began to happen.

As I was walking slowly down the street two young English girls stopped me and said, "I know this sounds strange but we are going to lunch and would like you to be our guest." Perhaps they also felt my vulnerability, as only women sometimes can. Unable to refuse such a kind offer, even though I was not hungry, I joined them for some mashed vegetables and told them of my recent quest.

When we hugged and said goodbye, I slowly walked down the main street of Alice Springs wondering where I was going to stay and what I was going to do next. I was now far too sick to again set up camp in the bush.

I stopped and stood gazing at a beautiful looking pair of hiking boots in the window of a shop. A woman came out of the shop and asked me if I liked those boots and I said yes, so she invited me inside to try them on. They fit perfectly so I asked her the price. She said $300, which was more money than I had so I apologised to her and said that I could not afford them.

This lovely woman then said, "That's all right, dear, you can have them."

I was taken aback and said, "No I cannot do that. You will get the sack."

She smiled and said: "No I won't. I own this shop."

I walked out of there wearing a brand new pair of very expensive boots and left my old worn ones with her.

Nothing like this had ever happened to me before and I wondered what had actually transformed me out in that desert.

Continuing on, and feeling sicker by the minute, I wondered if I perhaps did need to have the inner strength left to set up camp by the river. Suddenly a good-looking blonde woman, whom I had once briefly met in Cairns at the hotel where I had been working a long time ago, stopped me in the street and said: "Fancy seeing you here. You won't believe this but I am flying to Canberra for a week and I have to leave today. Would you like to look after my home for me while I am away?"

I was once again amazed by this synchronicity reinforced by such trusting generosity. This pretty woman hardly knew me but she trusted me to look after her home.

I recalled the words of a wise man who once said: "The meek shall inherit the Earth." Well, I was feeling meek, mild and vulnerable, so these women must have been able to somehow feel and see this, as only women can, and then offer me what I needed. Women are love because they can feel so deeply.

Life revealed some amazing things at that particular time, when I was not dulled by sugar, spices and too much food. My mind was empty and clear as I had not read a paper or seen the news for weeks. I felt I was discovering how miracles and manifestation really work naturally.

After deep contemplation, I realised that when I chose to fast in the outback to find the answers I had been searching for, I was able to experience what all the masters were attempting to teach. Not only my body, but my thinking mind was forced to fast. With no diversions to inspire thinking, no desires possible to fulfil, and no ego to protect, my judgemental and fearful mind no longer operated. I had sacrificed all control over my body and mind. I was forced to surrender to nature. Nature had become my teacher, not an enclosed meditation room in a temple.

The world no longer existed for me. All actions were an immediate response to meet the needs of my now surrendered, innocent state of being. With this sacrifice my vulnerable mind was reconnected into my heart awareness where all needs were magically being met. My heart was unknowingly connected to the hearts of others. The feminine hearts responded to my needs without the masculine censorship of analytical judgement.

I had just fasted away my judgemental mind by disconnecting to all worldly news, people, and spices.

One of the things that I was now aware of in the town of Alice Springs was the plight of the simple Aborigines who had come here straight from their isolated desert wanderings.

It brought tears to my eyes to see them drinking and fighting while stuffing themselves with the white man's diet of alcohol, pies, fried foods, tinned foods, coke, sugar, and everything that their poor bodies were not used to. They were slowly dying before my very eyes, and no one else seemed to see the obvious reason why. My heart went out to them and in that moment, I realised that they were a reflection of me and I was also killing myself with these foods, if only a bit more slowly than them.

I felt such overwhelming love for these bandaged, battered and dying Indigenous people. I could see halos hovering above their innocent, childlike heads.

I also noticed the exquisite beauty of the trees and the birds on their branches singing the songs of nature.

I was aware of the death that was happening to me as I re-entered the normal world. During the next week I lay in the woman's luxurious home hardly moving. I was suffering terribly with fevers caused by a parasite named giardia which a doctor told me I was infected with. Aching, vomiting, heavy sweating and diarrhoea, plus a high temperature, all assailed me for five long days until finally my body shook it off.

Feeling half-starved I then watched myself becoming like the poor Aborigines, from whom I had obviously learned nothing. I stuffed myself with all the foods that took me out of the wonderful state I had been in. I ate cakes, fried fish and chips, croissants, and take-away fast foods, until finally all the magic was gone and I was ordinary once again.

Perhaps in the future when I was stronger, I would also use this knowledge and experience to become "meek" once again by choosing what, when and how I would eat. The only real difference at that time between these "Black Angels" and myself was that I was not drinking alcohol and smoking tobacco.

This episode in my life reminded me of the story in a great book called Dune where the people on a faraway planet were kept dull and controlled by their rulers feeding them "spices" similar to our salt, pepper and sugar. A great book that is worth reading because even though it is about another planet in the universe, it is really about life on earth. Satire at its best.

I eventually regained the strength to ride the thousands of kilometres back to Queensland, where I quickly went back to working two jobs and saving more money for a time when I would have to continue the search, where, from experience, I would soon feel nervous and lost once again. I did vaguely wonder if my experience in Vietnam kept me in a state of fear, always prepared for the next fight. But at this stage of my life, I did not cling to such thoughts nor give them the strength of truth.

Where would I have to go next? Who or what would I find to teach me? Existence seemed to answer these longings and eventually guide me. I don't know how that works but it is amazing when it occurs. I only need to trust existence and surrender until the next wave reveals itself.

Kiran

Chapter 34 - India Again

Osho had by this time been kicked out of America. You cannot dress in long robes with a mala around your neck containing the picture of your guru, while also wearing large pistols on your hip and carrying machine guns, and then expect to stay in America. It was obvious to me from the beginning that they would be kicked out. Why could they, the Orange People, not see this? Drunk with worship? Feelings of being special perhaps.

Belief, once again, was like a loaded gun waiting to cause trouble.

I decided to go to the ashram in Poona for a brief visit while I picked up my Enfield motorbike to ride off and visit Ramana Maharshi's ashram in the south of India.

Four newfound male friends from Sydney, Australia, flew to Poona with me and this had enhanced and made my journey easier. They actually talked me into going with them as I knew the way.

The ashram had become a bit like my platoon in Vietnam in the sense that I did not fit in at all. All these followers were telling me that Osho was "enlightened". What in hell do they think they are talking about? If they have no experience of enlightenment and are not enlightened themselves, how can they possibly know that he is enlightened? Why even bother to mention the word? Like Christians or Muslims, have they just surrendered their uniqueness to someone apart from themselves. In this case, an outside "God Man"? Someone who they believe is separate and better than themselves? Had they sold out, just for a feeling, and to fit in with a crowd, just like the new breed of Christians do as they dance madly in their new churches?

I knew I was not one of them because I could not worship Osho and so I did not fit in. It was time to leave on my own again, but I planned to go to one last satsang with this unique man who had achieved and done so much. To me he was just a friend. Even if a sometimes crazy and foolishly brave one.

That night, my friends and I sat at the back of the great Buddha Hall just in case I had to escape from another long and boring satsang.

Buddha Hall was packed with people, shoulder to shoulder, like sardines in a tin.

When he finally appeared after a long delay, Osho was acting reluctant to come into the hall. He would walk in and then walk out. Everyone screamed his name to bring him back, until he finally relented and sat in his chair.

He gave a short, intelligent talk that night, interspersed with well told jokes followed by the crowd's loud laughter. At the end he stood up, the music played, and he once again directed us all in the dance with his waving hands.

Something came over me, some deep knowing, and I became very excited. Buddha Hall was covered by a huge, thick plastic type of circus tent and at the back where we were standing there was a cement pillar about 4 feet high that the base of the tent was shackled too. In my excitement I jumped up on to this pillar. This was not allowed of course, and a guard immediately came over and tried to pull me down. I simply twisted his arm and pushed him away, and he immediately gave up his futile quest. I then threw up my hands and began hitting the huge side of the tent behind me with my arms, in time with the music. This sent enormous ripples up and down the tent's side. Osho noticed this and laughed, pumping his hands even harder as he looked for a long time straight at me and the tent flowing in great waves. It was strange how he always recognised me and allowed me to do as I pleased.

I then excitedly began hitting my friends, standing in front of me, on their shoulders and yelling: "Look closely at this man for you will never see him again."

One of them turned to me and said, "Why will we never see

him again? Where is he going?"

I did not know where he was going but I knew without a doubt that we would never see him again. I have no idea how I knew this. I yelled again at my friends: "I don't know where he is going but we will never see him again."

As it came to pass, he died suddenly soon after and never came back to give satsang and so we never did see him alive again.

The end of an amazing era had come. And what an experiment it had been. Sannyasins had learned how to surrender in dance, how to forgive and forget quickly, and they had learned a lot about breaking the moral barriers of sexuality. They had also learned many aspects of self-awareness through the group experiences held in the ashram and all over the world. They attended seminars and workshops with titles such as Tantra, No Mind Meditation, Breath, Return to Childhood, etc, etc.

One of the breakthroughs that I had definitely noticed this enigma of a man achieving while he was alive was weakening our KNOWING minds. One day he would tell us, for example, that this something was "black". We would go to our rooms thinking: "Wow, I've got it and I feel great."

The very next day he would come out and say, no this something is definitely not "black", it is "white". He would often tell us the exact opposite and confuse our thinking minds.

And then, to confuse our minds even more, the very next day he would come out and say that what he previously said is neither black nor white. It is most definitely "no colour at all".

I am only using colour as an example of opposites.

This would keep our minds in a state of confusion and I imagine that this is exactly what he meant to do. The thinking, knowing mind, after all, can be boring and repetitive and one must not bore God, so to speak.

Once again, I soon left the ashram for a long period of time and rode off into the heart of India.

When I finally returned to the ashram, a woman I knew told me that I must come with her to visit a man named Kiran as he was enlightened. Not another one, I thought. I refused point blank and said I had had enough of so-called "enlightened beings", and their followers, for one lifetime. She was adamant and insisted, over and over, every day, so after a week of this annoyance and because we were good friends, I finally relented and drove her deep into the suburbs of Poona to Kiran's fancy house.

This man, Kiran, was a successful businessman who owned some sort of factory in Poona that employed a lot of people. He had been a sannyasin with Osho from the very beginning and Osho had picked him out for special treatment. One of the things he had ordered Kiran to do was a meditation called "Dynamic". This meditation was almost like training for a professional sport as it was very physical. People would jump up and down on one spot for fifteen minutes, doing fast chaotic breathing through the nose, then roll around while catharting and screaming for fifteen minutes and then jump up and down yelling "Who, who", with their arms held painfully in the air. At the end of an hour, they would be soaking wet from sweat and quite exhausted. The final part of this wild process was to sit for a long time in meditation.

Osho ordered Kiran to do this every day without fail for one year. I had done this meditation several times and to me this seemed impossible but Kiran did it even when he had the flu or was quite sick.

Finally, after many years, Kiran did what must be done by a follower with every master, guru, god or spiritual teacher – he left. Simply walked away never to return.

He retired to the large backyard of his home and sat alone for years. During this time, like most masters before him, he suffered greatly. Kiran's pain was in his body and it eventually became so great that the doctors rushed him to hospital and operated on him. They cut open his stomach, and the scar, which he later showed me, ran from high up on the centre of his rib cage all the way to his groin. They took his stomach out and inspected every organ and came to the conclusion that there was nothing medically wrong with him so they put it all back and stitched him up. When he finally went home the pain disappeared and he was in a different state of being and possessed the deep understanding that relaxation of the mind, after deep pain, can bring.

Soon thousands of people from all corners of the world began turning up for satsangs in his large backyard. It was there that he would give a talk and then answer any questions.

The day I was taken there, about three hundred people were sitting before this man who was dressed all in white. He was sitting on a slightly raised chair at the front of the crowd.

When I walked in, I stopped dead because I was amazed at how much he looked like my lovable Uncle Boof. I said, quietly addressing this word towards him in amazement: "Uncle." At the exact same time I spoke, looking at me, he said something in Hindi.

Slightly embarrassed, because I had interrupted and spoken spontaneously, I acknowledged him with a nod of my head and quickly sat on the floor.

He was answering questions and because of the accuracy of some of his answers I was sure he must have attended a school in the West, probably America I thought. I soon discovered that I was wrong because he had never left India. It seemed impossible that he could answer some of these questions without ever having had the experience of them. He had found his way inside to somewhere unusual, it was obvious, where all questions have an answer. Shades of Jenny O'Connor.

After the meeting was concluded he called me over and said, "Please join me and my wife Vinodini for dinner." During dinner I asked him what he had said to me when I walked in. He said he had called me "nephew". How amazing, I told him and Vinodini, as I had called him uncle at exactly the same moment!

At the end of dinner, he said that I must go back to my home, pack my belongings and move in with him and his family immediately.

Strangely enough I had no doubts whatsoever about this so I did as he asked.

When I returned, I was given the only room on the second

storey of his large home where he lived not only with his wife but also his grown children and grandchildren.

Each day, in the afternoons, he would hold satsangs and share his teachings. A strange thing that he soon did was to dress me in his own white clothes, place a chair next to him, and have me sit there during these meetings. This would make me very nervous and I would sweat, as all these people, who I did not know, would be looking at me and wondering who I was.

After the meetings people would sometimes ask me if I was also enlightened like Kiran. As usual I found this question to be absurd, as if they did not know what enlightenment was how could they even ask this question? If I answered yes, would they start worshipping me and immediately hold me up as somehow better than themselves?

One day I asked Kiran why he insisted on me sitting next to him all dressed in white during satsang? He laughed and said: "Practice for the future, Sahajo." This did come to pass.

After a month of these meetings, he told me to pack my bags because we were going to the south of India to see a famous guru named Sai Baba who could produce watches and jewellery, seemingly out of nowhere. This man had a huge ashram and a large worldwide following.

With my ex-model girlfriend of the time named Sadhana who had been Osho's door opener for Satsangs, plus a small, select following of Kiran's, we boarded a train bound for this famous man's ashram in the south.

I, as usual, loaded my motorbike on the train to travel with us.

After a night in a fancy hotel in Baba's large city, where we had disembarked from the old train, we hired two vehicles for the day and everyone set off for Sai Baba's large and packed ashram. I followed on my motorbike.

The cars were parked on the busy main road outside the high gates of the ashram while I parked my bike around the corner on a quiet side street and paid two young Indians to guard it for me.

We then made our way past many buildings inside Sai Baba's huge, and obviously costly, ashram. We finally came to a very large temple constructed of concrete. This man was already world famous and had a large following of both Indians and Westerners, all happy to donate money.

Magic, like sex, must draw the crowds.

As Kiran had been expected and was known, we were all ushered to a reserved row of seats. Kiran was seated in the aisle seat to await the great man's entry and introduction. I was seated next to him.

After a long delay Sai Baba finally entered. He did so with a flourish and was accompanied by a retinue of guards and assistants.

On a raised stage he spoke for a while. To me his message sounded like spiritual kindergarten after having already listened to Osho, Kiran, Jenny, and others.

Sai Baba then walked down the aisle, stopping at certain people and producing so-called "gifts from God" using quite impressive sleight of hand and magic.

He wore a large robe with long sleeves, and watches seemed to appear magically in his hands out of nowhere. He stopped next to Kiran, said something in Hindi, and produced a watch which he handed to him.

Soon the performance was over and we went outside. Kiran smiled and handed me the watch. It looked like a cheap Indian watch and I could not help but think that if God was producing watches, he would most likely produce Rolexes not cheap Indian ones.

A young Indian boy was looking at this watch with his eyes and mouth wide open, so I smiled and gave it to him. He was ecstatic. A holy watch.

Sai Baba did not impress me in the slightest and years later I heard that he was exposed for sexual encounters, with some young boys among his congregation of worshippers. This is enlightenment? Ah well, he was obviously enlightening people of their heavy purses.

Leaving the crowded ashram, we went out into the busy street and bought some fresh food from one of the many street stalls. We then took this back to the hire cars.

When we were near the cars two tall European men, one of whom was obviously French, approached us. These were two of the bodyguards I had noticed earlier who had walked into the temple behind their master. They were both at least 6 feet 4 inches tall and towered over us. The Frenchman was the largest and looked as if he weighed at least 105 kilograms, while his sidekick was thinner and probably only weighed about 95 kilos. When they were close the Frenchman said to Kiran, "You are a pretend master and you are not welcome here so take your sluts and followers and never return."

Sluts? Pretend master? What filthy mouthed, uncalled for insults. I began to move towards them and said: "What did you say?"

The beautiful Kiran smiled and said, "It is okay, Sahajo." Turning to the Frenchman, he said, "Thank you. We are going and will not be back. I loved seeing your master." These wise words placated them for the moment.

With that, Kiran joined the others in the car. He called my name and I told him to drive off and I would catch them up on my bike.

I was touched by how gentle and wise Kiran had quickly defused the situation. He had taken no exception whatsoever to what this foul-mouthed and aggressive man had said even though his wife Vinodini was with us. Following his example, I turned quickly, ignoring them, and walked around the corner to my motorbike.

These two men took it upon themselves to follow me, all the while abusing me and giving me further warnings. I was determined to follow Kiran's lead and just kept walking. For once I would control my temper, even though it was brought on by foul-mouthed threats which included the women of our party. I would let them learn their lessons elsewhere.

At my bike the Frenchman, probably because he thought I was scared of them by now, poked me with his finger and called me a weak little man.

What is it with these large bodyguards and their fingers? Why are big men often such over-confident bullies? Because they can get away with it? Because they are so big people are afraid of them and so they have never been challenged or had real fights? Because they run in packs? Anyway, I let it pass and did not take advantage of his mistake.

Once again, I was fed up with spiritual leaders and their followers. Why are jealous believers so sure that only their god, or guru, is the One and allowed to be worshipped? Fundamental believers will even stoop to killing, just to protect their gods and gurus and to spread only their beliefs. This "Saviour Mentality" has caused most wars in the past and certainly a lot of pain and suffering.

Anyway, the last thing I wanted was a confrontation with these two men, and I did not know how many of their friends were just around the corner so I tried to mount my fast bike and quickly escape.

Sadly, it became obvious that these two were going to take this further because they were now pushing me up against my bike. I hate being poked and pushed by amateurish fools, especially these who had so rudely insulted Kiran and the women with us including my girlfriend plus Kiran's wife.

The finger poking was becoming harder. My bike was pushed over. As I loved my bike and feeling like I finally had no choice, I grabbed the biggest man's finger and quickly bent it backwards. Unfortunately, his finger broke and he fell down screaming. I did not mean to break it. But he was big and obviously dangerous. I had allowed both of them to crowd me, so this had probably caused me to overreact.

His sidekick attacked immediately, and threw a swinging right arm haymaker of a punch at my head. I simply leaned back easily avoiding his telegraphed, cheap punch, and then hit him softly, with my punch only being thrown 3 inches, straight into his exposed nose. He fell down holding his nose and blood spurted out. He cried out in a funny accent, "You have broken my nose, you have broken my nose."

I leaned down close to him and said, "Your nose is not broken, you wimp, and you two can think yourselves lucky that I did not put you both in hospital for your dirty-mouthed bullying outbursts. Let this little lesson wake you up and make you better men."

With that I whipped up my bike and quickly departed.

I spent the night wishing I could have handled the situation as well as Kiran and not reverted to violence. When will I learn how to talk people out of fighting? Surely I could have somehow talked them out of hitting or poking me. I was sickened to the pit of my stomach. These men had simply shown me a mirror of myself. I was just a bully and a thug when I used my skills like that. Surely it was unnecessary. How could I have handled it differently? Did I become afraid when the abuse and the poking disturbed me? Did knocking over my prized bike upset me so much? I realised that I would ask myself many such questions over the next few days or I would get no peace.

Of course, I did not tell Kiran or anyone else what had happened because I felt ashamed. I moped around for days and the observant Kiran asked me more than once what was bothering me but I said "nothing" as my recent actions would have surely hurt this gentle, wise man.

We visited a couple of other spiritual places on the way back to Poona. There are spiritual places all over India.

Back in Poona I asked Kiran if he would come to Australia if I arranged satsangs for him there. He smiled and said, of course.

I spent the next two weeks gathering money from some of his wealthier followers and bought two tickets to Australia. One for the beautiful Vinodini and one for him. Business class no less, as I had collected a lot of donations.

It is interesting to note here that Kiran and Vinodini had been observant Hindus, as well as sannyasins, and had been strict vegetarians their entire life. They had never eaten meat, chicken or fish. Of course, neither had Osho.

A magical moment occurred before I left Kiran and flew to Australia to set up the satsangs. This experience has always remained with me. It is one of the many great lessons and revelations of my time with Kiran.

On my final day when I walked outside into Kiran's garden,

he was sitting alone. He had a look of divine ecstasy on his smiling face.

He was looking up into the sky. Without disturbing him I stood peering around the corner of the door. I followed his gaze and saw that he was looking at a jet-black raven that was slowly circling, high overhead. The raven kept slowly circling closer and closer to Kiran until it landed on the arm of his chair. He could have reached out and touched it. The raven looked at him for a while and then it flew away.

I then walked up to Kiran and asked him what had just happened with that raven. "Was it your pet?" I asked.

He said something like this. "Sahajo, when I was looking at that raven, I only saw beauty and was in a state of intense ecstasy. My mind did not arise and say to me, this is a raven, or this is black, or this is beautiful. Without the mind entering, the raven and I became as one and I became the raven."

It seemed to me that when the busy naming mind does not arise then this joyful communion with creatures, or even other humans, is possible. I call it seeing through the eyes of beauty and agape.

This was like what Linda Tellington-Jones was doing when she looked at, and communed with, creatures of any kind.

The very next day I left for Australia. Kiran and Vinodini followed a month later after everything was set up in Byron Bay where the satsangs were to be held in the old whaling station. This was a very large wooden building and could hold a thousand people. It was also where thousands of whales had once been killed for their blubber.

I would hire it again in the future for Garimo, and later the scientist Terence McKenna. I mistakenly thought that if I could not find, in myself, what I had so long searched for then perhaps I could bring awakening to everyone else through the words of these wise teachers.

I picked Kiran up at the Brisbane airport and we began the drive back to the Bay when something quite remarkable occurred.

At the Gold Coast, about an hour out of Brisbane, Vinodini

said that they had not eaten on the plane because there had been no vegetarian food available. She asked if we could please stop for something to eat.

I stopped in the main street of Surfers Paradise and we each ordered a small pizza. This is where the vegetarian story unfolded in its unique and life-teaching way.

I had been a strict vegetarian for about fifteen years at this time in my life while they, and this I again stress, had been lifelong vegetarians and meat had never passed their lips.

When we had eaten a couple of slices of our pizzas, I looked at Vinodini and unthinkingly said, "Vinodini, I thought you were a vegetarian?"

She looked at me, smiling, and asked in that Indian head-moving way with the single word, "Kay?"

I pointed, and said, "Your pizza has meat on it."

It had slices of salami embedded in it which neither of them had probably ever seen before. This poor, wonderful woman changed colour from Indian brown to almost grey, as she realised that meat had passed her lips after a lifetime of being a vegetarian. I realised that my unconscious remark had caused this and was instantly horrified.

Kiran, who had also been a vegetarian his entire life, then demonstrated what the power of great love, combined with wise intelligence, can actually do. This is love that easily stops the controlling and judgemental mind in its tracks. This feeling of love has nothing whatsoever to do with the monkey mind. Kiran's love for Vinodini would rescue her no matter what he had to do.

Smiling calmly at his wife, this phenomenal human being simply reached over and picked up a piece of her meat pizza and ate it. I could do nothing more than emulate the action of this great, loving teacher, so I too reached over and also ate a slice of her pizza as she knew that I was also a vegetarian. The colour returned to her face along with her gentle smile.

The rest of her pizza went uneaten.

This was one of the greatest demonstrations of love I had ever witnessed and I was touched deep in my heart to be with

Kiran, this wonderful human being whose love overcame thousands of years of Hindu belief in a single moment.

I settled Kiran into a nice house in Byron Bay and for the next few days many people came to sit with him. I would take my close friends in to see him, one at a time, to be personally introduced. Kiran would look at each of them and usually say only one short sentence after the introduction.

Now I had known these friends of mine for a long time so I knew their personalities and how they acted in life. Kiran could not possibly know these things, yet with his one sentence he would cut right to the chase and was always on the money.

For example, one friend was an ex-university professor who was now a therapist of sorts. Kiran took one look at him and said, "Ah! Shane, when are you going to give up your knowledge?"

Now "Shane the Brain" had a huge library and was forever reading books on growth, psychology, astrology and such. He once told me that he never reads novels. He had a great accumulation of knowledge and he occasionally used this knowledge to satisfy his somewhat royal tastes and desires.

Another female friend who is one of the toughest women I have ever met and, when she loses her temper, is quite scary, stepped into the room and Kiran immediately said, "Ah! Maria, when are you going to give up the fight?"

This simple sentence must have struck a deep chord for Maria fell down on the floor crying her eyes out.

Maria is Italian and often used her ancient Christian, Italian, socially trained mind to morally prejudge people about anything sexual and then attack them with her judgements.

My simple and caring Kiran decided, after that emotional performance, to stop speaking so abruptly and clearly to the people I introduced him to. He did not wish to upset anyone. A pity, I thought. We need to take our medicine however it comes.

During the satsangs, the whaling hall was packed to overflowing and Kiran, as usual, insisted on dressing me in white and sitting me next to him on the stage. When I protested, he said that he sometimes needed me to repeat the questions that were asked of him. He smiled at my obvious discomfort and patted me on the shoulder.

Maybe this is what he meant back in old India when he said, "practice for the future".

After the satsangs we would occasionally be confronted by the odd angry person, and some were obviously affected by drugs or alcohol. This, after all, was Byron Bay. I had finally learned my lesson in India and physical force was no longer an option in stopping these people. I diffused the situations with a joke or a sentence and I never had to use force again when protecting my beloved Kiran.

Before Kiran and the gentle Vinodini returned to India, I interviewed him in order to have a personal gift of his wisdom with me always. My search for my place in this world has always involved the challenge within relationship. I asked Kiran for an understanding of the value and purpose within this world of relationships.

These are some of the answers he gave to my questions.

INTERVIEW WITH KIRAN

SAHAJO: In the West, our connection between man and woman is conditioned by, and based upon the romantic myth of Tristan and Iseult the Fair. This seems to have been cunningly combined with Christianity's form of marriage with all its patriarchal rules and promises of love such as the "holy order" for the poor woman "To love honour and obey (the man) till death do us part". Plus, rules written as holy commandments such as "Thou shalt not commit adultery" or "Thou shalt not covet thy neighbour's wife".

This idea, and creation of love, is not working, as the high divorce rate plus pain in relationships proves. We seem to have lost our freedom and imprisoned ourselves in the mind.

And if we are not in this accepted form of so-called love, then we spend most of our energy searching for it, to find our one and only lost love, our soul mate (as Louise Hayes stresses) and so voluntarily imprison ourselves in this never-ending, impossible search.

Our songs, our movies, our books, all help create a longing and a desire in us for this golden dream.

Can you speak about this please? And also, about what love really is?

KIRAN: All relationships in the material world are based upon selfish fulfilment. Some selfish motive is hidden in that relationship to fulfil some desire or have some expectation fulfilled. It may be a relationship between husband and wife, between girlfriend and boyfriend, between mother and son, between father and son, between father and daughter.

In Sanskrit, it is said; "Atmanastu kamaya sarvam priyam bhavati". This means, everything is fine, everything is good, everything is beautiful, so far as it fulfils your ego. Otherwise, if there is no fulfilment of ego, you do not enjoy that relationship.

So the motive behind all relationships is the fulfilment of ego somewhere. And we give beautiful names, we decorate these relationships with the name of love, affection, and what not. And we imagine so many great things out of these relationships. As if we are going to get heaven out of these relationships. But it is always our experience that this hope, this desire is never fulfilled. And that is why all relationships end in more pain, more frustration, more jealousy, fights and misunderstandings. This is because the base of the relationship is wrong and where the foundation is wrong, the whole structure on that foundation cannot be a strong structure. The base is ego, the base is mind. The fulfilment of ego, the fulfilment of desire is the base. And anything which is based on mind does not bring the joy, pleasure, or the bliss which we expect from it. Because mind has no approach to these things. Mind has approach towards anger, jealousy, and these are the things which mind can only give you. Mind cannot give you joy. You can only get joy when there is no mind. Pleasure you get when there is no mind. Bliss you get when there is no mind.

This is our experience with so many things. Whenever we had a glimpse, a touch of joy, it was an experience when there was no mind present. So the fate of all relationships ... [we were interrupted by two women arriving and we resumed this interview after five minutes.]

KIRAN: Now your question is that the relationship between a

man and a woman, husband and wife, girlfriend and boyfriend, does not bring that fulfilment, does not give that fulfilment of love?

SAHAJO. Yes!

KIRAN: It does not take you to that space where you can just fly. A touch of joy, a celebration. And all relationships bring more sufferings, more pain, more frustrations, even though the expectation is totally different.

So we have to go a little deeper in understanding the mind. As I said before, the normal relationships which we have, are always through mind. And though we give the name of love, affection, benediction, to that relationship, if you go a little inside you can feel that you have some expectation, some desire, to fulfil through that relationship. It may be the fulfilment of sex, the physical fulfilment, or it may be the mental fulfilment of getting that love. When you are in that relationship you always play loving. It is not actually from the heart. Inside, there is something different. You make a show that you are making love, or that you are loving somebody. You go on telling somebody I love you, I love you, all the time. When you are in love why do you want to say, I love you? Why do you want to express, I love you? Because somewhere inside there is a guilt. There is some point where you know that you are not loving. You expect something different. You expect some joy, some pleasure through her or him. You expect that through her or him you will get that joy, that bliss, and you give that relationship the name of love yet inside it is having a selfish motive. So when you are in love, or you are in that relationship, when you meet each other, when you are together, each one is trying to exploit the other. And see how far he or she is satisfying that hidden desire, or how far he or she is satisfying the fulfilment of the ego.

The problem is that the instrument is wrong. You want to reach somewhere and you are using a wrong instrument. You want to get bliss, peace, and you are using mind to try to get there. You are using some selfish motive through mind to try to get there. Drop mind from that relationship. That means, becoming totally awake and aware in the moment when you are having that relationship. You are sitting with the boyfriend or girlfriend, touching, close, going into sex. Then if you remain aware, remain awake to that moment, there is no mind. There is only awareness, and when there is awareness, there is consciousness.

You are totally conscious. You are not unconscious. You are totally conscious, totally aware. And when there is awareness there is no mind. And when there is no mind there is no desire which you are going to fulfil, and fulfilment happens by itself. And in that moment of awareness when you are in that experience of sex, or anything, you get the same joy, the same bliss, the same ecstasy.

In the first experience of sex, at that time, there was no mind. Then a desire was created in the mind to repeat that same experience. But if you drop that mind and stop comparisons, making yourself totally available to the moment you are in, then you get the same bliss, the same joy, the same pleasure, out of that relationship.

So it is the base that you have to change. Change the base from the mind to the consciousness. This is the problem. Understand the whole phenomenon, that you cannot be loving when you are unconscious. Love can only happen when you are conscious because love is the quality of that consciousness, that existence. And to merge into that existence is a strong thirst in everybody's being. And we try to express this urge, this thirst, by using the word love.

The demand for merging into existence, into emptiness, is our search for love. Somewhere, you have to merge into that emptiness, and merging into that emptiness, you get that joy, pleasure, bliss. That is the thirst inside.

So out of all relationships, out of all experiences, we are searching for this. We are in search of that experience, that space, where we can merge into that emptiness. Where we can become one with that existence, of which you are also a part. The part wants to merge into that whole. This is the longing. At this longing you feel that you are in love. When that longing is so strong, we give to that longing the name of love. And we try to find that love in all relationships. But that is the longing and that longing can only be satisfied when you become totally conscious because that consciousness opens the door for merging into that emptiness. Mind has no way to give you any opening into this. On the contrary, mind becomes a hindrance. The ego becomes a hindrance for your merging yet we use the mind and the ego to try to reach to that love, to get that love. This is the contradiction. And that is why all relationships through mind, through ego, though it may be a craving, a desire for love, end in frustration, pain and agony.

And you break the relationship. You blame the partner. Somebody outside. That because of he or she you are not getting this joy. You are not getting that love. We blame somebody and we go on changing partners. The mind tells you, okay, fine, if she is not able to give you this love, change the partner. Go on changing the outside things. So we go on changing the place and the objects.

But the search is for this – that we want to merge into existence. The part wants to become one with the whole and if you search through mind you will never reach this. You just bang your head and come back and you get injured. More frustrated and more painful.

And this is the history, the story, of all relationships. Everywhere we have burnt our fingers because we don't understand that we are using the wrong instrument.

If you go with that total consciousness, with total awareness, and you go into that experience without mind, without any comparison happening in the mind. Without any old conditionings. Go back into that relationship with total awareness. You get it. Then it makes no difference whether you change your partner or whether you are with the same partner. It makes no difference.

SAHAJO: *Why not?*

KIRAN: *Because now you are not depending on somebody for your joy and pleasure. Because now you have found your own source inside. Somebody else only becomes an instrument, a medium for your awakening. To take you to that experience. Just like that.*

So once you start getting this from inside then you enjoy all relationships. It may be with anybody. Because then you are not depending on somebody else. You don't blame anybody. And if both of you can understand the base of the relationship, then the real ecstasy, the real joy, happens. And that is the real love. And that love is god. The love is that existence. The love is that fulfilment. Then you really experience the love.

SAHAJO: *So to put the blame outside is to be in the mind. And then all I'm trying to do is to get out of this relationship that depends on mind, and therefore expectation?*

KIRAN: You cannot get out of relationship unless you find your own source of joy inside you. All the time you think that you will get it outside somewhere and then that effort to get it from outside creates relationship. So, unless you reach to that source inside, you are not going to get it. You are not going to be free from all relationships. It is difficult. That is separation. You are separating yourself. You are going to the other extreme. Because you feel that you are not getting it somewhere outside, you totally cut off and break the relationship. Going deep into relationship, or breaking relationship is going from one extreme to the other. They are two extremes. You don't get to where you actually want to.

That is not the way. The way is in the middle where you become totally aware and conscious of what it is you actually want from this relationship. What exactly is my demand? What do I want to fulfil? Ask yourself, what do I search for in this relationship, what do I expect from my partner? Understand the whole thing. And when you understand that it is the search for joy, for bliss, the desire to merge, a longing to merge into that oneness. That is the longing, you see. And then you understand that the instrument which you are using is wrong.

When you are not reaching somewhere, you have to stop and think. And then when you see with total awareness that, god, somewhere you were making a mistake. Because then you see that it is impossible to reach through that. Impossible.

SAHAJO: But once relating in this way it is very difficult to let it go.

KIRAN: Then, you see, you are caught in that vicious circle and you go round and round in that same experience again and again. You don't understand the whole thing. You get more frustrated going round and round, again and again. You go on changing partners and then you become so helpless, so frustrated, because it is not happening.

You miss the basic thing. That you are using the wrong instrument. You have not to use your mind, your ego. Just be awake, aware, in each situation. And in any experience, not just sex. Any experience outside or inside. If you are in that experience with total awareness, it gives you a touch of that joy immediately.

SAHAJO: And an experience either comes, or it doesn't, but you are not concerned either way?

KIRAN: Not concerned. Then you have found your own source.

SAHAJO: Then the searching is finished?

KIRAN: It is finished. Then you can remain with your partner and your partner can also be with you. Then your relationship is totally in freedom. You respect your partner and give total freedom and full trust. These are the qualities of love. These are the qualities of existence. Once you are here in that love then you give total freedom to your partner. Total trust is there, and total respect. And you respect him or her as he or she may be. There are no expectations. The other is fine exactly as he or she is.

SAHAJO: And to give that I have to have it myself first. I have to trust and respect myself?

KIRAN: Yes. That is the first condition. Unless you have this, you cannot give to somebody. When you have it, what happens is that you share your joy which is already flowing from inside.
    In all relationships you share your joy. In normal relationships with the mind, you share your pain, and you fight. But in the relationship with love, with that awareness, you share your joy. You share your pleasure, you share your bliss. So there is no fight. All relationships are so beautiful and fine.
    You see, it's like a play, it's like a game you play. You are not touched inside.

SAHAJO: I feel that all experiences are drawing us towards this point of awareness and understanding. I can understand better now, and am more aware of this, just by listening to you. Are all experiences in life drawing us towards this point?

KIRAN: Yes. And there is a demand pulling us, in all experiences, towards that joy. That is the search.

SAHAJO: Until we have it in us?

KIRAN: You have the answer to your question. It is very vital. It concerns a lot of phases of relationship. And I can see now that people are frustrated with all the relationships.

SAHAJO: It is the basis of our whole structure of society. And in a so-called normal relationship we take on the role of having to protect the other, among other things, and this also leads to fighting and wars?

KIRAN: You are searching for security in each other. Some safety in each other.

SAHAJO: There is no such security.

KIRAN: You cannot get security with the mind. Mind itself is insecure. Because mind is itself an illusion. So mind always moves with the fear. And when you have relationship on the basis of mind you always move with fear. Fear of breaking that relationship.

SAHAJO: Will people still get married after they have found their own source?

KIRAN: Not necessarily. You see, marriage is a commitment, a social commitment. It is a part of society. Because human beings are social animals. And to keep this society in that, they have created this marriage as an institution. Because, you see, there are a lot of problems in the beginning. If there is no marriage, no commitment with one woman or one man, then who will look after the children?

In the beginning there was a group and they had group relationships so those children belonged to that group. Slowly, because of the mind, because of envy, because of anger, some stronger man in that group wanted the best woman. So for that woman there was a fight. And then for that group to remain together became a problem.

So slowly that group started dividing into smaller groups. Until commitment to one man, one woman, became that institution.

Once upon a time there were two really strange people who thought it would be cool to live together in the same house. Little did they know that this was actually physically impossible due to there not being enough space in the one house for the both of them. Luckily the end to the story does not involve anger, violence, destruction, physical death or financial ruin, only a change in sleeping arrangements leaving the two strange people to continue their lives without being emotionally damaged by love, but to continue to live their lives to the fullest without the drag of conditional love but to be able to see only the good in each other until all that remains is the love for each other which does not need to accept or make allowances for the other but just to be able to be in their own space and see the other with love.

[End of interview with Kiran.]

Ah! If only I could have really heard these words and understood them way back then, I would have saved myself a lot of unnecessary pain of the deepest kind. The pain of lost love through deep unconsciousness and wrong approach. The unconscious pain that can, and too often does, cause people, both men and women, to kill the object of their misplaced affection.

Our religious leaders have taught us in the wrong way, trying to control us morally to enhance their own power. Controlling people morally, from the outside, using laws, is not possible, as experience has proved time and again.

Kiran and Vinodini returned to India. Some of my friends would go and sit with him at his home in India and they would ask him to come back to Australia. They would then come back to Australia and tell me that he always answered that he would come only if Sahajo came and got him. I never did and he died a few years later. I shall always honour, love and respect Kiran and he is always with me.

Chapter 35 - Another Master

It is probably worth telling this story about another master whose teaching a female friend persuaded me to experience. It was always women who insisted that I meet these teachers or gurus. After some resistance I usually, albeit reluctantly, went along. This was all obviously meant to be in a life that was not always, consciously, of my own initial choosing.

Women are in the vast majority of those who search for truth, change and self- improvement, both physical and mental.

This Australian master's name was Barry Long and he was appearing in a large hall on the Gold Coast in Queensland.

When my friend and I arrived, the huge hall was packed with his followers and most were sitting quietly in meditation awaiting Barry's arrival.

He finally walked into the hall and took a seat on a raised stage.

He was an older man, definitely not good looking, with grey hair and dressed in plain clothes that I later learned he bought from Kmart. You could not say that Barry was a fashion icon.

To my surprise, when he addressed the audience, he said, "I am God, do not put your I on me."

This is how he always opened his meetings. He then told us to close our eyes in meditation and go into the black space inside and sink deep into it. After ten minutes or so of this intention, Barry then spoke about whatever he chose to speak about. It was usually something to do with spirituality in the form of truth and common sense.

That is the usual form of satsangs with these so-called enlightened people. I was not interested in that word, but my friend Rod assured me that Barry was. Rod meditated for hours every day and had done so for the past twenty years. Maybe he knew what enlightenment is, but it still seems impossible to know that someone else is enlightened when you are not enlightened yourself.

Why do people need to think that someone is better than them? So they can use the "enlightened one" as a form of worship, as humans do with all imagined gods of the mind?

After Barry finished speaking, he then asked the audience if anyone had any questions. He then answered every question quite intelligently. I was impressed. He had something.

When the meeting was over my friend took me out to the back of the stage to introduce me personally to Barry. He shook my hand and continued to impress me so I asked him if he would come to Cairns, in the far north of Australia, and hold meetings for my friends up there. He readily agreed to do this if I would organise the halls and the meetings, plus drive him, take care of him, and protect him.

Here I go again, I thought, but at least I did not have to fly overseas to find this master.

I hesitated for a moment and then agreed. The next day I flew to Cairns to make preparations.

In this hot, humid city, I quickly organised accommodation, a meeting hall, and arranged and distributed the advertising for Barry's talks.

When all was prepared and his meetings were booked out, Barry flew in with his wife, and the meetings began.

Barry was a very strict teacher and for him our eventual enlightenment was his only concern. I could not help but wonder here if he knew exactly what he was doing? I decided to watch him closely and find out. If what he was saying was actually true then some of these followers would become enlightened. I hoped, with tongue in cheek, that they would not float away after "attaining" such lightness.

After the series of meetings in the hall were concluded, I organised a private, entire day of a small group-teaching with Barry for only ten lucky, or unlucky, people.

When the group began, he had us all sit together on hard,

cold iron chairs with no cushions. He spent the day addressing us as if we were in an army. Nod off for even a moment and he was on you hard and fast. He often said to individuals, "Sit up straight and pay attention. Don't you wish to be enlightened?"

I thought, if they are sitting here wishing for enlightenment then they are not going to be present.

After six hours of this army-like regimentation my anger finally got the better of me and I stood up while he was speaking and told him that I was leaving. He immediately commanded, in a loud voice, for me to come out the back with him where he sat me on another of the hard iron chairs, which I had now grown to hate. He sat on one opposite me and pulled it up close, knee to knee, then leaned forward, just inches from my face and said, "Now what is this nonsense about you leaving?"

I felt like having a little fun by then so I leaned in even closer to him and answered, "Barry, it makes absolutely no difference to my eventual enlightenment whether I sit here with you on these uncomfortable chairs all day or not."

As he obviously believed in enlightenment, to his credit he leaned back in his chair and said, "That's fair enough, Darrell. Pick me up at five o'clock when this is over." Darrell is the name given to me at birth.

I walked out the front door and exactly at that moment my curious nephew pulled up on my motorbike and carried me away. Synchronicity, I thought. This was meant to be.

Barry never mentioned that moment again while I was with him though we spent many days alone together, walking and talking. He wrote many books including one filled with his poetry. Barry also fancied himself as a bit of an expert on sexual matters and made tapes instructing men and women on how to make love. These tapes sold worldwide and he often held large meetings in overseas countries. His many followers came from all corners of the world to sit with him. To his credit he was the only "Australian" guru or master in the world at the time. I suppose it takes a certain courage to announce oneself as he did.

Barry, like other so-called masters, had some pretty good

words, advice and teachings to share, and his words often came back to me in the stillness, when I was camped alone in far-off places. One of the interesting things he said was, "We are here to learn what not to do in life." There is truth in that as I had already discovered.

Now Barry always said that man must be true to one woman and one woman only. And he also preached that man must be truthful.

One day, before a meeting on the Gold Coast I became quite angry with Barry because I did not think he was practising what he was preaching. I had learned that he was having sex with someone else in his following as well as with his wife so I confronted him about this dishonesty. To his credit, once again, he surprised me and everyone else in that packed meeting on the Gold Coast.

On stage he announced that he had been questioned about the significance of truth in relation to sex and living an authentic life. Because honesty was a foundation of his teachings, Barry realised that he needed to immediately announce to his shocked audience, and me, that he was not only making love with his wife but also with five other women here in this congregation. I was shocked into open-eyed amazement. Five women – not just one – and he had just revealed it to the world because all these meetings were taped.

His confession caused me to look around the hall and wonder who these five women were as there were a lot of beautiful women in the audience among his devoted followers. One woman had changed colour and looked down so I knew that she was one of them. Such a beauty and she was sitting next to her devoted partner. She actually told her partner later that she was one of "Barry's women".

A woman in the audience (obviously not one of the five women) burst into tears, jumped up, and asked him how he could do this when he always said that a man had to be devoted to and be true to only one woman?

I smiled and wondered how he would get out of this predicament that he had created, but Barry answered without a moment's hesitation.

He said, "I am a master. It is okay for a master to do this. It is not okay for any of these men here to do this as they are still unconscious."

Nice comeback, yet somehow this did not sit well with me and I got up and walked out of the meeting. I never returned or saw Barry again.

For the next few years, he was devoted to his five women, but also took the opportunity to enlighten other women during his seminars. He actually believed he was enlightening them through the sexual act. But then many other master teachers believed this as well. So even in the state of declared enlightenment, temptation and desire obviously still existed and are also forceful masters.

Barry finally got the big call with prostate cancer. Facing death changed his thinking. Barry returned to the devotion of only one woman. It took cancer for him to realise that the deepest love comes through the devotion of one man and one woman. He had finally realised his own teachings after all those years. In his own words, "I died into love". What also amazed me was that the other five women continued being devoted supporters and students of Barry. He had "taken each of these women on". He had taken them on fully. Their emotions, their jealousies, fear, lack of trust in men. Everything and anything that arose in them. I suppose it is true that most men could not do this.

The wisdom within his teachings still returns to me at appropriate times and I now have no judgement of him. He was a good man doing the best he could. He had told me that he had experienced some sort of satori, while suffering excruciating pain for days. This may well have brought some sort of deep understanding that he had to share as a teacher and finally realise for himself.

His books are worth reading and can bring relief and understanding when it is most needed.

It is worth noting again that many of the masters I have met, including Byron Katie, Kiran, Krishnamurti and Osho, all suffered terrible pain for long periods of time before something in them awoke and they began teaching. Suffering is a master teacher. It can create a profound surrender of the human mind and activate the inspiration and mastery of authentic teachings. Accepting this suffering with grace and head bowed thankfulness soon dulls its pain and allows the wisdom and realisations to arise. As ancient scriptures say, "You must die to be born again." To imagine you are dying can achieve the same result.

36 - The Familiar Road

I soon earned some extra money and decided to go to San Francisco where, I was told, there was some type of teacher or master on every corner. I wondered if the threat of imminent death there, through the many earthquakes, made people try harder to realise what constitutes the spiritual path and spiritual connection.

As usual, I also had a woman friend, not a lover, who lived there and had invited me to sit with her guru. My friend was German and her guru was a famous New York Jew named Andrew Goldberg. He had been a follower of Papaji (Poonjaji) in India and, like other followers of Papaji, he had also declared himself enlightened. He now had many followers worldwide and I soon discovered that he was making lots of money.

My friend picked me up at the San Francisco airport and I was surprised to see that her beautiful blonde hair was gone. She had shaved her head. I asked her why she had done this.

"Her guru had ordered her to".

She then told me that her and about ten of Goldberg's "inner circle", the specially chosen, lived in a special house of his. They all worked at different jobs in normal society. She was only a cleaner, but the others were specialists such as carpenters and teachers.

Then she proudly told me that the chosen people in this house did not do what the Christians do which was to give ten per cent of their wealth and wages to the church as a tithing. Nope, they believed they were better than mere Christians so they gave their chosen enlightened master the princely sum of twenty per cent of their wages each week. Jesus wept!

Living together in this one house they were only allowed to have sex with someone who also lived in the house.

Another sect, with controlling regulations, obviously.

On top of this, my poor German friend informed me that she had been instructed by Andrew Goldberg himself that she was to remain celibate for one year. No sex allowed for her with anyone. I did not fancy her as a lover but I was sorely tempted to have sex with her as a way to break this belittling covenant.

This all seemed more than a little strange to me and I wanted to walk away, but I was willing to sit with Goldberg, as this woman had been a good friend of mine in India and had nursed me when I was very ill.

So she took me to a little hall that was packed to overflowing with people. We all had to pay $20 entry fee but I like to be different. I did not trust what this man was doing with money, so I paid nothing and just boldly walked in, accompanied by one of the "chosen ones". Mere helpers guarding the door were obviously not going to question me when I walked in with such authority.

We were all instructed to sit in the hall and meditate until the great man appeared. Because I was with one of the "chosen ones" I was given a comfortable seat up front.

Strangely, for some inexplicable reason, when I closed my eyes to meditate with the rest of the people in the hall, all I could see and think of was Adolf Hitler. I had never thought of him before and I usually did not think of anything much when meditating. That is not the point of meditation and I had been well trained in ignoring thoughts yet Adolph with his funny moustache stayed with me for ten frustrating minutes, fading away and then rising from the insistent ashes of my mind once again.

I swear on my life that what happened next is true. This is synchronicity in action.

Andrew Goldberg strutted in. I had never seen him before and was amazed to see that he was a short man with short hair and wearing an exact copy of Hitler's moustache.

This was becoming stranger by the minute.

Andrew Goldberg took his seat on the raised stage and for the next ten long minutes he spoke only about Hitler. His long tirade was infused with a deep hate and many anger-filled judgements about Adolph Hitler.

I wondered how he could possibly know some of the things he said about Hitler because he had not been there with Hitler so long ago. It was patently obvious that he was only repeating what he had read and heard. Preaching hatred and not forgiveness did not seem to be the work of a so-called enlightened man who people happily gave twenty per cent of their earnings.

I wondered if this was a deep ingrained Jewish trait and this hate affected not only him but also the whole Jewish race?

His tedious tirade made me think about this cultural dilemma. Hatred is a terrible burden to carry as it mirrors back the hate to the one who is hating. Hitler is long dead. When will the Jewish race find forgiveness? Possibly after they eventually act unmercifully to another imagined race of other people, because hateful memories create present hate? We shall see.

After ten minutes of this emotional release of judgement and anger he thankfully and finally asked if anyone had any questions. I certainly had none for him though I was sorely tempted to tell him that Hitler was a distant relation of mine, which of course he is not.

When the Hitler impersonator in the form of Goldberg left the hall, I stayed behind with my friend and the rest of the inner circle after everyone else had also left. All the occupants of what they called the "safe house" were here. Not "safe" from Andrew Goldberg, I thought.

We all gathered in a circle. The inner circle? On my friend's recommendation, they invited me to come and live in their house.

I thought it was high time to play a little game of wake-up so I asked them why they lived in this "safe house" and why they paid Goldberg twenty per cent of their hard- earned wages?

The leader, a tall man, said it was because "the master" had a target for them and that target was enlightenment.

So, smilingly, I asked them what this target of enlightenment looked like?

Quite a long silence followed and then the leader said that they did not know what it looked like, as only the master knew that.

I then said, tongue in cheek, "So you are like archers with a bow and arrow aiming at a target you cannot see so, obviously, you will never hit it."

Complete silence followed my outrageous statement and I could see the leader's face changing colour. Finally, he pulled himself together and said: "You cannot live in our house."

"Oh, bugger and tragedy," I answered, then turned and walked outside where I waited for my friend.

On the drive home I tried to talk her out of living in that house and following this man's strange orders, but I am not a teacher and cannot, frustratingly enough, seem to help anyone escape from strange beliefs. I realised that each person has their own harsh lessons to learn in their own way, as experience and suffering are the master teachers. She chose to stay in the house, pay him twenty per cent of her hard-earned wages, remain a disciple, never make love, and I never saw her again.

However, I heard years later that Andrew Goldberg's team of the chosen ones finally rebelled against his extreme control. Power and control was HIS temptation and it overcame him. His followers must have realised that his egoic control was not a pathway to enlightenment. They confronted him and then departed. Perhaps in the shock and despair of failure, Andrew shut down his teaching organisation and hid away in retreat. We each learn in our own way.

Over the next week I went and sat with different advertised masters and enlightened ones of my choice, thinking that there must be someone real in this exciting city.

There was one lady called "The Golden Angel" who was supposedly channelling one of God's angels. She also had a lot of followers. I was once again amazed at the simple gullibility and desperation of people as I listened to her impossible kindergarten type of nonsense. I wondered why people needed to fit in with a group and a common belief, while totally suspending logic? Loneliness, desire, and no spiritual connection, perhaps.

Another woman I met, who was very famous and wealthy from her teaching (ah! the money-spinning spiritual business), was channelling someone called Rama, or some such name. This Rama was supposed to be an enlightened man who had actually been alive ten thousand years ago and had now, magically, entered this woman, Kaght. Rama would take over her body and mind during each presentation in order to bring enlightenment to the many followers and the world.

Don't laugh! This woman had followers here who had come from all over the world. Kaght became extremely wealthy as she demanded large sums of money for the pleasure of being with Rama, and hopefully, instant enlightenment. She also persuaded her inner group of wealthy believers to invest in yachts and race horses for her pleasure.

When she was channelling Rama her voice changed into that of a man. It was quite an act. Worthy of Hollywood. Rama spoke around the world. Many thousands of people were enamoured with the experience. But it was Kaght who became enamoured with the money and possessions that spiritual teachings can attract. Her desire for wealth eventually became her downfall.

I saw one more of these strange spiritual ravers in San Francisco, but I cannot even bother to remember his name, or his game.

One thing all these people had in common was that they were all boring and I had difficulty in staying awake to listen to their impossible ramblings. I wondered if there was something in the air in San Francisco which was affecting everyone.

It was certainly time for me to depart before I became afflicted with this air. I hoped I had at least been cured of sitting with what I thought were strange, rich impersonators, and their many gullible followers.

Chapter 37 - The Unexpected McKenna

I left San Francisco the next day and returned to Esalen Institute in Big Sur where I went back to work in the garden as a work scholar 2 for one of Esalen's famed teachers. A wonderful and kind man named John Soper. He and I hiked many days in the Big Sur Mountains and I appreciated his gentle teaching, imparted to me, unconsciously, through our hikes and conversations.

One day John told me that a man named Terence McKenna was coming to Esalen as a resident teacher for a week, to talk only to Esalen staff and work scholar 2s. I reminded John that I had had more than enough of unknown teachers, but "Dear John" insisted that I accompany him to at least one evening lecture as this man was anything but boring.

"Okay," I agreed, "I will do it for you but if he is another boring 'believer' spouting enlightenment, then I am walking out."

John just laughed and said, "Do as you must."

That night a man, over 6 feet tall, walked into another packed room, took his seat, and began talking. No one had been asked to meditate or keep silent as we waited for him.

For two seemingly short hours I was enthralled by everything this amazing man, and respected scientist, said. Not even a moment's boredom touched me.

For the rest of the week, I was there every night and his vision and knowledge continued to astound me. At the end of the week, when Terence was leaving, John introduced me to him and I immediately asked him if he would come to Australia and go on a speaking tour.

"Sure," he said, "if you organise it."

I took his phone numbers and email address and said that

I would definitely do this and give him all the details of his Australian tour in the near future.

I felt strongly that McKenna's teachings were a bridge between East and West and would help people combine the teachings of both the East and West and become more whole and grounded. A combination of science and the spiritual.

I soon said my goodbyes to John, my friends at Esalen, and the wonderful hot tubs where everyone bathed together naked in the supposedly healing waters.

I had no wish to organise a tour for McKenna myself but I knew a promoter in Australia who I contacted on my return. Fortunately, this professional promoter knew of him, had read all his books, and was keen to organise the entire tour. All I had to do was introduce Terence at every gathering, and be his bodyguard. Terence had asked me to do this for him. He had been told that I was a good bodyguard and minder. The spiritual grapevine at work. I was pretty good at most things I did but only for a short time as I always had to move on.

The tour was organised quite quickly and I rang Terence in Hawaii, where he was living, told him the good news, and booked him a seat on a plane.

I had no idea at the time that this man had an almost cult-like following in Australia and around the world. When people in Europe, England and elsewhere overseas heard that he was coming to Australia to tour, they immediately jumped on planes and flew over to listen to him.

At the time I had this idea that the gurus and masters around the world were mainly holding satsangs, which had a distinct Indian flavour. India was the spiritual capital of the world and India was actually doing a good job by keeping the spiritual somewhat alive for the rest of the world, where it was dying. But there was obviously something missing. The bridge between the East and the West. I thought, once again, that McKenna could supply this bridge. He did not meditate or talk of gods. He was a scientist through and through with one of the greatest minds of the time.

I was told that he was with Timothy Leary when Leary died

and Timothy had taken off his cloak and handed it to McKenna telling him that he must now carry on "the great work".

Timothy Leary was born in 1920 and forged a career as a noted psychology professor and researcher at a university in San Francisco. During the 1960s he became a highly controversial advocate of psychedelic drugs, which he said would raise consciousness.

He experimented with mushrooms and eventually LSD. He became famous and a media icon with his much-quoted line, "Turn on, tune in, drop out."

In my usually naive way, I knew nothing about McKenna and his experiments with drugs, when he arrived in Australia. All I knew was that he had a brilliant scientific mind and his talks were riveting and full of interesting information. I had not read any of his books about his experiments with DMT, ayahuasca, LSD, ketamine, marijuana and god knows what else.

I picked him up at the Brisbane airport and took him to a small cabin near Byron Bay, which was to be his home while he was in Australia.

Each morning when I picked him up, instead of doing what most people do when awakening, which is to have a cup of coffee, Terence would pull out a massive ten-paper joint and smoke the bloody lot on his own. How he could function in a normal manner after this was beyond my comprehension yet function he did and he did it well. Mary Jane, marijuana, obviously affects everyone differently.

I once asked him what DMT was and he said that a dose fits on the head of a pin and when you take it you go straight to god – whatever that was meant to mean. I asked him if you could die from taking it and he said: "Only if you can die from amazement."

I, however, did not wish to experiment with any of these new drugs as that was not my path.

However, one new drug, ayahuasca, is legal in the Catholic Church in many countries, and the Catholics have a powerful influence. Australia is one of these countries, so I did try it. I vomited, and vomited, and then suffered from extreme dysentery. So much for the Catholic Church and their recommendations.

I should have known better. Drugs are obviously not the way for me, but those like Terence – and there are many of them – are explorers in the realm of the human psyche and they discover new territory outside of the normal. Brave people! Good luck, and I have no judgements of them as I thought that this "normal" world we live in definitely could do with some urgent changes.

Together, McKenna and I organised the old whaling hall in the Bay in the way that he wanted it set out. It had two massive halls that had once been used for cutting up whales. One of the halls was where he would sit on stage and talk for an hour or two, and the other we set up as a dance hall with high quality disco sound equipment. After his talk he would walk to this second hall and direct the dancing "doof" – the great pulsating dance with the constant deep beat that had swept the world. He would do this by "rapping" to house music, with his own unique message, while everybody danced.

The large crowd who attended surprised me. They were lined up all the way down the street and around the corner, waiting to pay to get in. I realised in surprise, yet again, that Terence had a huge following even in Australia.

On the night of his first appearance, I picked him up early and he directed me to take him down to the nearby beach where he lit up one of his ten-paper joints donated to him by one of his followers. This "dope" was always of the very best quality and even one puff of it would have me stoned immaculate. I had tried it once before at his insistence.

On the beach, however, he forcefully insisted that I have a couple of puffs as I had to be in the same zone as he was if I was to introduce him.

Now I had three foolscap pages of written details about the history of this fascinating man and I warned him that if I had even one puff, I would find it almost impossible to do his introduction properly. He told me not to worry about that, so I reluctantly had two puffs while under his fixed gaze and direction. "Hold the smoke longer in your lungs," he commanded when I tried to blow it out quickly.

When we arrived at the hall I was totally stoned and this was only after two puffs of his monster joint. He had smoked the rest alone. Eventually, I managed to walk onto the stage to a huge cheer, armed with my papers and my heavy cloak of paranoia.

I began to read the impressive history of Terence McKenna. After two paragraphs and many hesitations, I just gave up and threw all the papers into the air and shouted: "Bugger it. Here's – Terence."

Everybody screamed in approval at my spontaneous action and short introduction.

With that I got off the stage as quickly as possible in my spaced-out condition, and he began his long lecture while I hid behind the curtain.

What was impressive about his talk was the pure professionalism of a truly great speech from a truly great mind. It had a beginning, a middle, and then an end that tied up perfectly with the beginning. How he could do this after smoking the entire, huge, exceptionally strong joint, was beyond my understanding. It was again obvious to me that everyone is different and drugs affect everyone differently.

After the talk I escorted Terence down a long narrow corridor to the large adjoining music hall. I was almost feeling normal by this time. Halfway down the corridor an extremely agitated man, who was obviously drunk, began abusing Terence at the top of his voice about the horror of drugs.

I was walking behind Terence, protecting his back, so I grabbed his arm and pushed past him saying, "I will handle this."

Terence pulled me back, placed me behind him, and said: "It's all right, I will take care of it."

Now Terence McKenna was not a strong physical specimen of a man and I doubted if he ever exercised at all, let alone practised any form of self-defence, so I once again had a huge surge of Mary Jane induced paranoia and imagined Terence flat on his back when this large drunk hit him. I knew that I could not protect him from behind in this narrow corridor.

Terence McKenna, I read in his astrology chart, has something like seven planets in Scorpio. He can deliver psychic sentences more deadly than bullets. The music was playing quite loudly and because I was standing behind him, I did not hear what he said to this tall man in front of him. I would have dearly loved to know what he said although it was also probably how he delivered it that had an effect.

Jenny O'Connor and the Nine once told me that in the future, psychic power would be much stronger than physical power. Here in front of me an eye-opening example of this was aptly demonstrated.

McKenna said all of two sentences and the big drunk fell apart, suddenly crying his eyes out. I was amazed, and also thankful that I did not have to use primitive physical force. McKenna was my psychic bodyguard.

The dancing and Terence's rapping in a deep voice went off with no further hitches and by the end of a long evening he had made some serious money. He deserved it, doing what he does. Even his rapping, in his strong voice in tune with the music, was delivering a message of change.

One day later I drove him to the home, in the far-off hills, of a friend of mine who is also famous worldwide as a speaker, musician, and leader of ayahuasca meetings. McKenna had heard of him.

His large home was built in the middle of a rainforest and we sat down on the high veranda after I made the formal introductions.

During the course of the day many large hash joints were smoked. I think I may have gotten slightly stoned from the second-hand smoke in the air. I abstained from participating because I was the driver.

Our host was a good storyteller with a good heart, and also a speaker and musician of some note.

Terence was also a great storyteller and he told us about a trip he took to South America where he took a strong drug in the rainforest. He was standing in the rainforest in a drug-induced state when hundreds of butterflies came and alighted on him. He showed us photos of this extraordinary phenomenon.

After a few more of these stories and the amazing things that had happened to him while on drugs, I spoke for the first time. "Don't you both know that the experience that you have on any drug is always a part of you and you can reproduce it at will, without the drug, by using your memory of the feeling?"

This seemed obvious to me but my statement went over like a lead balloon and was met with complete silence and long stares. This may have seemed to them as if I was trying to take their drug-based experiences away from them but I was just trying to save Terence. I somehow knew without a shadow of a doubt that if Terence kept using as many drugs as he was, then he would soon join Timothy Leary in the grave. The body can only take so much.

I felt very strongly about this and was not happy with their superior stares, looking at me as if I were an idiot. So I forcefully directed their attention to a stump in the rainforest about 50 metres from where we sat. "Concentrate on the stump," I instructed them, "and we shall all remember the butterflies."

As we sat staring at that black stump, about a dozen butterflies fluttered down and landed on it.

"Do you see what I mean?" I asked.

They never answered and never acknowledged that this even happened. For the rest of the afternoon, they completely ignored me and I never said another word. They were about as ready to give up their drugs as meditators are ready to give up their quest for enlightenment, or the religious are ready to live without their beliefs about gods.

And I am obviously not good at saving people's lives.

Terence and I left about three o'clock in the afternoon and he asked me to drive him to Nimbin, where drugs are smoked freely in the streets and in the cafes. Nimbin is like a place from another planet with an entirely different set of rules to the rest of Australia.

I had driven there many times before and knew the way very well. However, the second-hand smoke of the long day must have affected me more than I realised and we ended up in a town in the opposite direction. This amazed both McKenna and especially me.

However, my time was not wasted because on the long drive I asked Terence to share his wisdom with me on different topics. One of the questions I asked was, "Terence, please speak about religious beliefs."

This man does not waste words and he turned to me and simply answered: "BELIEF MAKES YOU LESS THAN HUMAN"."

That's all he said on the subject, this scientist of many words. Straight to the point. I thought about his answer, every now and then, for years after. I suppose when suicide bombers blow up innocent women and children in marketplaces, this is certainly "less than human" and they certainly do it because of their beliefs – these primitive beliefs implanted by their spiritual leaders.

And, also, when the Christian governments kill millions of innocent civilians in their seemingly never-ending wars, and drop large bombs in foreign countries, this is also far less than human.

Both of these outrageous, unconscious acts of gross inhumanity and irresponsible stupidity must certainly be fuelled and governed by the many conflicting belief systems present in the world.

Personally, after thinking about his statement for a long time, I may have toned it down a bit and said: "Belief can, and often does, make people act in ways that are less than human." But I have no planets in Scorpio like Terence, if that actually makes a difference, which in his case it certainly seems to.

On that drive he was explaining to me about ordinary people and I asked him what he meant by ordinary people because I do not find people ordinary at all.

He looked at me and said: "Well, you are ordinary."

I did not think I am ordinary and this statement annoyed me a little because I thought that it arose from comparison and ego so I said: "Well I may be a bit naïve, and even a little stupid, to find myself sitting with someone like you in my car, but I am not ordinary."

Thankfully he did not take offence at my slight, or else he may have also had me crying with just two sentences. He just laughed.

Perhaps he meant ordinary in not being world famous like him. I have never wished to be famous. That takes a certain kind of bravery.

He continued to talk and told me about future developments of science. He said that soon man would have a little pill with a camera in it and we would swallow this and it would even heal what needed healing.

He also said that in the future man would be able to make cars from a scoop of river mud.

One of the ideas that stuck in my head, from something he said, was that if I died right then and he put me in ice and froze my body, then in the future I could be unfrozen and brought back to life. What great vision and imagination this man had.

I asked him one more question: "Terence, what can I do in this life?"

He answered immediately, "Write the poem that cannot be written." This sounded like a Zen koan to me; nevertheless, I have pondered on it occasionally.

Not many years later Terence rang me from Hawaii and said that he was dying from, of all things, a brain tumour. I urged him to remember what he had told me on our drive to nowhere, and to freeze his dead body and have it brought back to life in the future because a mind like his was worth saving. He was adamant that he would not do this and I could not understand why, after what he had told me about the future.

This great man died soon after. He was not that old. He had indulged in too many drugs, as I had warned him, which he remembered during our conversation, but that was his sacrifice. Terence McKenna left this world with some valuable insights in the books he wrote, plus information he gained while high on drugs. Just like Timothy Leary, Stanislav Grof, and many others.

This amusing story on our road trip will also remain with me.

He said that if you are having a debate with a "religious believer" and you are getting nowhere, then you must finally ask the believer if there is anything you can possibly say that may change his belief about his god. If he says no, then there is only one last thing you can do.

He said, "You pull back your coat and reach into a deep pocket."

Now I was sure he was going to say that you pull out a 38 pistol and shoot the religious fanatic. But no, of course he did not say this as he was a gentle man.

He said: "You reach deep into your pocket and pull out a huge ten-paper joint and light it up, for there is absolutely no use in any further debate or discussion."

Something else he told me was that in ancient England when tobacco was first discovered it was illegal to smoke. If you were caught you were beheaded, so people would meet in secret locations and smoke it and get high just like they do now on marijuana. However, he said that when tobacco was legalised for the masses it no longer had the mystical effect of making you high and became simply commonplace. He then said that when marijuana was eventually legalised, and he was sure that it definitely would be, then people would no longer get stoned from it in the way that they do now.

He was also very sure that the use of drugs had lifted and quickened the pace of the growth of human consciousness. I thought he was probably right about this from my own experiences. Yet I knew even then that the pursuit of pleasure or further experiences using drugs of any type was of no interest to me. Intelligence gained from everyday experiences was my pathway. For me, living experience is everlasting but the effect from drugs always ends as the mind slowly fades back down into the present. For Terence, repeating the use of mind-altering substances seemed to be the only way to maintain that high, until the physical body could no longer handle it.

Ram Das (Richard Albert who experimented with Timothy Leary) once told me, at Esalen Institute, that when you take drugs, you go straight to heaven but then you are booted out.

Chapter 38 - Garimo

Life slowly returned to normal after Terence McKenna left. My beautiful daughter was grown up by this time and going to university so I decided to go bush again and sit by myself in isolation for a few weeks. I did not fast this time, or meditate. The deep Australian bush has always brought me peace just by being surrounded by it. Its vastness and the endless natural phenomena revealing itself, if you have the eye to see, is a meditation in itself.

After three weeks of total isolation from man, from TVs and radios, papers and books, I drove across the Blue Mountains and into the city of Sydney.

Strangely, yet again, a female friend of mine insisted that I come to a church in the upmarket suburb of Vaucluse and listen to a woman named Garimo, an American from San Francisco, who was appearing there. They must breed them there, I thought. It was the last thing I felt like doing. I felt that I had been with enough so-called enlightened beings and anyway, I had already met Garimo at Papaji's home in Lucknow. But my friend was very insistent. I have found in the past that women's feelings and intuition can be quite accurate so I relented, out of curiosity, even though I was trying to closely guard, and hang onto, the feeling I had picked up in the desert.

The huge building, in an exclusive school in Vaucluse, was packed. I discovered that Garimo now had many adoring followers worldwide and many had flown here to be with her. In fact, she was one of the more popular spiritual leaders of that time in the world.

A sign at the door read, DONATION $20. If it had only read $20 I would have paid, but donation obviously means you have a choice. As this sign was not really being honest, I felt it was my duty to ignore it even though four people were standing there with eagle eyes making sure that everyone paid.

I was told by my friend that this blonde, quite attractive woman, Garimo, is a Gemini and has seven planets in Gemini. Shades of Terence McKenna, I smiled, with so many planets in the one sign.

Anyway, this means in astrological terms that she could talk the legs off a table and talk she surely could. Wonderful words sprinkled with wise teachings. Unfortunately she was promoting the usual Eastern, alluring, goal which could be found in one's distant future – the holy grail of enlightenment.

Yet she was certainly a master of the spoken word and she appeared to radiate love through her well-chosen words. Quite impressive and, thank goodness, not boring.

At the end of the satsang the packed hall of people was told to sit in silence until Garimo had left. After a few minutes her assistant, her right-hand woman, stood up and made a few announcements.

There were perhaps a thousand people in this building and I was sitting quite far back. The woman pointed in my direction and asked for the person who she was pointing at to come outside and meet Garimo. I looked behind me to see who she was pointing at. She said, "No, no, you." I pointed my finger at myself and mouthed, me.

"Yes," she said, "you, come forward."

I stood up and joined her. She took me outside to a car where Garimo was standing waiting.

Garimo said, after meeting me graciously, that she would like me to be her driver and protect her from over-zealous people while she was here in Australia. I thought about this for a moment, and quickly realised that once again this was my ever-repeating destiny as it was following a pattern, and so I agreed.

I wondered if Garimo had forgotten that she had welcomed me into her master's home (Poonjaji-Papaji) in Lucknow, India, so long ago. I had also sat with her once in Sedona, in America. At that sitting my mate from Esalen, Jenny's husband Russell, was sitting next to me and he raised his hand to ask her a question. She looked over and said, "Oh, look at that face. I am in love."

Russell thought that she was talking to him and tried to say something else but she waved him away and said, "No, no, not you, the man sitting next to you." She was pointing at me, to my surprise. I cannot even remember what she said to me then. Strangely enough, I had forgotten that meeting until I saw her standing here.

So now, here I was, about to become her driver and bodyguard. I wondered why this kept happening to me. What does she have to teach me that I may pass on to others? And what do I have to teach her? There always seems to be an exchange between myself and these famous spiritual teachers.

For the next week I ate breakfast with her every day and we spent a lot of time together. She certainly was a fascinating person with snippets of wisdom to share.

She asked me to accompany her on the rest of her tour in Australia, which was to take in Melbourne, Perth, and finally Byron Bay. There was, as usual, no pay offered for my service, even though I was not a devotee of hers. She did pay all expenses for the trip, however. This at least gave me a free holiday and some interesting company.

Once again this is where destiny had landed me so I had to surrender to what is. Kiran had taught me that "isness is our business".

I remember something which had happened when I took Kiran to see Barry Long when Barry was addressing a packed hall of people. I was whispering to him, complaining about Barry. He turned to me smiling and said, "Sahajo, don't you know that everything is perfect. If existence did not want him up there, then he would not be there." Now that's a demonstration of faith in life and the directions it takes. Surrender in action.

During the tour with Garimo there were often excited people who rushed at her, mainly to hug her or get closer, and part of my job was to stop them, which I easily did as they were not dangerous.

A couple of times, though, men became angry with her and these I stopped forcibly without actually having to use the deadly physical force that I was capable of. I must have learned something about violence from Terence McKenna and Kiran.

One man who was advancing angrily towards her got the shock of his life. I was standing on a veranda about 20 feet high and Garimo was below me walking to her car while I held back the surging crowd. This man came racing at her, loudly yelling obscenities. I immediately leapt, without thinking, over the railing of the veranda where I was standing and landed right in front of him. He got such a fright from my sudden appearance that he buckled at the knees and stood there, wobbling and smiling stupidly. I did not have to touch him. He was happy to shut up and quickly walk away unharmed.

After Melbourne she held a live-in retreat in a nearby, pretty little town that was later burnt to the ground and totally wiped out by huge bushfires. This retreat was held during a full moon and lasted for five days. I told her and her staff that this might be a mistake as some people tend to go a bit mad during full moons in Australia.

Unfortunately, this madness came to pass and I was kept busy. Most of the outbursts were minor and easy to handle but one person kept writing notes to Garimo threatening her with violence. We did not know who was writing these progressively violent messages but I could tell that they were serious and written by someone who was "full moon mad".

Garimo was staying in a little house about a kilometre from the retreat hall. She stayed alone, which she always preferred, as her husband was off in other parts of the world getting himself into trouble that would soon be revealed.

On the night of the full moon, I felt that something was going to occur so I crawled, unseen, under her house, which was raised a few feet off the ground. There I lay silently without moving for hours. Sure enough, at midnight a car pulled up down the road. I recognised it as being owned by one of the people attending the retreat. I had taken note of everything to do with this retreat, even all the cars.

I quickly crawled out from under her house and using the surrounding bush as cover I approached the car from the rear.

When this young man opened the car door and began to step out, I grabbed him, spun him around, and pressed him back against his vehicle.

He almost fainted from shock, and quickly admitted, on questioning, that he had indeed written those notes threatening her. I asked him why he did this and he said it was because she was ignoring him.

Ah! The fanatic disciple's delicate ego on full display.

I took him back to the satsang hall where Garimo's staff were summoned by one of her employees. They wanted to call the police but this poor young man was just moon-mad and by this time in tears and very apologetic. He said that he had no intention of hurting her. He had stolen one of her shoes and was going to give it back to her by placing it on the steps of her house.

I, being in charge of security, insisted that we did not need the police and I drove the young fellow back to his family's home in Melbourne with his sincere promise that he would not return. The rest of the retreat went off without any further madness now that the full moon was waning.

As we continued travelling together, concerns began to arise in my mind. When we were in Perth, I brought up the wording of the $20 sign at the door of her satsangs, telling her that the word "donation" should be removed as it is not a donation when it is a demand. The word never was removed as this had something to do with the fact that she did not have to pay taxes on her earnings. It seems that donations are treated like they are in all religions and are not subject to taxes.

She certainly earned a lot of money on her trips to Australia and paid no tax whatsoever on these large earnings.
Religious madness, in another of its many forms, where religious leaders, unlike honest hard workers, pay no tax. Little wonder that the Catholic Church is so wealthy.

One night during satsang in Perth she spent hours telling

her multitude of gathered followers that they must empty their minds of everything, all thoughts. They must enter into the nothing, etc, etc. Then at the end of this long discourse she dropped in the words: "The day that I became enlightened."

On the drive back to her house I felt I had to point out to her that by saying these words about her enlightenment she had immediately filled up the emptied minds of her followers with the thought, and endless expectation, of their eventual enlightenment because they immediately wished to become enlightened just like her.

In my opinion this was surely a slight form of induced schizophrenia, as the minds of her followers were being pulled in two opposite directions, and I told her so.

She looked at me and sighed, saying, "Yes, I know, Sahajo, but I must try."

I did not really know what this meant and I felt that she had not addressed what I had told her.

Yet I was still somewhat entranced and a little in love with her way with words and what she was trying to achieve. This would not change until her next visit to Australia a year later, when her husband came to sit on the stage with her. Two "enlightened masters"?

Of all the teachers or masters that I had served and spent time with, Garimo became the only one who would not let me interview her or ask her any questions for my writings. I would find out why later. If anyone wished to know anything about her, she told me, they could buy her books and tapes.

Yet at the beginning, I, like thousands of others, was entranced by her words and her radiating of what looked like love from a loving soul, but I did not worship her. In all my encounters with the many named masters I had been close to, I had never bowed down to a master as all my friends did.
Not with Osho, Kiran, Jenny, Dick, Joseph, Peter, et al.

When I was younger, I did move into a house with a Baptist group and I did let them stick my head under the water, hoping to feel something. I felt nothing whatsoever and so I packed up and left, much to their consternation. Misery indeed likes company.

Anyway, at the beginning of our time together I decided, okay I will give this a try with Garimo. Everyone else was doing it so let me see if I could actually feel anything. She called me up to the front of a large gathering and said to me that when the disciple is ready, the master comes. In this case the mistress?

I let myself be totally open and I waited to feel something, anything. As with the Baptist minister, I felt absolutely nothing. What to do! I am what I am and no relief was felt in bowing down and making a declaration to one who says that they are enlightened.

She did, however, honour me by giving me a letter to read out to the first gathering of the Rainbow Tribe in Australia. I stood in the middle of their big, sacred circle, with their magic hat (which they pass around the large circle for people to place money in), at their gathering in the deep bush, inland and far from civilisation.

The things I do? Honestly, this has been a long, ever winding road.

The next year Garimo returned. Yet, contrary to expectations, she didn't bring her husband.

I was contacted by her and asked to perform the same service on her Australian tour, and so once again, I spent a lot of time with her.

I was no longer enamoured by her divine presence, though I was still a little in love with her. I began to notice small things such as the fact that she swallowed twenty-seven different types of vitamin tablets in the morning. And she now ate meat because she had read a book about hunter gatherers of 40,000 years ago who all ate meat, and how good it is for us today. That thought totally dismissed human evolution.

Yet, she was one of the few spiritual leaders who also gave attention to physical wellbeing instead of focusing only on the spiritual dimension. I only wished that Jenny of the Nine, and Osho, would have cared for their health as Garimo did.

I was further surprised that Garimo did not approve of me drinking coffee. She said, "That stuff will make you high."

"That is why I drink it," I replied, tongue in cheek, as she sipped her decaf latte.

I also became concerned about her absolute commitment to making as much money out of these tours as she possibly could. When I mentioned this in a roundabout, subtle way, she said that she needed the money because she helped prisoners in the American jails. Must be costly help as she was taking in an awful lot of money. Thousands of people paying $20 a head to attend an hour-long satsang every day, and then paying $500 for an exclusive, booked-out weekend retreat with her.

Many of her books, tapes and other paraphernalia were also sold for high prices at all these gatherings.

I was silently critical about the large amounts of money coming in, while none of her many helpers and staff were being paid a penny. She only paid her personal assistant, who worked long hours, a small wage. We all served the master for free in the name of enlightenment, or in my case, in the name of learning. Perhaps I was learning what not to do, rather quickly.

Throughout the tour I could not help but take notice of the everyday life behind the scenes of satsang. I became surprised at the way this lady was displaying and showering large amounts love at her satsangs and retreats, yet when she was with her staff or a few followers, outside of satsang, she would dismiss people with a shortness, laced sometimes with what looked to me like anger. She seemed to have no concern whatsoever for people's feelings and one of her helpers left in tears one day.

Surprisingly, she continued to mention her enlightenment in satsangs even after my warning in Perth about confusing the minds of her followers.

A teacher I am not, and neither was she a master.

One day she asked me to drive her to the airport on the Gold Coast because her husband was flying in to meet her. Now it was common knowledge among her staff that her husband, Ben, had been having an affair for years with a young 24-year-old student of his. The young lady attended his three-year-long self-awareness course which cost a fortune. I guess everyone desires to be enlightened or go to heaven and will pay a fortune to do so. Ben had managed to keep this sordid affair a secret from Garimo for a long time. She had only recently found out about it because the young lady was suing him.

On the drive to the airport, she said to me, even though I had known about it for a long time, "I may as well tell you this. My husband has been having an affair with a young student of his for years. I am no longer with him and we shall be getting a divorce."

I asked her how many times she had been married and she told me twice. I quickly said, in order to try and cheer her up, "Not to worry, third time lucky."

She replied, "No, Sahajo, I shall spend the rest of this life alone with no man."

Fair enough, I thought, she at least knows what she is doing and what she wants. She is wise enough, at least, when it comes to the sort of needy love she must have been involved in.

When the large, overweight husband, nicknamed "The Monkey", came through the gates, she rushed up and threw herself in his arms and they kissed passionately. My mouth hung open in shock.

I drove these two self-promoting "enlightened" people back to their very expensive lodgings. Ben seemed to always be speaking about his enlightenment and I suspected that Papaji may well have been rolling over in his grave when he kept hearing this self-promotion.

One day Ben pulled out a small box that measured about 5 inches square. I had never seen something like this before.
This box was used to smoke maryjane while apparently keeping any actual smoke from going into the lungs.

For the rest of the tour Garimo and Ben appeared to be madly in love, or at least in drug-induced lust. What about the coffee high? Are only some highs spiritually legal? Not that I have anything against marijuana.

So this is a closer look at enlightenment. Good lord! And it became progressively worse.

Ben would smoke pretty much all day through his little box. At satsangs he would sit on the stage with Garimo and give advice. He was always "shit-faced" with a big marijuana smile on his face. I wondered why everyone in the audience could not see this as it was so obvious.

The house they had rented for the many weeks they were here, had been rented by Brandy Bires, another so-called enlightened soul from California, who was also adept at relieving people of their hard-earned cash. This old wooden house on the beach in Byron cost the princely sum of $5000 a week, they proudly told me. When we entered it for the first time, the rubbish disposal under the house had a strong smell of old fish coming from it. I said, that is no problem, I will clean it out with some antiseptic. Upstairs, the floor was a little dusty. No worries, I said, I will sweep it.

No! No! No, they said sternly. We are paying $5000 a week for this house and we shall ring the agent and make him come and clean it.

I found this situation to be ridiculous. The cost was outrageous, and they had been offered a beautiful modern house in the hills by one of her followers, for free. Were they trying to maintain their enlightenment and social status? They had to be by the beach and in the same house that Brandy Bires had rented.

The high cost of accommodation required a high cost from their students. Quite a few of the poor, desperate people, who thought they loved and worshipped this pair, would beg to be allowed to come to the retreats, but did not have the full $500. They would always be told that, for their own good, they needed to come up with the full amount. No exceptions. And definitely no discounts on the $20 fee for each satsang.

And it got worse.

The retreats became a farce. They were no longer "live-in" events as the participants' accommodation cost the "enlightened ones" too much money. The intensive retreats were now held in a small, old, hot community hall in a little town called Bangalow. The participants, and they were in their hundreds, were instructed to go home at night after each day's "retreat" and to remain in silence. During the day, Ben and Garimo would grace them with their enlightened presence for an hour in the morning and an hour in the afternoon. The rest of the time the staff would show them movies about growth or have them do age-old exercises borrowed from Esalen or Osho.

One day, Ben asked me to drive him to Nimbin. He told me on the way that he wanted to buy some hash. I warned him that Nimbin had cameras in the street and if he bought hash we could very well be pulled up by the police and arrested on the drive out of there. This did not deter him.

On the entire drive to Nimbin, which took over an hour, this man talked non-stop about himself. I wondered if this was why Garimo was with him as she is worshipped and adored by everyone else and being with him is how she gets a break from this tiresome worship? Could well be the reason.

One of the stories he told me on this drive was about a time in California when he was a large dope grower. The police came after he harvested a successful crop and took him to jail along with his crop, already dried and jammed into a large hessian bag.

He told me that when he was in the jail he created a diversion and Garimo came in and took the dope while the cops were occupied with him. Therefore, the police had to release him and he was never charged because they had no evidence. Why tell me this? How could this fool be so self-absorbed? He should be protecting his wife, not telling me egotistical stories like this. I doubted whether they were even true.

It was at this point that I had to stop and look at myself in the mirror. What was this man teaching me about myself? Jenny had taught me that all life experience is a mirror so if I am judging him then I must be judging something about myself. What? I suppose I have my own ego, perhaps well hidden from myself.

It was patently obvious that he wished to be seen as some sort of hero as well as a spiritual leader.

Did I take pride in being the bodyguard of famous people? And I probably also imagined myself as someone who knows a lot. So I should be grateful for this boring opportunity to take a deeper look inside my clouded mirror. Here was his lesson, in the flesh, and I was grateful for it.

Thankfully, after finally arriving in Nimbin Ben got out of the car, walked straight into the building called the Hemp Embassy and asked the man behind the counter for some hash. The man looked at him as if he was an idiot and by this time I thought that he was as I had just told him that asking for hash at the Hemp Embassy would get him nothing but a suspicious look.

Ben didn't even hesitate and continued down the street by himself, and I refused to follow this fool. Sometime later he came back with a block of hash as big as my fist.

Now I cannot judge anyone for smoking pot. Some great men before him have taken this path. Terence McKenna, Timothy Leary, Veeresh of the Humaniversity in Amsterdam, to name just a few. I don't care that he injects Human Growth Hormone, which is legal in America but not Australia, smokes hash, and takes who knows what else. Good luck to him. But how he goes about his business and who he thinks he is, did not impress me in the least.

When this enlightened couple finally left Byron for America, I took a friend of mine to their house to clean up their mess. We were both shocked to find that this self-consumed, egotistical man had left his used Human Growth Hormone needles in an open plastic rubbish bag for the cleaner or me to dispose of.

I had to ask myself again, what was Garimo, who seemed like an intelligent woman, doing with this man? I could not come up with any satisfactory answer.

When I asked her again for an interview, she again said that if people wanted to know about her, they could buy her books or come to her satsangs and retreats. The compassion of spiritual abundance doesn't seem to thrive in the pursuit of monetary abundance.

Later that same year I was informed by her staff that people in America had marched in front of their home when they had heard about Ben's affair with his young student. This young woman, obviously taken advantage of by a spiritual leader, had finally hired a solicitor and was suing him and his large company, which runs his worldwide groups and satsangs.

Now this man, if nothing else, is cunning. My father would have called him "as cunning as a shit-house rat". To avoid being sued for a lot of money he declared to the world, both on the internet and in his satsangs, that he had an incurable cancer and had been given six months to live, with no cure possible.

His company was disbanded and he went into retreat, out west in America where incurable cancer patients are taken care of before they die. He posted pictures of himself in a hospital bed, dying, with no chance of recovery.

The good-hearted victim of this supposed spiritual master dropped all charges against this poor dying man.

After the charges had been dropped, he made a miraculous recovery, jumped up out of his deathbed, recreated his company, and started bringing enlightenment to people all over the world again – for a higher price of course.

And, like the good businesswoman Brandy Bires, he could charge even more now, because he had the added sales pitch of having cured himself of incurable cancer. Obviously, he never had incurable cancer. Incurable, as it suggests, surely means incurable.

So, even though I may have been hoping for a positive message from Garimo to pass on, my negative one should save you much. Monetarily, mentally and emotionally.

The following is this particular guru's bodyguard's message from experience. If spiritual leaders mention, or boast, to you about their own personal enlightenment, as so many do these days in the age of Aquarius, quickly grab your wallet and run a mile in the opposite direction or these enlightened ones will swiftly relieve you of your purse.

Men like Kiran never had to say that they were enlightened and never did. He was a good man, with a message. And, like Kiran, the honest others such as Papaji and Jenny did not have a sign at the entrance to their front door that said "Donation $20", with four people there to make sure you paid the $20 or entry strictly forbidden. Donation indeed!

These days many spiritual masters like Garimo and Ben have websites, and on digital sites they beg for money, always for very good causes of one sort or another of course. Yet, looking beneath the surface, the main cause for their begging is their fat bank accounts and lavish lifestyles.

Ah, the spiritual business. It pays better than oil.

Yet, even in her clinging to be worshipped by others, or the comforts that money could shower upon her and her husband, Garimo still occasionally imparted knowledge that was impressive and of value in the quest for awareness and consciousness. How does she do this? I think one of her secrets, like Jenny and the Nine, is that she surrenders to someone or something else. In her case she has surrendered to Papaji as her master, someone more than herself. This surrender helps her get out of the way of her own ego and with an empty mind, messages of wisdom come through.

Kiran once said, "Sahajo, don't you know that everything is perfect, otherwise existence would not have him up there." He was talking about Barry Long, but he may well have been talking about Garimo or the many others in this Aquarian age who offer a symphony of wisdom on stage.

Ah! The many plays of consciousness. Learning what not to do is a fine teacher. How much more unlearning did I have to do?

The further lessons that I needed were about to come from elsewhere. Existence is perfect because it always seems to give me what I need, not what I think I want in a demanding, begging or prayerful voice. Existence is, it seems, only interested in the evolution of consciousness and awakening into presence. I had been discovering that we are here for a purpose. On the path to awareness, intelligence is set free from the primitive laws and beliefs that presently bind societies, and a new way of living slowly reveals itself when the ever-thinking mind is shown how to sometimes rest.

However, suffering is one of the greatest heralds of change and is no friend of the ego as my ego would soon find out.

I continued to sit with and listen to many so-called masters. Masters such as John Rawer, who seemed to me like a madman travelling around the world collecting disciples while he sat and stared silently, with unblinking eyes wide open, at everyone, as if his eyes had something to actually give or silently transmit. Yet he must be giving something to those that think they need that something that he gives. He gave me nothing but a head-shaking disgust at the madness of it all.

Though Byron Katie, another famous spiritual leader, is definitely worth pursuing if that is what you need at the time. I would say she definitely underwent a spiritually transforming experience that changed her from a fat, sick, angry, pill-popping person, into a thin, loving, pill-less, spiritual being who has helped many people in difficulties all over the world. She is real, and I highly recommend her and her work if you are often overwhelmed by anger, judgement or jealousy. She has also written worthwhile books and instructions on living her work. I watched her transform people, on stage, from angry or jealous ravers into calm and peaceful beings all in fifteen minutes. I found this wonderful and amazing.

I also travelled through the outback with a young up-and-coming guru named Vartman, who had many followers. He had been told about me and asked me to show him outback Australia one day, as I sat minding my own business drinking coffee in a small cafe in Byron Bay. I love the quiet of the outback so I readily agreed as he was paying all expenses for this trip. No wages of course.

Every night while I set up camp he sat and played a strange game of building empires on his computer. He, who could also deliver a statement like a whip, yet needed some whipping himself. I did this for him eventually, with my tongue, when one day he got carried away with himself. I told him, among other things, that his sexuality needed some attention and he may find out what it was if he visited Byron Katie and Da Free John. Da Free John was a popular guru who was focused on sex and relationship. After my constructive critique of his character, Vintman said, "I thought I was the guru here." I answered, "Everyone gets their fifteen minutes."

There were others hardly worth mentioning, such as Peter Caddy, from the famous Scottish garden Findhorn, who was immersed in thoughts of a Christian god, and also young women. When I had had enough of his continual talks of God and God's work, I asked him, while he was sitting in the back seat of my car, to explain the word god to me, which I pointed out that he often used.

He was clueless, and in a knowing tone he merely said: "I will have to talk to you later, Darrell." He never did. He did not know any god and had no answer to my simple question.

The road seemed long and often just a waste of time, but I had a strong feeling that the lessons I needed for changes I desired in myself were on the way, harsh and painful even as they sometimes were and had been in my long life.

Having recently written and thought so much about enlightenment and what it could possibly mean, this wisdom was soon revealed. Synchronicity took action again when I randomly picked up a book and read these words:

> *"Enlightenment is a destructive process. It has nothing to do with becoming better or being happier. Enlightenment is the crumbling away of untruth. It's seeing through the façade of pretence. It's the complete eradication of everything we imagined to be true."*
> Adyashanti

I did not even know who this Adyashanti was, but that was an interesting explanation of my life journey.

Chapter 39 - Desire

"For most of us, belief has greater meaning than actuality. The understanding of what is, does not require belief; on the contrary, belief, idea, prejudice, is a definite hindrance to understanding. But we prefer our beliefs, our dogmas; they warm us, they promise, they encourage. If we understood the way of our beliefs and why we cling to them, one of the major causes of antagonism would disappear."

J Krishnamurti

I always like to think that each dawning year will bring a magical breakthrough and the pain, suffering and disappointments I face will be over, or at least subside. So much for my thinking. Seeing one's own mirror reflecting mind can be embarrassing. Like being caught with my pants down. It does, however, support the teaching that peace is found in "no mind".

After my guru guarding stints, my mind was confused about the path to self-realisation. Desire, and the need for another, always seem to rise up along the path leading to inner peace. Most of the spiritual leaders had proven that to me as they were confronted with the darkness of the world offering the desire-filled fruits of fame. Control over others, sexual pleasures, money, power and drugs, and being worshipped, are earthly offerings easily manifesting in the creative consciousness of mastery. The temptations are veiled behind the status of being spiritual. So I was not surprised when I found myself in a wild and hot love affair.

I easily managed to persuade myself that I had finally found the girl of my dreams and would at last be married for the first time in my life and maybe live a "normal" life like other people. My fear of marriage was instantly forgotten in this romantic hot bubble with its basis in sex and powerful lust on a daily basis.

We travelled, we made love, we laughed, we played and we improved our lives together. We did this by exercising, plus watching what we ate, having fun, and living in the wonderful pristine bush lands of vast Australia.

When I met this woman, she was addicted to tobacco, speed, plus legal drugs in the form of pills. But with care and love she finally gave up all these things. Her crippling afflictions lifted and her skin glowed. We continued to make love a lot.

Personally, I do not think that any drug is in itself harmful. This includes tobacco, alcohol, marijuana, opium, ecstasy, et al. What is harmful, however, is becoming an addict who uses any of these drugs even once, then has to have them every day. Addicts cannot smoke one cigarette or joint a day. No, they have even one beer and then must have twenty. These types of people are in the minority but everyone else has to suffer the draconian and useless laws we are forced to live under because of a minority of poor hopeless addicts. Unfortunately, my new friend had that addictive personality.

At the time I saw Trish as an extremely beautiful woman. But she was smoking a packet of cigarettes a day mixed with marijuana and pills and her back was giving up.

After six months in our loving relationship, she had become strong and healthy once again and her back no longer hurt her.

When we walked into a crowded room both men and women would turn to look at her. She was 5 feet 9 inches tall with witch-like curly black hair.

Trish was a white woman from South Africa who was living in a small town in Western Australia. I met her at a pool party and could not take my eyes off her and her very small bikini.

However, I noticed that she was chain smoking and she talked a lot – a real lot. I thought at the time that this woman could talk under 3 feet of wet cement with marbles in her mouth. Because I have never wanted to partner a smoker, I left the party early and did not see her again for a year.

One cannot avoid one's destiny, it would seem, and I met her again one fateful day at another swimming pool, complete with that string bikini.

After talking for a while, we realised that we had a lot in common and so she eventually left her job, sold her home and we set off for a new life in the far north of Western Australia. At the time she was in a bad way from the overuse of drugs and tobacco and had to leave her job anyway. She found it difficult to walk because her kidneys and back were giving her hell.

She, as addicts tend to do, blamed everything but the tobacco, alcohol and drugs, yet they were the obvious cause of her worsening condition.

Trish desperately wanted to give up the planet's deadliest drug and biggest killer, which is tobacco, but she could not do it in her home town in WA. Not only tobacco was available there, but also the other illegal drugs that her many friends would persuade her to partake of. I realised the urgency of her situation as a deadly new drug had just appeared called "ice". I wondered what chance did the poor girl have, even though she readily admitted that she was an addict?

Trish, like many other people in our society, had an addictive personality. Giving up anything was a challenge of massive proportions. I felt really sorry for her and my love for her seemed to grow with this sympathy. Feeling sorry for someone added to the complicated array of loving feelings I was burdened with.

I convinced her to do a three-day vision quest with me in a deep, isolated part of the Australian bush where nobody ever goes. To do this I told her that we must give up coffee, sugar, processed foods, pot, drugs, alcohol and tobacco, in order to have any chance of success. After many weeks of trying, she finally managed to do all this. No more picking up dirty cigarette butts in the street and then hiding away to smoke them.

I finally took her to a pristine river to fast, after we had successfully been off everything for weeks. It was easy for me as I was not addicted to anything, and especially not legal prescription pills.

I set her tent up a mile away from my tent and we began our three-day quest, drinking only water, eating no food, reading no books, and listening to no radios.

Occasionally I would sneak close to Trish without her seeing me, just to check that she was okay.

On the third day of the quest, I gathered a lot of firewood. While lifting the last large log I tore the ligaments in my right shoulder and the pain was excruciating. My pain was almost unbearable combined with my hunger. I had also managed to contract dengue fever from a mosquito in a city I had been in, and my bones were aching as if they were broken. This is probably why I tore the ligaments in my shoulder.

On the fourth day I went and got Trish. We lit the fire and shared our experiences. Wonderful, I thought, we did it together! I was proud of her.

That night I was sick from the dengue fever and also in agony from my torn shoulder. Both would take well over twelve long months to heal. In our tent I lay still and quiet and tried to breathe through the pain and sickness. I took no pain killers.

Dengue fever makes the entire body ache and I was also feeling nauseous and had a pounding headache. Doctors call dengue the "bone breaking disease".

On the next day I somehow managed to painfully pack up our camp and we headed back into town. Once out on the main road, Trish asked me to stop the car. She confronted me with a frown and a harsh voice, informing me that I did not love her. I was shocked as this was the furthest thing from my experience so I asked her why she thought this. She said that when she had finished her vision quest, she expected me to cuddle her when we went to bed and to make love, and I did not.

Ah! The lesson of expectations, desire, and the trouble they cause.

I looked at her and a cold feeling ran through my body because I realised that something in her was amiss. I did not know what it was and thought it may be lack of food, lack of drugs, or lack of tobacco, so I grabbed her hand and explained that I was sick and in pain but I still loved her dearly. She said no, that was definitely not true, and stressed that I did not love her. I realised, with some shock, that she was talking about herself. Life is a mirror of our reflective mind, so I wondered if she could not love any man for long. The Spoiled Princess Syndrome was also staring me in the face. Or was she a mirror of me and the fact that I had not been successful in relationships?

I knew our relationship was in trouble. Over the next weeks she spent hours telling me what she thought were my shortcomings, and many faults, after she had a few alcoholic drinks.

Seeing clearly the obvious outcome of her criticisms, I tried to explain to her that she was setting herself up to leave me but she disagreed and said she would never leave me.

I wish people could see themselves better. An old saying from Robbie Burns, the Scottish poet, came back to my mind. "Would the God the gift he gie us, to see ourselves as others see us." But this was never going to happen with Trish.

We travelled to Africa together but unfortunately Trish started smoking pot laced with tobacco again when we returned to her old town in South Africa.

Almost every day she would find fault in me with such trivial things like I did my washing with the washing machine not totally full, or I walked around in my shorty pyjamas in the mornings and other people may see me, and so on and on.

She told me that I was a "spoilt brat" because I would not stay in a haunted church on the first night when we drove to an isolated town in Africa. On the long drive to this town of her sister, Trish told me terrible stories about how black people were tortured and killed in the cellar and all white men who stayed there came away bloody and beaten the next day by these dead people.

All these judgements of me were minor, petty and undeserving so I asked her if she hated her father. It seems that women who hate their fathers or ex-husbands usually, but unconsciously, eventually project this hatred onto any male they are with. I tried to persuade her that forgiveness could set her free from this hatred. Forgiveness sets us all free from hatred, in fact.

The "spoilt brat" mantra that she constantly voiced, never affected me. Life, in this case, was like a magic mirror as things that you angrily label and judge other people with are usually what you are yourself, or what you need to learn about yourself.

Finally, back in Australia Trish wanted to go home for her birthday and demanded that I drive the thousands of kilometres back to her hometown to celebrate it.

We had a few arguments when I tried to explain to her that it was a dangerous and sad move for her to return to her addict friends so soon because she was not yet strong enough. She angrily denied this and swore she would never smoke again now we were back in Australia. She continued her drinking, and constant assaults of me so I offered to fly her home and be done with her. She would not hear of this. There was to be no easy escape for me and the lessons I was about to face, so I packed up and began the long drive to her home.

I again warned her that this was not a good idea.

Trish adamantly assured me again that she was done with tobacco, drugs, and her druggie friends. She was with me now, she said, and free of all this. She promised that she would never smoke again. I wondered, later, why people so easily break their promises, even the ones they make to themselves?

Though I still had dengue fever and my damaged shoulder throbbed with pain, we still made love every day, impossible and crazy as it now seems. I surprised myself by still being able to do this, or even wishing to. But not having orgasms, which is based on Tantric lovemaking, conserves a man's energy. This saved me.

Arriving home, we bought a caravan and decided to live in the hills, about five miles out of her town, on the farm of a friend of mine, far from her old addict friends. We loved our little home on wheels and decided to go on a tour around Australia after she visited her family who also lived in the little town. All seemed forgiven and she was happy and loving once again.

Each day she would go off alone to visit her family. I would stay behind and work around the caravan, clearing land and growing vegetables.

One day she came home and excitedly told me that she had a new best friend who she wished me to meet. Her friend was a

55-year-old lesbian who had been a heroin addict for thirty years but had now given up heroin.

I had developed a saying that I often used with Trish when she told me of her past lovers or sexual escapades, as this was information that I did not want to hear. "Too much information" I would often say. And I told her that this was too much information about her heroin reformed new lesbian friend but she still insisted that I meet her.

She turned out to be a sweet woman/man who was more than friendly. Meeting her was like meeting a long-lost friend and she insisted on lots of hugs. I came to see that she was like this with everyone she met. I have never been able to totally trust people like this as they make the best cons.

I noticed that she rolled cigarettes, combining tobacco with marijuana, every fifteen minutes, and her hands shook when she lit them. I noticed that Trish shared these joints with her even after she had promised me that she would not. Within a week Trish was rolling her own tobacco, with dope supplied by the lesbian. I could not believe that this was the beautiful woman whom I thought I loved.

Her sweet lesbian friend was a jack-of-all-trades just like a competent man. Painting, fixing, building, but then she finally hurt her shoulder and neck. A case of too many drugs, I thought, and her body was slowing her down as a warning. It was a warning she did not seem equipped to ever listen too.

One day when visiting our caravan home, she pulled out a huge bag of marijuana, about a pound, and a large bag of ecstasy tablets, and told me that this is how she was surviving and paying her rent because she could not perform her usual hard trades.

This, again, was far too much information for me as I am paranoid about drug dealers. I knew immediately that Trish was lost.

I looked at Trish and knew without a doubt that her previous pain and crippled body would soon return tenfold.

That night Trish invited me to join her at a party at "the orgy house" in the hills. I asked her what kind of place this was and she said it was a place where a lot of married people went and had orgies together, by freely swapping partners. I explained to her that I was not interested in orgies and then she told me that she was not either. I wondered why she wished to go?

It was obvious that Trish was growing more remote over the following weeks. She spent more and more time staying at her lesbian friend's house. I asked her to spend more time with me and told her that I was paranoid around her new best friend because of the drugs she was selling. I warned her that the police could, and probably would, eventually arrest both of them. This just made Trish angry. She stomped out of our van and I did not see her for three long days.

The type of needy, jealous love I began to experience was exceptionally draining.

When she finally returned home, she informed me that we were no longer a couple. However, she would continue to live with me but on her own terms with no sex because I was not strong enough in my head. She was right about that because worry for her had already drained me and I no longer knew what I could do to help her.

I didn't agree with her interpretation of me, however, because "not strong enough in the head" was definitely her issue when it came to drugs and tobacco. However, I did need to face the fact that I was becoming obsessed with her and her problems.

New Year's Eve arrived and she told me that she was spending it with her lesbian friend and that they were going to a party and were going to sleep in the van together.

By the 3rd of January I had not seen or heard from her, and I was getting very little sleep thinking about where she was, what she may be doing, and if she was okay. When I finally was able to contact her on her mobile phone, she informed me that she had been partying for three days at the orgy house.

Again, all this was too much information for me and I lay awake for days and nights, wrestling with what to do about it. My mind was driving me mad as visions of all these people performing terrible sexual acts with my beloved kept attacking me. For three long nights every time I finally dropped off to sleep,

I would awaken almost immediately with visions of the poor woman in Vietnam being electrocuted, with her thumbs tied with wire and pulled up towards the ceiling as she stood screaming in pain on her tiptoes. The horrified looks on the faces of her children would flash in front of me.

I would awaken, sweating profusely, as this was combined with a vision of Trish being held down on the floor and both men and women raping her.

Part of me wanted to take my bowie knife and a shotgun that belonged to the farm and rush to "save her". I sharpened my knife for three long days and nights and cleaned and nursed the double-barrelled shotgun.

The wounded macho man in me wanted desperately to save his woman even at the cost of life if that was necessary.

Strange synchronicity, as usual, had been occurring this entire time. Every time I turned the TV on there was a story about someone's lover being murdered by his or her partner.

At least one person a week dies in Australia from the hand of a jealous lover, the newsreader said.

I thought, in my turmoil, that I could easily raise those figures and increase the average by a lot.

My disturbed mind went around in circles as I tried to figure out what action to take. From past experience, I knew that I could heal from a false broken heart if I did not see or talk to the beloved for six months. I just had to face the burning pain and the relentless mind. And I needed to do this alone so I could not talk about the lost love to anyone.

If, on the other hand, I murdered my beloved, and her lovers, then I would never heal and I knew that suffering brought about by my own hand would be my constant companion until life's end and probably beyond, as existence is not limited by time. I recalled the wise saying, "As it was in the beginning is now and ever shall be"; so this suffering for killing would probably become my eternal companion. So please think of this before killing your lover for not loving you and for having sex with someone else, even a woman or multiple partners. I told myself these things.

I had to continually acknowledge the fact that I was only suffering this terrible jealousy because of what religion and society had taught me about love and women.

Finally, after almost unbearable suffering, I knew I must eventually walk away and let my ongoing pain, caused by the pursuit of pleasure and desire, heal what needed to be healed, and then be released from me, never to arise again. These logical thoughts were keeping me from acting the hero and saviour. Strangely enough this situation was curing me of constantly thinking about the wrongs done to a woman in Vietnam so long ago even though it sometimes still seemed like yesterday.

Even with everything I had discovered in my travels, this mammoth struggle was harder than any mental or physical fight I had ever been in.

I realised that I was also doing battle with my ingrained male, superior macho side that thinks that women are weaker than men and must be protected. There was also that ingrained Christian sexual side in me that cannot bear the thought of someone else having sex with my beloved. Why? In case she enjoyed it more.

I remembered that males use religion to control women sexually all over the world. The Christians do it with teachings of guilt and sin, while the Muslims cover their women from head to foot, in black, even in hot weather, and treat them as far lesser beings than men, perhaps so that they do not have to face what I was now facing. Maybe this is what they are afraid of and why they have these strange teachings of control even though this eventually drives these believers to distraction and inhuman actions, even wars.

And why must men possess women? Because women are stronger, sexually, than them?

I was fighting for understanding as I wallowed in my deepest hell.

After this trip through a special kind of hell I finally realised that Trish had chosen exactly what she was doing and could not care less about me and what I was feeling. I had no right whatsoever to hurt anyone or to even encourage my mind to imagine that she needed rescuing by big, brave, macho me. She was a grown, strong, single-minded woman and she had the human right for freedom of choice.

I recalled that she had once told me, when we were making love, that one of her dreams was to be in an orgy. That shocked me then as it shocks me now, though I had successfully put it out of my mind for a time. "Too much information." Unfortunately, we eventually live our dreams and desires before we can wake up from them. I had been living mine with her and this was my wake-up call.

I had tried to escape from her, and this future outcome, during the time I was with her, because I knew it was going to end badly. But existence, or consciousness, call it what you will – though it is better to give it no name – had conspired to keep me in this relationship and to finally break me, together with the hidden teachings residing deep inside. I needed this lesson, shocking as it was. Every time I had tried to escape from Trish and this disconcerting madness of false, pleasure-driven love, a message would be presented telling me without any shadow of a doubt that I could not leave this woman at that time. It would be the written word I randomly read or something that had been told to me by a psychic man informing me that I must not leave her even though he had never met her. This was all amazing synchronicity that left me with my mouth hanging open in wonder and amazement. I was being cleansed in a shocking but effective way. Perhaps in the only way I could be made new.

It felt like my heart, desperately clinging to dreams, in league with my ego, was being smashed into a million pieces.

I can handle a woman I am with running off with another man, as I have had to do in the past, but this was way beyond anything that I, or anyone else, should ever have to suffer. Trisha had run off with a lesbian who had also been a heroin addict for thirty long years and then moved into the orgy house with her. Really? My fiancée! On death's tail? And I could not save her?

Strangely enough I knew it was not personal and that there was no one to blame, especially not Trish. This was my own personal lesson and it could not have been scripted any better in a movie. There was some hidden energy working for me here. Eventually I felt touched as I recognised its evolutionary progress.

It became obvious that love that turns to sorrow for the other is not love at all. It was a chance for me to be aware of myself and all that passed through my mind in this dangerous state I found myself in.

And I remembered some of Kiran's words on love and realised that I had been looking in the wrong place for love. Yet, still, it was as he had said: "Don't you know that everything is perfect." So it was perfect for me, this deep imagined suffering.

I knew that I had to continue to face this heartache without alcohol, pills or drugs of any kind. Understanding only comes through awareness and the deep feeling of pain, not the numbing of it.

I cried. Perhaps I was crying for the death of thousands of years of the way we have been taught to love.

I lost weight and dropped quickly from 85 to 70 kilograms. My body was still racked with pain from my wounded right shoulder, combined with the dengue fever, but now it seemed much worse with my imagined love gone. Some nights my chest hurt so much that I thought I was about to die. I dealt with death on many levels over the next months. Sometimes I dearly wished to die and be done with this extreme physical and mental pain of life that was being visited upon me.

Every morning when I awoke my first thought was of Trish. My last thought at night was always about her and most of my thoughts during the long endless days were also about her. How long would this special relationship take for me to be rid of my teachings and my deep unconscious longings? Too long it seemed to me at the time.

Now it may seem like I am complaining but the truth of the matter is that underneath all this pain I began to realise that I was being changed forever from the inside out. Some things in me needed changing and this was evolution's way of doing it, so everything was indeed perfect, horrible as it seemed. Evolution was with me and working for me. I was simply evolution in progress. Could I surrender to that?

My face had changed rapidly. When I looked in the mirror, I was beginning to see an old man with large bags under his eyes. I had never had bags under my eyes before and never looked over fifty. My hair was quickly turning grey, it seemed.

A part of me that was once the confident, easy-going, take what he could, competent lover, was slowly dying and disappearing.

I became so damaged by the deadly and dangerous process I was undergoing that I finally contacted an Indian Vedic astrologer. Following that, I had a reading from one of the world's leading psychics, closely followed by a reading from the leading Western astrologer in the district.

Strangely enough they all told me much the same thing.

The Vedic man told me that I was undergoing a process that only came once in a thousand years and during it I would continue to suffer greatly and think I was dying but to hang on because this too would pass and on the other side was wisdom and I would be able to share this wisdom.

The Western astrologer told me something amazingly similar.

The psychic, when she saw me, kept grabbing her heart and telling me I had a lot of "filing" work to be done and it was going to hurt like hell. She kept doubling over as if in pain, as if she somehow felt my present and future pain.

I was lucky to be given this information by these three people because it made this process a little easier although easier does not at all explain this dying of the old in me. It simply made it bearable as I once again realised this had to happen to cure me of false dreams and the endless pursuit of romance and pleasure.

I bowed my head and thanked the fact that this had come to pass because my unconscious longings were becoming conscious, rising to the surface, and fading away.

What continued to hurt me deeply was the fact that I did not have to be psychic to see Trish's future, and just as with the woman in Vietnam, I could not save her from this future. People must face their future according to their current choices. I could not be their saviour.

One of the awful things that I could see in Trish's future was the eventual shape of her face as it changed and swelled. Obviously, her teeth would gradually turn a different colour, and the clear eyes she possessed would cloud over and take on an often-haunted look, as she continued to take her next smoke, or drug and alcoholic drink. Her beautiful body would eventually turn on her and she would become crippled.

I have watched my mates die like this, those who took this route to ease their unforgiving nightmares of Vietnam and what they may have done over there.

I wondered what Trish may have experienced in South Africa to affect her nervous system like this?

It was obvious that I had to leave town as soon as possible to help this process of forgetting Trish and be shed of it.
The pain was still rising every time I saw her in town, smoking with her friends. At night it reintroduced the nightmares of the electrocuted woman and her three poor little children, together with Trish being overcome by people raping her.

I also sometimes wondered if they – the children in Vietnam – were still alive and how what happened to their mother affected them?

The day before I left, I went to see my still beautiful Trish for the last time. I explained to her that I was grateful for what had happened and she was in no way to blame, as life, through her, was allowing me to clear away many hidden, unconscious things inside myself. I told her I just loved her and could not control, quite yet, this feeling I had for her. Obviously, I was still in the cleansing process as this was, of course, not the love I ever wished to be in.

I explained to Trish the distress I had been going through, imagining what she had been doing in what I saw as the house of horrors, the orgy house, and my concern about the assortment of drugs she was taking, and especially the tobacco she was once again chain smoking. I told her to please think of AIDS and other common sexual diseases possibly found in an orgy house in this day and age. And, of course, I told her of the cost to her body of tobacco and too many drugs. She would eventually have to pay the "Ferryman", I warned her.

She surprised me by saying that she would rather sleep with a black man than with a lesbian and that Shirl, the lesbian, was just her friend. She also explained to me that she was not really interested in orgies and she had slept with no one in that house though she spent a lot of time there, as she "loved the people".

Once again, she was giving me "too much information" and it was obviously not the truth. People eventually live their dreams and she would live hers when it suited her. I realised that she is a tough person. Tougher than me that's for damn sure.

In defence of her "black man" statement: when I was in South Africa there was no white family who I spoke to who did not have a story of rape, robbery or death from attacks by black gangs. There are something like fifty million black people in South Africa.

Yet these attacks are definitely not the entire fault of these poor black people. The slums in South Africa run as far as the eye can see. And, unbelievably, they seem more primitive and poor than the slums of India. These poor black people live, if you can call it that, in tiny one-room tin shacks, that are unbearably hot in the harsh African summers. And these people, often, do not have enough money to eat.

The few million whites are generally wealthy, or at least well off, and the blacks have nothing, so of course they turn to robbery just to exist. A pity the whites do not share the South African wealth with them but who gives their money away, especially when these whites are so afraid, and have to be alert every day.

Trish finally and angrily told me that she was a fifty-year-old independent, autonomous woman who did not need anyone to save her.

I looked up the word autonomous in the dictionary when I got in my car and it means "self-governing".

It seemed that everyone was innocent of the dreams and visions I was being subjected to and it was only my conditioned male mind that had been driving me mad.

Yet I still had the uncomfortable knowing in my logical mind that if she continued to use ecstasy and other drugs she would eventually relax into the ways of these partner swappers and orgy lovers. Why else would she hang out with them and tell me how much she now loved them, and that she no longer judged them?

I may well be naive but not entirely stupid.

Trish had still not lived "her dream" of being in an orgy and people usually live out their dreams to eventually get over them. Just like me now.

I realised from what her family members told me that she was a tough, intelligent (in some things) South African and would do exactly as she wanted, when she wanted.

The pain of lost love would hurt me for the next year. It took longer than six months this time to be rid of memories, pain and heartache. This experience had been a deep one on many levels.

It became obvious to me that all the wisdom in the Bible, the Koran, and all the other religious holy books comes from our personal experience and history, not from what some imagined God, Allah or Elephant God is purported to have said. Do I "throw the baby out with the bathwater" by abandoning all histories hard-earned wisdom hidden in these holy books because I do not like or approve of religion. One may read between the lines in these holy scriptures by ignoring all references to gods and their often unholy orders, which are only the orders of power-hungry men.

Trish did not have much money at this time and she told me she had signed up to do a university course but had no computer. I took her to the nearby Mac shop and bought her a going-away present, a Mac Air computer for $1500, and told her that this was less than little payment for what she had cured me of and taught me about myself and love. Even though she did not know what she had done.

Her still beautiful eyes lit up, as she loved the new Mac. Obviously better than she loved me.

She hugged me and I said goodbye and drove off.

On the outskirts of town, I was waved down by Shirl, her new best friend, who she now lived with. She told me that Trish had told her that she would never have sex with a lesbian and that she tells anyone who will listen that I am the best lover she has ever had and she has had a number of lovers.

This information did nothing to ease my pain but it did show me that it does not matter how great a sexual lover you may be, it is still not enough to keep your lover with you, unless you really know what love is. There are many other requirements, such as being clear and conscious in yourself for a start, and not prone to the affliction of attachment to the other or how you think you want them to act. It seems strange that I know all these things and have been taught by masters, yet I had to have hidden things virtually burnt out of me through rough experience. Suffering – the great teacher.

Yet at the time this was once again "too much information" from Shirl and would only make it harder to forget Trish.

According to the astrologer and the psychic, my journey through suffering was not due to end for many months yet, so existence was obviously set on giving me all the unfortunate information I still needed to prove their predictions correct. How right they were. It was now necessary for me to be alone, with no friends, no lover, no home, and no family. What a grim process I still had to go through as the shocking and painful burning of hidden beliefs slowly ran its course.

My faith remained blind, unknowing, but nevertheless convinced of the existence of Something other than, but also a part of me. I did not use thought to think of myself as separate from this "Unknown." It was just evolution working on itself. Not a religious inspired contract between two unequal partners – master and slave.

Sometime later, and far away, when I was walking alone in the isolated bush, and my thoughts became exhausted, I could feel this unthinkable something inside me and all around me. This was life found in emptiness and in everything.

On leaving home and Trish, I did not know where I could

actually go but as usual some unseen force seemed to reach out and direct me. Life is like that if I watch carefully. It is coming towards us to give us what we need, not what we think we want. Life knew better about these things than I did in my present condition, and I surrendered to this. I bowed to the mystery.

A year later my professor friend sent me this poem by the ancient Sufi poet named Rumi who once lived in Konya, Turkey, from 1207 to 1273. It sums up what I was experiencing at this time.

The soul is a newly skinned hide, bloody and gross
Work on it with manual discipline, and the bitter tanning acid of grief
And you'll become lovely and very strong
If you can't do this work yourself, don't worry
You don't even have to make a decision, one way or another
The Friend, who knows a lot more than you do, will bring difficulties, and grief, and sickness
As medicine, as happiness, as the essence of the moment when you're beaten
When you hear checkmate,
And can finally say, with Hallaj's voice
I trust you to kill me.

Rumi

This is an ancient description of surrender, and partly of my experience, that he wrote about so long ago. I imagine what he means by killing is to kill what is necessary in us such as a swollen ego, or even a romantic misconception. Not the death of our bodies.

Isn't life sometimes wonderful and amazing, in between when it is being a suffering inducing teacher.

Arriving back from isolation in the deep outback, I was still raw when a friend I had not seen for years told me about a German man named Karl Renz who had had experience in a foreign army. He later lost the love of his life and his millions of dollars, and all in a couple of crazy days. This sacrificial suffering caused him to have an amazing awakening experience. His life changed dramatically and he began to travel the world talking to and teaching people about life and true spirituality.

My friend said that this man was the last stop on the spiritual trail and usually only the desperate or the broken-hearted came to sit with him.

That certainly sounded like me at the moment – both desperate and broken-hearted.

We agreed to fly to Thailand together as I felt too weak and unsure of myself to travel overseas alone. We planned to sit with this man who was about to talk and teach for two weeks on the island of Koh Samui in the south-east of Thailand.

It seemed to me that mankind's evolution is often being honed and prepared quicker in the darkest times. Vietnam tipped me off to that fact. Shock treats and awakens. Or destroys if the situation is unbearable.

With my weight now less by another 10 kilograms I felt as weak as a kitten. The pain in my bones from Dengue fever, especially when I tried to bend over, had made it really difficult to even tie my shoes. I, like this Karl Renz, felt that I had now lost everything including my body.

I booked and paid for my ticket, then my friend changed his mind. In my weak state I really wanted to go with him for support, but life, as usual, does what it likes, not what I want. Yet again I was meant to be alone and so I was to fly back to a country in Asia where I have never really felt safe, for long, since Vietnam.

My shattered ego, intense humility, broken physical body and broken heart would all have to be my sole companions on this journey.

As often happens when something is happening in my life, it is reflected in the world and the media. I read in the newspapers and heard on the TV that partners were still being murdered daily by their jilted and jealous lovers – the killers, and sometimes thoughtless maimers, being both men and women. One woman, who was now famous, had been burnt beyond recognition by a jealous woman who had seen her talking to her boyfriend. It was not only men affected by our ancient teachings.

Jilted lovers please take this assurance for yourself and know that there is "consciousness" guiding you through difficult evolution.

Again, I iterate that if you kill your partner in a misguided loving frenzy, this heart pain will increase a thousandfold and it will last all this lifetime and perhaps forever. But take heart and moral strength, and all fear, jealousies, hate, and all other unconscious chains shall disappear and you will eventually be brand new.

I was also obviously reassuring myself here even though all thoughts of killing or retribution had long ago left me.

Chapter 40 - Thailand

It seems that when I finally surrender to what is, synchronicity takes over. Destiny produces a myriad of tricks to get me to where it wants me. My life search for truth and to bring necessary changes to this world has been never ending and full of miracles, even here in Thailand.

My first stop was in the northern Thai city of Chiang Mai. Seemingly out of nowhere I had been contacted by a man in this city who had been meditating for the past thirty years for up to eight hours a day. His house, of all the places that it could possibly be in this huge city of millions, was perched on a hill a few hundred yards from my old Buddhist wat and temple where I had become a monk so long ago. I was stunned but the message seemed clear. I was to again spend time in my old temple. I did not know why.

Every day for the next two weeks I visited the wat, the temple.

My old master had become a revered, saint-like figure. His home was sealed off and there was a very lifelike statue of him sitting crossed-legged, next to his old cabin. Inside were pictures of him over the years. He was a really lovely man who treated me with respect and caring. To see him like this touched me and took me back to his feet and to my old Buddhist shack, which was still sitting nearly next door to his.

The wat grounds were much smaller back in the 70s. The grounds were now huge and there were two temples, not one. Westerners were these days practising sitting and walking meditations in the courtyards, or anywhere they pleased, and the men and women were mixing. It was almost like a Buddhist Vipassana, a music dancing doof in extra slow motion.

They also did not meditate for nineteen hours a day or only

eat in the mornings like I had to. I looked in their rooms and saw books, TVs, food and computers. How things had changed. They also meditated for only eight hours a day, though this is still a long time for a Westerner.

Each day I came and sat in the temple and meditated, contemplated, and sometimes physically wandered the wat and remembered something about each of my special places of so long ago.

Yet it once again became obvious that I was definitely being directed, guided and supported by this strange relationship between myself and the "unknown". One of the ways it often did this was with writings that seemed to appear from nowhere.

The spiritual meandering into my past life here didn't last long, however. I woke up at four o'clock on the morning of the full moon feeling tired and utterly sick of this quest of meditating every day in my old temple, still seeking something elusive within and without, perhaps like people seeking enlightenment. Damn it, I thought. Enough of this. I shall take my ex-lover's advice and go and have fun. I planned to go out and find a beautiful, kind, loving Thai woman, as they seemed to be here in abundance.

I sat up feeling my heart's longing. I looked down and next to my bed was a book that had not been there when I went to sleep. A book that was just sitting there, waiting for me to randomly pick it up, as so often had happened to me in my past when I was questioning my situation.

I picked up this book and opened it with a flick of my wrist. It fell open at page 22, and this, to my absolute amazement, is what I read:

How many friendships have you forged, how many loves have you formed, how many threads of attachment have you tied – and every time you have lost everything. All that came into your hands was anguish and sorrow. But still you did not wake up. Still you hoped you would find someone somewhere else. Let me search a little more, just a little more. Hope never dies. Experiences tell us that we will not find what we want, but hope goes on winning over experience. Hope goes on weaving new dreams. Only the person who awakens from hope awakens from the world and becomes free. No, there is no soul mate here, and the inner flower of the heart never blossoms here. It can only blossom with the touch of the divine. Seasons will come and go but this flower within you will never blossom, never. It can only blossom when the season of the divine comes. That is its spring; the rest is autumn. You can wait as long as you like, but sooner or later you will have to return. An intelligent person returns sooner, a foolish person takes longer. An intelligent person learns after even a little experience, a foolish person makes the same mistakes again and again – and gradually becomes used to these mistakes. He repeats them more and more, he becomes skilled at this. Wake up! Don't repeat your mistakes. All the Sages are guiding you through this door.

> *Remembering the divine,*
> *The serpent of time and the creepers of sorrow*
> *Do not bother you.*
> *Hence, embrace the divine*
> *Leave the net of the world behind.*

How was this possible, I wondered? Such amazing synchronicity? After reading this I realised that I must stop looking for love and obviously must persevere with my writing and continue to do long meditations in the temple with my friend Vimal, the marathon meditator.

I saw my own desire and longing, so decided, here and now, to do no seeking outside in the world with all its tempting flavours. I would continue seeking alone and inside myself. The message was absolutely clear and I was meant to do, and really could do, nothing else but this.

I had to rise beyond my usual needy self. I had been shown that more was possible only in this direction. The divine is experienced in how one is seeing. Perhaps a reflection of the tune that the divine is playing.

Each day my friend came and sat with me in meditation.

I would sit for an hour while he would sit for four hours, then he would go home for lunch and then sit for another four hours in the afternoon.

I took to watching him because his eight hours of meditation fascinated me. I wondered why and how he sat for so long without moving, every day, for the past thirty years? I asked him why he sat. Did he think he was going to become enlightened, the East's form of heaven or paradise? "Why? Why do you sit for so long?" I asked.

He answered, "Because while I am sitting, I am happy and in bliss."

Now call me a pessimist, but I know that sitting for four hours without moving will definitely create many experiences but not permanent happiness and endless bliss. Oh no, sir! Life itself is not like that. It is not endless bliss.

And when happiness, bliss, or in fact any feeling or thought has disappeared, then I have experienced this as meditation in this very wat.

So that is why I took to closely watching him. I noticed that he sat with his body straight for about fifteen minutes then his head would begin to nod off as in sleep. He would jerk it back up the first couple of times, but by the third or fourth time his head would be hanging down to his chest, his back bent, his body hunched forward at his hips, and he would no longer be sitting straight.

He had learned how to sleep while sitting. Amazing! I had never been able to do that. I bet it does take thirty years of practice. Whenever I fell asleep while meditating I would fall backwards and that would wake me with a jolt.

I recall Papaji saying that it is all very well meditating in the quiet of a cave but true meditation is your ability to remain in the centre of the cyclone of life in an aware meditative state. Papaji would take people into the busy centre of the city and say, "Meditate here".

I looked at his slumped body and hanging neck and could see, because I have been a body worker, that he would have a sore back, a sore neck, and at least one sore shoulder. I touched his neck and lower back and told him he was most likely sore in these two places. He told me that he could not lift his arm above his shoulder, and his lower back and neck had been giving him so much pain for so many years that he could no longer work.

Barry Long had told Vimal to stop meditating. Even though he thought of Barry as enlightened, and his master at the time, he did not stop. It would be a waste of time for me to ask him to stop. He still meditates for long hours to this very day.

We would eat together when we were not in the temple. He was a handsome man and all the beautiful Thai ladies who were single flocked around him. Because I was his friend, they soon also showered their love and laughter on me.

These Thai people are so open, friendly and beautiful. And they seem happy even though they work long hours for little money. Five per cent, even if that many, of the Thai people have about ninety-five per cent of all the money in Thailand. These five percent are the powerful generals, the politicians, the royalty, and some businessmen. The rest are mainly poor except for a few of the middle class like doctors and lawyers. Yet these extremely poor people appear to be happy. Most would like a better life, especially in Australia, so in their eyes my friend and I might become potential lovers and husbands, who could take them to this imagined better life. I don't know about that because people in Australia do not look as happy as they did and when you take people away from their friends and family to a strange country they sometimes fade away and their smiles fade away with them.

I remembered the written message that I had discovered next to my bed. It revealed to me that it is not possible for me to partake in any of these games of love. So I gave these lovely women generous tips and a few little presents because I admired and respected them. I enjoyed making them, and myself, happier by helping them to laugh. This was an easy kind of love to experience free from romance, promises and sex. The mysterious book was perfect for me in this intense time of learning.

One of Vimal's beautiful lady friends was an avid runner so I taught her the finer points of running. How to breathe in time with your steps and how to totally relax while running. These lovely people experience gratefulness so easily, even for such a little gift as this, and she showered me with hugs and a free meal at her café.

Pursuing enlightenment through meditation is like pursuing heaven before you are physically dead. As I have noted before, a lot of Catholic sannyasins, who now hate Catholicism, have merely replaced the thought of heaven with the thought of eventual enlightenment, yet it is the same mindset of something unknown to be gained somewhere in the future.

I suppose that at least enlightenment can be imagined to be gained in this lifetime, unlike heaven or paradise, which can only be found when you are dead, so it is a small step in the right direction. Yet these types of pursuits are all based on longing and desire and would be best forgotten.

"Best we forget", not "Lest we forget".

However, I continued to respect and appreciate some aspects of Buddhism even with its limitations. Buddhism is by far the best of a bad bunch of religions that dominate the world today. Buddhists are more tolerant and peaceful, more sexually free without judgement, and less warlike in their way of living. Yet like all the world's religions that I have studied (including Hinduism), the women are looked upon as less than men. Is this because all the current gods we worship are said to be male. Surely it is time we stopped worshipping gods, or even goddesses, that separate us from ourselves and are outside somewhere, living in magical kingdoms. Isn't it time that we started including all life as part of ourselves?

Even at this time, in my old wat there was a sign in one of the temples that clearly stated: "Women are not allowed". There were no places in the wat where men were not allowed to go. There was a strict rule for males that they may not touch or look at women. Are women considered dangerous, or just less than men?

However, there are no rules for women that state they cannot touch, look at, or have sex with men. In this case I wondered who was better off – the men or the women.

Women are still not allowed to become monks or wear maroon robes, even today, and they can look at men as much as they like.

Again I suspected that men are sexually weaker and in fear of the sexual strength of women?

Do holy men gather power to themselves by imagining that they can control both men and women sexually? An impossible task eventually.

Why do current religions create the separation of men and women? This further enhances religious orientated duality. I imagine that this actually separates us from the ecstatic feeling of the divine.

One of the underlying and great truths that I had learned during my travels is this one statement that I had heard repeated over and over.

It was this: "There is only one self".

The understanding of the truth of this "one self", which therefore must include all possible gods, all humans, all things, all universes, is the beginning of the death of religious taught separation and the current ingrained way of thinking that causes all the unnecessary separation, war, terrorism, greed and pain.

I had finally had enough of these temples, and all religions, so it was time to fly to Ko Samui and see who this Karl Renz is and what he has to teach.

Before parting from my friend, we went on many long walks in the streets of Thailand. He impressed me greatly with some well-chosen words of wisdom, and I realised he was a beautiful human being who, I am sure, will eventually find a very special place inside himself. He will share this with many, for he will one day find everything he longs for and then need search no more. Then he will not have the time or the need to meditate for so long every day. Indeed, I suspect his every moment of waking life will be a meditation, after so many years of practice.

Through him I did realise that there are many paths to where evolution/life/self is slowly leading us and all these many paths are heading towards the same place.

I also hoped my meditating friend would eventually turn away from his current master, one of many from his past. This current master's name was Bubba, an old man who lives in India who is also a believer in hours of meditation every day. To give up Bubba and meditation my friend would simply have to give up his beliefs about meditation and enlightened humans. Until then he would probably keep returning to sit with Bubba in meditation for long hours every year.

Look with the eyes and see the divine in everything.

Chapter 41 - Karl Renz ... The Authentic Truth

After listening to Karl Renz for two weeks I realised that this man deserved a chapter heading all to himself. Let's see if I can explain why.

I flew onto the island of Koh Samui on the east coast of southern Thailand and booked into a five-star hotel. I quickly discovered that I did not like the pretension of a five-star hotel so after one night there I booked into a cheap bungalow costing only $10 a night and perched right on the beach. I waited quietly each day in solitude until finally it was the day to sit with Karl Renz. I did not have the energy or inclination to be with anyone else from any country, especially after what I had read by my bed in Chiang Mai.

Existence had been quietly hammering me into a different person, having given me not only deep heartache, which sometimes felt like it alone could kill me, but also this bone-breaking dengue fever and torn shoulder ligament. The pain was still lingering from both sources – heart pain and physical pain.

I was gradually being softened into someone else as my ego felt like it was slowly being strangled to death.

Some days my vision faded, I would nearly pass out, and then I had to quickly sit on the floor or the footpath depending where I was when this vertigo struck. The feeling of having no control over passing out, as my breath faded, sometimes invited the spectre of death. I had faced death many times before and the thought of it quickly passed as I no longer feared it. Better death passing than me. I would sit and breathe deeply with a stilled and calmed mind, not bothered by the thought of dying.

And so this was my present life in Thailand, and I had passed this way many times before. A lonely, almost defeated monk on a spiritual mission? Good Lord!

I finally got to sit with Karl who was surrounded by other people who had travelled from all corners of the globe to be with him.

This man, more than anyone, including Osho, took away any tiny shreds of hope I still had left in me. In saying this I would also scream from the rooftops that Karl is in touch with something spiritual that is pure and honest, even though it is relentless in forcing the mind to give up its constant knowing.

The bare truth is often extremely difficult to take and it does not agree with normal, hope-filled thinking. The normal people here left after one day. Some of us stayed for two weeks and loved this man and his teaching until he left, even as disturbing as it sometimes could be.

One poor woman had to be taken away to a hospital after ten days when she became emotionally out of control and could not be calmed down.

I'm sure that my praise of him would cause Karl to laugh at me and immediately show me some folly for thinking about him in this way. If I quoted what he said (and he told me not to) he would abuse me, intelligently and relentlessly. He gave the mind no place to settle and then he broke into this beautiful speaking realm where he actually almost revealed how to align with and recognise god (for want of a better word). He was always quick to point out that god is not some being that is separate from our self. He was using this term, god, merely as a spiritual symbol even though it has become grossly over-used.

Ah! The wonderful map of truths he revealed over the two weeks I was with him.

The truth is so powerful it needs no guardians.

I would love to quote what he spoke as I wrote down a lot of it, but it is enough that I recommend you to find out for yourself if you become desperate like me. He is the last signpost on the road to consciousness, then it is up to ourselves.
He leaves nothing of himself to worship or even hold on to – a truly impersonal transcendent being.

If you have a yearning to see Mooji, the big black man who

is a spiritual phenomenon in the world at the moment and is currently speaking to huge crowds in Portugal, then put Karl on your itinerary, as he is usually nearby, also speaking in Portugal. His speaking may well lay the icing on your cake of discovery and awakening in this short life.

That is, briefly, what I got from sitting with Karl Renz in person. I cannot praise him enough as he has not sold out the truth for money, worship, fame, power, sex, or any of the other trinkets that have turned many a gifted sage's head. I told him this at the end and urged him to stay strong and resist the urge for a new Porsche or millions of dollars. He laughed and said that the truth was what was important to him and he would never sell out. Good on him.

Beyond right and wrong
There is a field
I will meet you there.
                    Rumi

Chapter 42 - Belief

Belief is more dangerous than a gun. Belief is what pulls the trigger.

When I finally returned to Australia I decided to fly to Broome, in the far north-west of the country. I knew that something from my deep subconscious mind was still slowly rising to the surface to be seen, understood and dealt with. Until I became aware of these emotion-filled thoughts, I would find no peace. I felt that maybe these feelings still had something to do with my sexual judgements and longing for women, together with somehow thinking of them as innocents who needed saving.

After weeks in Broome, I was no longer thinking of Trish day and night. Thoughts of her and the orgy house and her lesbian house mate sometimes crossed my mind but I could now consciously stop these thoughts and overcome them.

Yet even then, after all this pain and spiritual therapy, life obviously had other ideas and further lessons for me.

I went to the weekend markets in Broome. As I wandered through the many colourful stalls, shopping, and thinking of nothing in particular, an attractive woman with her skirt pulled high above her knees stopped me and engaged in conversation. She was rolling a cigarette, and I noticed that it was the same brand of tobacco that Trish used.

Looking very closely at her I said: "You are obviously a Libran." Her mouth dropped open and she asked me how I could possibly know this. I told her that she reminded me of my old girlfriend who was also a Libran.

We talked for thirty minutes or more and she asked me if I would like to meet her after the markets ended. It was clear to me that I must do this because of the obvious mystical message so I said sure and invited her to drop by my rented apartment.

When she arrived, she produced a "local joint" as she described it, and told me that some of the best dope in Australia is grown up here because of the tropical heat. I told her that I had not smoked for a long time and did not wish to, so she smoked alone but continually blew the second-hand smoke in my face. I am not sure if I got a high simply from being near her, or if the second-hand smoke was so strong that it affected me, but I soon felt stoned, and then found myself in the mystical realm.

This pretty Libran named Jen then told me that she was forty-nine and her birthday was on the 13th of October. This uncanny synchronicity once again stunned me into sharp awareness simply because Trish is forty-nine and her 50th birthday is also on the 13th of October.

How could events like this be possible? Who or what is doing this to me? Obviously, it wouldn't allow me any peace until the inner job was done. Was I being given the opportunity to live this story of love all over again and do it differently this time, or at least heal my obsessive memories of Trish?

After a while Jen told me that she was feeling quite sexual. She told me that she had made love with a few men in Broome over the years and all she had wanted to do was to have fun, but the men always fell in love with her.

Of course, this falling in love often happens to men and women if they cannot possess the lover they are with. We want what we cannot have.

My mum had a saying: "Treat them mean and keep them keen."

So once again here was this stoned, attractive woman inviting me to make love with her.

I stopped and looked long and hard at her.

According to my famous astrologer friend Varij, a Libran's ruling planet is Venus, and Venus is the planet of love, so Librans are often very good at making love and then breaking hearts when they leave. Here was a classic example revealing itself. Well, I had already found this out the hard way with Trish and

I had finally learned "what not to do". Life is a relentless teacher, often teaching through suffering, as had been the case for me. If I did not learn the lesson it returned, with a dunce cap, for another round of teaching.

Staring for a long time at Jen with my eyes half closed, because this is how I see people in a deeper way, I finally spoke.

"We are here together for a deeper reason than sex. It is more mystical. Right now, looking at you, I see neither man nor woman. If I had to describe you, I would say you were a crone, even though that suggests female. Yet you are really neither to me so I cannot make love with you."

I then asked her if I could touch her shoulders. Her mouth was hanging open in shock, or wonder, at what I had just said, but she managed to nod her head.

Having looked at her so closely she seemed sick to me and I wanted to find out what I was seeing. I felt her shoulders and could feel the grit and the tension in them, but it was more than this. I felt her throat and there I felt the beginnings of a small lump. She began coughing violently from my touch and suddenly spat out a lot of green spit.

Now I had already learned many times that I could not help people by telling them anything about their bad habits but in this case, I had at least moved some sickness in her throat due to my awareness.

I asked her how long she had been smoking and she told me since she was fifteen.

I then told her the truth of what I could easily see.

"If you keep smoking you are going to develop cancer of the throat and it will kill you. Do you understand that?"

Tears came to her eyes and she said, "Yes."

I asked her if she could quit. She assured me that she could now, so I showed her a way to smoke that could help her quit.

I told her that it quite simply entails opening your mouth when you take a puff of smoke and letting half the smoke escape, then sucking in fresh air to cool the remaining smoke. Then you breathe the smoke in, but only drawing it to the bottom of your throat, not into your lungs. This allows a person to gradually give up the tobacco and the lungs may clear. One then needs to cut back smoking from a packet to three cigarettes a day, having one after each meal when the longing is strongest, eventually giving up entirely.

I realise that many people find life terribly difficult, as it sometimes can be, and their nervous system is probably soothed by "spin", another name for tobacco because it makes your head spin when you first start smoking. I can no longer judge smokers for smoking this poison. In my own family both my mother and sister, plus her son, had all died from cancer, thanks to cigarettes.

I told Jen all this and demonstrated the technique I had described to her until she said she had got it. Then I asked her if she could really give up cigs. She assured me that she could and she would. I told her to go home and lie with the sun on her throat and let this all sink in and I would see her next week at the markets.

The next week I found her at the markets and asked her how she was going with the smokes. She said "great" and then proceeded to roll one. I watched her closely to see if she was smoking like I taught her. Nope, she was taking it all the way down into her lungs. I again looked closely at her and saw her future self. Very soon she would become an old, sick, grey-haired woman with all her beauty destroyed through tobacco and drugs. I sadly wished her luck and quickly took my leave.

I was saddened again by the fact that I was not a successful teacher or even much of a helper.

When I finally returned to the south, Trish rang and asked me to have coffee with her. Her new gay best friend was with her. Trish looked dishevelled, with swollen eyes, and told me that she had been crying for two weeks. She said that she did not know why.

I knew that one of the things that hurt me so much was the fact that I could also see Trish's future and I could do nothing to help her. Trish and her friend were still living together. Two addicts supporting each other's addictions.

I could not resist asking her if she still smoked. She said yes and took her "rollies" out of a bag and began rolling. So did her mate. Both added marijuana. She then said that she was going to give up cigs and all drugs – soon.

As I took my leave, my last words to her were these: "If you now live with a smoker and hang out with other addicts, as you have chosen to do, then it is almost impossible for you to quit anything."

She hung her head and said, "I know."

This made me happy because if she knows this, then she has a chance.

My heart did another somersault and it went out to her. I told her that I could not take any more pain watching her destroy herself and wished her luck as I said goodbye. It was up to her now. I clearly could not help her and so was obviously not meant to. Perhaps it was just like I could not help the woman in Vietnam who had her thumbs tied to the ceiling. I suppose that she too was autonomous and had chosen her path. This understanding seemed to lift a weight I had been carrying for such a long time.

The thoughts of Trish and the pain the relationship brought have passed and finally gone. In any love affair gone wrong, so long as one does not write, speak to, or see the lost lover, the pain always passes. The one I once thought I could not possibly live without was finally lived without quite easily and for that I am thankful.

A special poem brought to light my experience of love lost.

There is always some
Madness in love
But there is also always
Some reason in madness.
Even the
Darkest night
Will end
And the sun
Will rise.
                      Victor Hugo

I am aware that I have experienced, many times, the folly of desire and the seemingly never-ending quest for repeated pleasure. Desire and thinking go hand in hand. Joy eventually replaces these longings, when joy is free from thought and desire. Desire, of course, can be almost overpowering and the experience of it seems to have been one of my best teachers.

Of course, thought, like memory, is necessary for daily living such as work and communication, but a break from thinking and life, through feeling, is nectar indeed.

But a feeling of oneness with the mysterious "All" and everything is divine nectar.

> *"The earth is ours; it is not English, Russian or American, nor does it belong to any ideological group. We are human beings, not Hindus, Buddhists, Christians or Muslims. All these divisions have to go, including the latest, the Communist, if we are to bring about a totally different economic-social structure. It must start with you and me."*
>
> J Krishnamurti

Chapter 43 - Journey's End

This life, as it has unfolded, has never been my conscious choice alone to do as I wished. If I had had total freedom of choice I would have chosen not to go to Vietnam and I would have chosen to live a normal life as a husband and an accountant, home safely at night with the family I loved, happy in a normal society, or at least in a society that I accepted as normal. This normal life was not meant to be for me. It was impossible for me to conform to, even though I tried many times with all my might. The migraine headaches, the back and neck pains, the strange moods, the heartbreaks and the broken body would not let up until I listened, learned and followed the signs that were always presented to me.

I tried many times to live a normal life. Within days of leaving my safe normal home, all my pain would seemingly magically disappear; even though surgeons had once assured me without a shadow of a doubt that I needed an operation on my neck so as not to become a cripple for the rest of my life.

I realise now that some surgeons operate because that's how they make their big money.

So I finally realise from my life experience that something? someone? existence? the friend? consciousness? any or all of which are also me (the who or what does not matter), has been leading to what consciousness needs to become through me, and the information I needed to impart.

The unseen cannot be named. Obviously, if I was meant to name it and tell you it's rules, then it would have declared itself with a loud voice and shown itself to me. It has not. So, the unheard, or the unseen, is not meant to be spoken and described, by me or anyone else – by no priest, mullah or holy man, telling us who and what each of their gods is and their special rules.

Faith can only be blind and this allows life to remain the surprising mystery that it is. Surely it is enough to finally realise that we are not separate from existence or any imagined god and then take immediate responsibility for ourselves and all our actions. Reclaim our morals that have been taken from us through misinformation.

I have always felt that I was put here on this remarkable earth for a specific purpose and by the end of this book I would hope that this purpose that has driven me is satisfied and fulfilled and I am allowed to rest.

And so, in this final chapter, I shall attempt to get out of my own way and let my hand do the writing, and let be written what needs to be said for a world obviously in dire need – a world with its religiously fuelled wars; with huge bombs that add to human suffering and chaos, plus cause earthquakes, tsunamis and climate change; with man's monetary imbalances caused by greed and lack of morals. Our fading religions, which create religious fanatics who murder innocents because of impossible and improbable beliefs and godly instructions, have helped to almost extinguish freedom and love in mankind. Beyond religious beliefs there is an earth where fanatics murdering themselves and innocent people en masse no longer occurs; where love and acceptance are the norm.

I wrote a poem when I was very young, before I met any of the incredible people that I have met, or learned any of the things on the road to pass on to you. This is the poem, as I remember it, though that broken psychiatrist from my chequered past burnt the original.

THE JAILER

And the Jailer came
With a simple key
And with a simple twist
He set man free.

*Now there are no
Priests or Nuns
To trap our minds
And teach our sons*

*For Gods have been
Since the World began
Shaped, twisted, invented
By power hungry man.*

*But the Jailer
Has come
And set our minds at ease
Now we need not kill others
Our Imams, priests
Or governments to please.*

*For God is not an Idol
A man, the Sun, or Moon.
God may be felt in silence
Beyond thought
In a breath, a bird
Or even a pristine lagoon.*

What was I seeing in this poem? The final death of all the current gods and primitive beliefs of our times. This death of false and primitive gods has always happened in the distant past in all our ancient civilisations. It happened in Rome, in Greece, and in other ancient cultures, and the signs were always the same when it was occurring.

The following were some of the ancient signs found in our history.

Most of the money, and therefore energy, in these ancient civilisations was, at the time of their monumental collapses, being spent on armies, wars, police and security. Farmers could no longer make a decent living as all the money was directed into the cities and into the pockets of the wealthy businessmen, the ruling politicians, generals and religious leaders. Another sign of these crumbling civilisations was that the gay populations multiplied and then entered politics and became part of the ruling class. These things are all happening today.

Also, great arenas were built and the populations became entranced with what happened in these arenas, just as we now do with our sports today.

Orgies became commonplace as people turned to rampant sex, devoid of love, to escape the upheavals and chaos of their failing beliefs and lost morals.

And last, but certainly not least, the young and the wise no longer believed in the old gods with all their silly, man-made, corrupted demands and laws.

All these signs are happening in the world right now. Change is upon us and we have a choice whether to enter this change consciously, with awareness, or in chaos and upheaval as our forefathers have always been forced to do.

Urgently and quickly, we must stop separating ourselves, through belief, from any man-made "thought gods". At the present time we have at least a thousand of these gods spread throughout the world. Probably more. We must experience a new way of satisfying our spiritual selves that is not through thoughts of separation from gods.

One of the teachings that I must mention again, which all the gurus and even some priests agreed on, was this, "There is only One Self".

So, if we do not separate ourselves from gods, each other, nature, the universe, or anything else such as carved idols, then we will not be so warlike and afraid. Fear precedes wars. Fear is often followed by intense anger.

Otherwise, we shall continue to live in a world of damaging duality – duality created by beliefs that take away our responsibility for our own individual actions.

Belief that takes away our morals and puts them in the hands of false gods, prophets or priests. We are, and must be, responsible for our own actions and morals. Jesus did not die to take away our sins and no priest can forgive us.

Know at this time, without a shadow of a doubt, that you and I are responsible for everything we do and there is no god in this universe, or any other universe, who can forgive us for our sins or wrong actions. We, and we alone, are forced to pay for our murders, rapes, and all wrong actions.

Remember Jack and my other mates who killed in anger, whether for revenge, country or priest, and the shocking price they paid for you. Surely they paid this price for all of us to learn from and we must not let them have died in vain.

Remember the man who was recognised by his peers as one of the greatest minds of his era. Remember one of the statements he made. This man was Terence McKenna. When I asked him to talk to me about religious belief, he simply said, directly and clearly, "Belief makes you less than human." Think about that and the truth of it that you can read about in the papers or see on the news every day. And any belief, whether it is cloaked in enlightenment, meditation, communism, paradise, or heavens, will take one away from the ability to really feel in the present moment.

And the beautiful Kiran; when I asked him what one needed in this world, he answered: "You only need three things, Sahajo: surrender, acceptance and awareness."

This is how I interpret Kiran's words. Surrender (to something more than ourselves but which we are a part of not separate from). Acceptance (of whatever befalls us as it is evolution in progress). Awareness (of every little thing that we do, including our thoughts and movements). In writing this explanation, I would add that Kiran's words need no interpretation but your own.

Another venerable and well-loved master that I must mention is my Buddhist master from the temple in Thailand. This is the gift he taught me. If you lie tossing and turning during the night and cannot sleep, perhaps because you have eaten too much, been forced to take strong prescription drugs, or are worried about your work, etc, and the monkey mind seems impossible to still, then do this. As your stomach rises and falls with each breath, simply acknowledge each thought, neither being for or against it. While repeatedly doing this, also note a different part of your body between the rising and falling of each breath and, with breath awareness, let that part relax. No single thought can then overcome you and no thought lasts for long. All thoughts pass away.

All the masters I have mentioned in this book had something to reveal at some time, even when they were showing us what not to do by their misguided actions. I hope the writing of their teachings can help each person towards badly needed change, until, finally, when we realise "what not to do", a new consciousness shall pop into place, just like "The 100th monkey" story from old Japan.

The masters I met were all subjected to more temptations than the average man. Temptations such as money (endless amounts of it), power, worship, or sex with any of their followers. We really cannot throw away all of their teachings just because some may have finally surrendered to one of these powerful temptations. How do we know that we could have done any better? There is good to be found in everything and everyone. The masters all had something to give, initially, or they would not have been on the world stage. I suspect this includes religions when they first arose, before they were sullied by time and men, and their time was over.

As the loving Kiran informed me with a smile when we were sitting with the Australian guru, Barry Long, and I was complaining about Barry: "Don't you know that everything is perfect or he would not be up there?" The negative teaches us as does the positive.

Kiran was also quite adamant about one other thing and that was this: "Isness is our Business."

Isness, as I understand it, is always happening in the present, not the future or the past. Perhaps one experiences this isness when one has no concepts or judgements at all, and the mind is at rest, in waiting. Then we may experience Isness spontaneously.

In this way the fact that everything is perfect is true for we are always learning and growing from our experiences, especially the painful ones. This is consciousness at work as we slowly learn what not to do, and how to live, by watching ourselves and our thoughts with sharp and acute awareness.

Let's look, for example, at how this brutal, heartless, modern phenomenon of the death cult, ISIS, can possibly have anything to do with being perfect as it unfolded? What are they teaching us? Obviously that religion no longer serves us, as primitive beheadings, murder and killing are not spiritual, human or religious.

This ISIS mob are Muslims. Yet because of their depraved, murderous and merciless actions many Muslims have abandoned their Muslim faith and its ancient beliefs. These individuals abandoning Islam have woken up to their own misdirected belief because of ISIS's bloodthirsty killings of people who simply do not believe in their god, their Allah. Infidels, all of us, they scream, and all infidels must have their heads cut off for being "unbelievers", as their interpretation of their holy book proclaims.

This is not human behaviour guided by love or common decency!

Evolution spares nobody on its relentless path to the transformation of humanity. Evolution takes no prisoners. Mankind is evolution in progress.

And it is not only the Muslim religion being abandoned.

The Christian leaders in the countries of the West encourage their ruling politicians to use tons and tons of bombs in war, all for the good and in defence of us and our Christian beliefs and way of life. We are taught to imagine ourselves as the good guys and saviours. Yet these bombs also kill the innocent, including the little children. How do we know this? Just in the Vietnam War, for example, the combined Western and communist armies killed two million civilians, mostly by misdirected bombings and misdirected bullets.

We, the soldiers, are taught to beat bombings by digging fancy, double tiered trenches in which we can survive any bombing, whereas civilians know nothing of these bomb-proof trenches.

In Vietnam perhaps only 80,000 soldiers from all sides lost their lives while two million innocent civilians lost theirs. This has always been a consequence of useless and misguided wars.

I once lay in a trench in Vietnam and watched American B52 bombers bomb an entire mountain. The massive mountain seemed to lift and I was positive that no one, neither animal, bird, civilian nor soldier, could possibly survive. The whole country shook, even my trench shuddered and this was far, far away from the bombing. When the bombing ceased, we went in to clean up and got cleaned up ourselves as the wily Vietcong rose up from their trenches and caves, re-laid our stolen mines and cut us down.

America is doing a disgusting, useless and shocking job of acting as "Sheriff of the World" with its John Wayne and Christian mentality. America should pack up and go home from other countries before they create more death, turmoil and upheaval in all the countries they bomb and are invading right up to this day. Go home and forget god. Let countries grow in their own way by their own mistakes. That's how they will wake up and should be allowed to do so.

And of course, America had a major hand in creating the Taliban (they armed and supported them against Russia). They then armed ISIS against Syria, and also armed any number of other terrorist organisations who they continue to support and create, if it suits them, to overthrow a country's leader who is not selling them oil or supporting their money ideology.

Wars keep America wealthy because they sell their fancy war inventions all around the world.

I think one of the greatest crimes and tragedies of modern man is that we now spend at least seventy per cent of all our energy, and that includes money, on armies, weaponry, police forces and security, and this seventy per cent is quickly growing. If this energy was spent intelligently, on science, right education, inventions, health, farming, the homeless, the sick, the handicapped quadriplegics and the dying, then an awful lot of the pain and death in the world would be gone, and we could finally all live in peace and love, working less and enjoying more, without the taught guilt and laws of false gods that have merely created the opposite of what they claim as they sacrifice our freedoms.

All these religious leaders are quick to take responsibility for any good things in our world and do not see that they are the cause of all the bad together with nearly all the wars as well.

I have no doubt, and never have, that a new consciousness, through evolution, will gradually win this struggle that is taking place in our world, whatever the seeming cost.

In the past twenty years or more, the children of Australia have grown up in a country where it is a law that they are not to be beaten or hit by adults or parents. This has allowed them a freedom that is now producing great art and free thought. This freedom from domination may allow them to look at a flower, a tree, or a raven, just as Kiran looked. Kiran did not examine the Raven with all the knowledge and information he had been taught of it. Kiran did not look, and with a trained mind see a raven or even a bird. No, his mind was empty and this allowed him to see the amazing beauty of a bird. In this seeing he could feel the bird. The bird felt this, and though wild and untamed, it came and landed next to him on his chair.

To look at anyone or anything with ecstatic feeling is surely one of the beautiful outcomes of a freed mind and one of the places where consciousness is leading. To look at a tree, its leaves, its flowerings, with no mind, enables us to truly feel it, the miracle of it, and the earth from which all this is born. In this form of seeing and feeling we are feeling ourselves and the miracle of us.

Let us for just a moment be aware of "the helper" without knowing who or what it is. Drop, even for a moment, all the teachings, all the hollow fear raised warnings of all the priests, the imams, or any so-called religious leader.

No god has ever spoken out loud or we would have all heard that voice. And if their gods were actually all powerful and real, they would have stopped all these wars and this killing of innocent children. Peace is obviously not in the hands of gods. Peace rests in us.

When mankind is at peace, free from thought, time disappears and this includes the past, the future, and even the present.

When we are free of beliefs of a future paradise or heaven and can actually feel, then no word such as god need ever be spoken or taught us again. God, if I may use that name, is like loving. Loving is a feeling, not a thought. What we have for so long called god, therefore, can only be felt, not named or thought.

One thing is surely patently obvious to us all. This sacred world, our earth, can be made a better place than what we are presently making it. It does not like all the many bombs being dropped on its surface and is more sensitive than we realise. We arise from it and therefore are it. Would we treat ourselves in the way we treat the earth? No!

We are also, at present, the outcome of all our religious teachings and misplaced beliefs, but this is rapidly changing as intelligence grows and we become more sensitive. The meek and the sensitive shall indeed inherit the world.

As for me, I now look like a samurai sword fighter who never learned to duck. I have had cancer removed from most parts of my body. I have long scars down both sides of my face and my mouth. I have had cancer removed from both eyes and one eye will not heal. Both eyes have had skin grafts. Cancer still keeps popping up and some doctors have finally admitted that the Agent Orange, sprayed on me in Vietnam, has obviously caused this.

I have now been diagnosed with an incurable cancer called follicular non-Hodgkin's lymphoma.

But now I have been cut enough and am not having any more operations or even R-Chop chemotherapy. Why bother if the medical establishment tell me that my condition is incurable?

I am now proud to look like my old sergeant, my great mate and great man, good old Jack. Cancerous melanomas have recently been diagnosed on my legs and I have refused to let them cut off large parts of my leg. I am sick of being cut and will heal them in other ways or go back to my beloved earth, which of course I also am.

I hold no hatred of the Americans who did this to me. They were only trying their best at the time even though their best was far from good enough and is still not good enough. To sacrifice one person for the good of the many, as they do, is no excuse for killing anyone.

These are just the wages of any war and there are many innocent children, people and soldiers, with scars much worse than mine.

I am happy to say that I hold no hatred of anyone (or any country) that has passed through my life. Hatred only hurts the hater as do the festering sores of anger, mistrust and lies. Suffering has helped burn these sores from me. I no longer cry about my sufferings nor imagine that they have some negative hold over me. Evolution works in mysterious ways to achieve what it needs to and I am merely its servant.

What is to come, I do not know, but in surrender life is the mystery that it is, and as it should be for me.

Surrender – Acceptance – Awareness. These are the words I think of, to help me live life.

I also think of these words, and their first letters, when I honour my dead mates on Anzac Day. I salute and say SAA (surrender – acceptance – awareness), not SIR.

Love may survive in humans only in freedom, without rules, without contracts, without judgements, without demands and without religions creating hells, heavens, devils and gods which all create fear through false judgements. Actual loving, once felt and experienced, can only continue to exist in freedom, innocence and trust and devoid of any judgements.

I have sometimes thought that I was writing this book to rewrite myself. In a way I must have because of the love I have found. One aspect of this I shall now try to pass on to you, the reader.

I have in the last few years had the pleasure and calmness of experiencing this type of love with a woman who I have had a 35-year connection with. We began as friends and then became lovers. We tried having relationship but it was only when we both reached this state of freedom with no rules, no judgement, no blame, that we have been able to always meet each other with open arms and heart. When I look at her it is like seeing all women and I never recognise her face as only Maria because true feeling is not limited by any thought. In her presence I feel the satisfaction of pure appreciation. Of course we can have differences of opinion and slight misunderstandings but finally I get to experience that the foundation is solid so any wobble is just a flutter.

When truly loving somebody, you do not recognise the face you are looking at, no matter how long you have known that person. In feeling there is no thought and the eyes see with wonder as they look with beauty. This type of loving is also healing. This type of making love is also naturally tantric with no effort or knowledge needed of ancient tantric teachings. It may also be experienced without sexual penetration and sometimes takes the breath away in a gasping agape manner. All the senses are awakened and remain awakened as long as any thought is ignored. And thought will try and exert itself both positively and negatively. Ignore all thought, all knowing, and this becomes the powerful meditation of loving.

Good luck to you all on this journey of life and awakening. In the end it will all be okay. If it is not okay then it is not the end. And it can be the end in any second, with understanding and intelligence.

I greatly admire the human spirit. How it carries on through all the disruption in today's world.

I suddenly remembered this. Spending days in Bombay searching for a man named Ramesh S. Balsekar. When I finally found him, he invited me into his apartment and spent many days talking to me. Strangely enough, last week a woman at a sale handed me a book and refused to take any money for it.

It is called The Final Truth and it is written by Ramesh S. Balsekar. I opened it at page 34 and this is part of what I read:

"The external worship of a form of God is prescribed only for those

whose psyches have not been sufficiently purified and whose intuitive intelligence has not been adequately awakened. Such worship of an object created by themselves as a concept may give the worshippers a certain amount of satisfaction and peace of mind but it is a futile process from the point of view of experiencing one's true nature.

The God who is fit to be worshipped by the highly evolved intellects is one which supports the entire phenomenal creation as its substance…. A God beyond all concepts, infinite and temporal … A God which, like the flavour in food, is within every sentient being and therefore needs no seeking … A god who cannot be comprehended because He transcends the mind and five senses of cognition.

Universal Consciousness is to be worshipped by one's own personal consciousness, not by offering flowers or food, nor by lighting incense and waving lights. It should be worshipped without any effort, by self-realisation alone, by the supreme meditation in the continuous, unbroken awareness of the within, the indwelling presence. This worship needs no effort because there is nothing to be attained which one does not already possess."

And finally from the preface of his book:

"Truth cannot be described or explained. Truth is "What Is" and the acceptance of it. Every word that is uttered concerning Truth can only be a pointer towards it. The understanding of Truth cannot be achieved. It can only happen. … And when it comes, it cannot be accepted unless the mind is empty of the "me" and the heart is full of Love."

Perhaps I may rest now if this life's mission is over.

Ha ha! With that writing of what I thought were my final words I again picked up a book and opened it at a random page and this is what I read. Ah! Life, again revealing itself with its divine synchronicity in its unexplainable mystery where miracles abound. This is the man who I first met on this journey of life.

And this is what I randomly read:

> "When you call yourself an Indian or a Muslim or a Christian or a European, or anything else, you are being violent. Do you see why it is violent? Because you are separating yourself from the rest of mankind. When you separate yourself by belief, by nationality, by tradition, it breeds violence. So a man who is seeking to understand violence does not belong to any country, to any religion, to any political party or partial system: he is concerned with the total understanding of mankind."
>
> J Krishnamurti

I would just add that not only are we separating ourselves from mankind with these thoughts Krisnamurti mentioned, we are also separating ourselves from all other life forms, plus the earth, the universe, and all existence.

Go now without gods as life is miracle enough.